History of Telecommunications Technology

An Annotated Bibliography

Christopher H. Sterling
and George Shiers

The Scarecrow Press, Inc.
Lanham, Maryland, and London
2000

SCARECROW PRESS, INC.

Published in the United States of America
by Scarecrow Press, Inc.
4720 Boston Way, Lanham, Maryland 20706
http://www.scarecrowpress.com

4 Pleydell Gardens, Folkestone
Kent CT20 2DN, England

Copyright © 2000 by Christopher H. Sterling and George Shiers

First edition, by George Shiers, published in 1972 by Scarecrow Press, Metuchen, N.J.

All rights reserved. No part of this publication may be reproduced, stored in a retrieval system, or transmitted in any form or by any means, electronic, mechanical, photocopying, recording, or otherwise, without the prior permission of the publisher.

British Library Cataloguing in Publication Information Available

Library of Congress Cataloging-in-Publication Data

Sterling, Christopher H., 1943–
  History of telecommunications technology : an annotated bibliography / Christopher H. Sterling and George Shiers.
    p. cm.
  Includes bibliographical references and index.
  ISBN 0-8108-3781-1 (alk. paper)
  1. Telecommunications—History—Bibliography. I. Shiers, George. II. Title.

Z5834.T4 S74   2000
[TK5102.2]
016.621382—dc21   00-024823

⊖™ The paper used in this publication meets the minimum requirements of American National Standard for Information Sciences—Permanence of Paper for Printed Library Materials, ANSI/NISO Z39.48–1992. Manufactured in the United States of America.

Contents

INTRODUCTION vii

1 GENERAL REFERENCE WORKS. 1
    A. Bibliographies *1*
    B. Dictionaries and Encyclopedias *6*
    C. Directories of Archives, Libraries, and Museums *7*
    D. Chronologies *8*
    E. Statistics *11*
    F. Atlases *14*

2 SERIAL PUBLICATIONS 15
    A. Guides to Periodicals *15*
    B. Indexes to Periodical Literature *17*
    C. Periodicals—General *18*
    D. Periodicals—Electrical and Telecommunications *23*

3 GENERAL SURVEYS. 33
    A. Electricity and Electrical Engineering *33*
    B. The Process of Invention and Patents *35*
    C. Telecommunications *37*
    D. Electronic Media *42*
    E. Foreign Telecommunications Systems *43*
    F. International Telecommunications *45*
    G. Military Telecommunications *49*
    H. Maritime and Naval Telecommunications *54*

4 INSTITUTIONAL AND COMPANY HISTORIES 57
  A. Reference Resources   58
  B. Government Bodies   59
  C. Electronics/Telecommunications Industry Surveys   64
  D. Societies and Institutions   67
  E. Specific Companies—U.S.   71
  F. Specific Companies—Foreign   83

5 BIOGRAPHY: INVENTORS, SCIENTISTS, AND ENGINEERS 93
  A. General Biographical Reference   93
  B. Collected Telecommunications Biography   95
  C. Individuals   98

6 TELEGRAPHY 123
  A. Contemporary Books on Telegraphy (1823–1895)   123
  B. Later Telegraphy Works (1896–1950)   134
  C. Modern Histories of U.S. Land Telegraphy   136
  D. Foreign Telegraphy Systems   138
  E. Contemporary Books on Submarine Telegraphy (1855–1905)   143
  F. Modern Histories of Submarine Telegraphy and Telephony   149
  G. Vintage Telegraphy Collectibles   152

7 TELEPHONY 153
  A. Contemporary Books on Telephony (1877–1907)   153
  B. Later Telephony Works (1908–1950)   158
  C. Modern Histories of U.S. Telephony   161
  D. U.S. Regional Telephony Histories   164
  E. Foreign Telephony Systems   166
  F. Submarine Telephony   170
  G. Video Telephony and Teleconferencing   171
  H. Vintage Telephony   172

8 ELECTROMAGNETIC WAVES 175
  A. Modern Histories   175
  B. Observations and Experiments to 1900   176
  C. Developments after 1901   177

    D. Microwaves   *179*
    E. Spectrum Management   *180*

9 RADIO 183
    A. Contemporary Works on Wireless (1892–1908)   *183*
    B. Wireless and Radio (1909–1925)   *188*
    C. Survey Histories   *195*
    D. Radio and Broadcasting after 1925   *199*
    E. Circuits and Patents   *202*
    F. Vintage Radios   *204*

10 ELECTROACOUSTICS AND RECORDING 207
    A. General Surveys   *207*
    B. Phonographs   *209*
    C. Electronic/Magnetic Audio Recording   *212*
    D. Motion Picture Sound   *214*
    E. Stereo Sound and Recording   *217*
    F. Video Recording   *218*
    G. Video Disc and Cassette Recorders   *220*

11 ELECTRON TUBES AND SOLID STATE DEVICES.................................. 223
    A. General Surveys   *223*
    B. Developments to 1920   *225*
    C. Developments after 1920   *226*
    D. Cathode-Ray Tubes   *228*
    E. Microwave Tubes   *229*
    F. Solid-State Devices   *230*

12 TELEVISION 233
    A. Contemporary Books on Television (1924–1939)   *233*
    B. Other Historical Books and Articles   *240*
    C. Historical Surveys   *245*
    D. Color Television   *248*
    E. Vintage Televisions   *250*

13 NEWER MEDIA TECHNOLOGIES............. 251
    A. Facsimile   *251*
    B. General New Media Surveys   *255*
    C. Cable Television   *258*

D. Teletext and Videotex   *260*
   E. Remote Control and Interactive Media   *260*
   F. Digital Audio/Radio   *261*
   G. High-Definition/Digital Television   *263*

14 TRANSMISSION: MOBILE, SATELLITE, FIBER OPTIC, AND THE INTERNET 267
   A. General Surveys   *267*
   B. Mobile Communications   *269*
   C. Satellite Communications   *270*
   D. Fiber Optic Systems   *276*
   E. The Internet   *276*

15 TELECOMMUNICATIONS HISTORY ON THE INTERNET 279
   A. Reference   *279*
   B. General Surveys   *280*
   C. Institutional and Company History   *281*
   D. Biography   *283*
   E. Telegraphy   *284*
   F. Telephony   *285*
   G. Radio   *285*
   H. Electroacoustics and Recording   *286*
   I. Electron Tubes and Solid-State Devices   *286*
   J. Television   *287*
   K. Newer Media Technologies   *287*
   L. Transmission: Mobile, Satellite, Fiber Optic, and the Internet   *288*

NAME INDEX................................. 289

TITLE INDEX................................. 303

ABOUT THE AUTHORS....................... 335

Introduction

This book contains more than 2,500 annotated bibliographic entries that trace the history of major telecommunications technologies over the past 175 years. Restricted to works in English, and thus focusing substantially on work done in and about Britain, the United States and allied countries, it is designed to be a road map to an expanding historical literature about a fast-changing field.

No such work can be absolutely comprehensive—what you hold is a selective guide to most of the important historical material. It substantially revises a valuable book first published a generation ago and long out of print.

How This Book Came To Be

More than a quarter-century ago, in 1972, Scarecrow Press published George Shiers' *BIBLIOGRAPHY OF THE HISTORY OF ELECTRONICS,* a 336-page guide listing some 1,800 published sources and annotating most. It grew out of the author's already impressive career in electronics and authorship about electronic drafting as well as the history of communications electronics.

George Shiers was a freelance technical writer and part-time college teacher in Santa Barbara, California. Born in Coventry, England, in 1908, Shiers was educated in London and worked in Britain's electrical industry before World War II. After wartime service with the military he worked in the electronics business, moving to the United States in 1948. He continued employment in electronics for another decade, then turned to freelance writing and teaching. Among his eight books were long-standard texts on electronic drafting published in the 1960s, the 1972 book on which this bibliography is based (entry 1-017), and several historical anthologies on telecommunications (6-120, 7-091, and 9-080) on which both of us worked. As he focused more on historical works in his later years, he authored a series of important scholarly articles on electrical communications history in standard journals (see, for example, 5-075, and 12-072). Virtually all are cited here—most appeared after the 1972 book was published. Shiers was a life member of the Institution of Electrical and Electronic Engineers (IEEE) and the Royal Television Society, as well as a member of the Society of Motion Picture and Television Engineers (SMPTE). When he died in April 1983 at age 74, he and his wife May (a childhood friend whom he married in 1936 and who assisted in all of his research) had been married for 46 years (May Shiers died in January 1990). By the time he died, there were only 100 copies of the 1972 bibliography remaining and it went out of print shortly thereafter. Never replaced, it has been widely cited in the years since, and begged for updating.

Based on his own bibliographic publications (1-020 and 1-021), the undersigned first suggested such an update in mid-1997. An impressive generation of new historical material had appeared and needed to be integrated into Shiers' original book, which remained valuable, though increasingly dated. At the same time, I felt some subject areas could be excluded this time around, as there are now excellent bibliographic guides to that material.

What *Is* Included . . .

Most of the telecommunications historical citations contained in Shiers' original book are retained here. Some items now of lesser importance, including materials clearly superceded, have been dropped to make space for the vast amount of new research published in the generation since the 1972 book appeared, as well as the many more recent subjects not covered there. The focus has been sharpened to include only material directly relevant to the history of telecommunications technology.

Among subjects *added* or *expanded* into their own sections here are: directories of archives and museums (chapter 1-C), chronologies (1-D), historical statistics (1-E), atlases (1-F), foreign and international telecommunications systems (3-E and 3-F), the role of telecommunications in the military (3-G) and in naval and maritime commerce (3-H), and works on collecting and appreciation of vintage telegraph (6-F), telephone (7-H), radio (9-F), and television (12-E) equipment.

Likewise I have added sections or chapters covering services that either did not exist in 1972, or then had histories too short to warrant inclusion. Added here, they include video recording, video discs and cassettes (chapter 10-F and 10-G); newer media including cable, fax, and teletext/videotex, digital radio and television (chapter 13); and newer modes of telecommunications transmission including mobile and satellite services, fiber optic systems, and the Internet (chapter 14). Another indicator of changing times is a guide to some of the more useful—and hopefully lasting—sites concerning telecommunications history to be found on the Internet (chapter 15).

As a guide for collectors, a special effort has been made to construct as complete a chronological listing as possible of the pioneering English-language books on different telecommunications services. These include the first three-quarters of a century of telegraph books (1823–1895, section 6-A); the first half century of works on submarine telegraphy (1855–1905, section 6-

E); the first three decades of books on telephony (1877–1907, section 7-A); and the first 15 years of works on wireless and radio (1892–1908, section 9-A). A more selective listing on television's first dozen years of publication (1924–1936, section 12-A) is drawn from the far more complete survey found in Shiers (1-018). These sections vary in the period covered as, after the specified cut-off dates, most relevant books were of a practical or how-to nature, offering little of historical significance. These sections appear in chronological rather than alphabetical order by author to better indicate development of the relevant technology. (The majority of the book appears in the usual bibliographic order by author or title.)

Each entry (with the exception of the periodical listings in chapter 2) has a unique number to which the index refers. That number indicates the chapter and entry number in this fashion:

```
   ┌─ Chapter: Entry number ─┐
   └─────→ 11:123 ◄──────────┘
```

Entries thought to be especially important or useful are noted with from one (*) to three (***) asterisks to highlight that status. A conscious attempt has been made to record later reprints of older material, and to provide considerable cross-referencing to related material listed elsewhere.

A few last minute additions after the book was assembled and indexed are evident by the use of letters after the number.

. . . And What Is Not

In the generation since George Shiers originally assembled his 1972 book (back in the days when such work involved typewriters and paper reference slips), the literature on all aspects of electronics has greatly expanded. Luckily, various new modes of bibliographic control have also appeared so that where Shiers stood virtually alone in the early 1970s, a generation later we have many places—and ways—to find historical resources, a number cited in this volume, especially in chapters 1 and 2. That availability, plus the huge expansion of historical material overall (and the topics added in this edition, as noted just above), has led to some tough choices of what to include in order to keep this volume to manageable size.

Specifically, entries carried over from the original book have been refocused primarily on telecommunications and closely allied technologies (defined to include recording). Put another way, several broad subjects found in the original volume have *not* been carried over to this new edition for lack of space and because each has been more than adequately detailed elsewhere. These include general electricity and electronics,[1] radio astronomy,[2] industrial electronic tubes and industrial electronics,[3] most computer references,[4] radar,[5] and the large field of solid state electronics[6] (though a selection of useful resources on the latter *are* included at the end of chapter 11).

Other exclusions will become obvious. The work is restricted to material in English. This is not to say that historical material in other languages is not important (there is a vast literature on the telegraph and telephone in French and German, for example)—it simply recognizes that no single book can annotate everything and remain relatively easy to handle.[7]

x   *Introduction*

One other difference from the 1972 book will be evident—with the important exception of the five special sections on pioneering contemporary works on the telegraph (6-A), submarine telegraph (6-E), telephone (7-A), radio (9-A), and television (12-A), already noted, all entries appear in author (or title if no author) order rather than chronologically. This is in keeping with other references, library catalogs, and on-line resources, and should assist users in locating material more quickly.

Acknowledgments

First and foremost, I am indebted to the late George and May Shiers. It was my acquisition of a review copy of the original bibliography in late 1972 that eventually led me to track down the Shiers in their Santa Barbara, California, home which George had largely built with his own hands from the time they arrived in 1953. Over the decade from when I first corresponded and spoke with them by telephone in 1973 to George's death in 1983, we worked on several projects together (including a number of series of reprinted books for Arno Press, then a *New York Times* publishing subsidiary), all the time sharing our common interest in the literature of telecommunications history.

With great patience, George taught me more about the technical historical literature of the field than anyone else. Many of the books and files from his library now reside in mine and helped to form the core for this project. Just before George's death, I promised them both I would see through to publication his then-incomplete magnum opus bibliography on the development of television to 1940. That promise took the efforts of many people—and 13 years—before it was fulfilled (see 1-018).

A considerable file (two-decades' worth) of specialized book dealers' catalogues has proved most useful in chasing down variant editions, pagination, and other details. In this regard I am grateful to the substantial bibliographic work evident in the catalogues of Frank Bequaert (Rainy Day Books, Fitzwilliam, NH), Len Kelly (L.V. Kelly-Books, Tiverton, England), James Kreuzer (New Wireless Pioneers, Elma, NY), and the Old Authors Bookshop (Morrisburg, Ontario). Bequaert and Kelly also took time from book selling to carefully review parts of this volume and made many valuable suggestions.

Several other authorities also reviewed all or parts of this book which is vastly improved thanks to their collective labors. They include Dr. Bernard Finn (Curator) and Elliot Sivowitch (Museum Specialist) of the National Museum of American History's Division of Electricity (Smithsonian Institution), and Dr. James K. Bracken of the reference division of Ohio State University's main library and my partner in several other bibliographic ventures (see 1-001, 1-020, 1-021). The curator of the BT Museum in London, Neil Johannessen, was very helpful with both citations and publications (see 7-135).

Likewise, several specialized book collectors have been of great help, including Robert Voss whose knowledge of—and willingness to share—bibliographic data and name of other authorities on telegraph history

Introduction

was a huge help. He also put me in touch with telegraph authority, dealer, and collector Robert Dalton Harris who provided extensive detailed comments, additional material, and two of his own valuable publications (6-247, 6-248), saving me from several errors. Steve Prigozy also helped to pin down several telegraph citations. My thanks to both Joe Knight and Jim Kreuzer for the former's "A Bibliography of Known Wireless Books," (revision 10a, 1997). Thanks to Len Kelly who put us in touch, collector Douglas Penisten was kind enough to review and comment on several chapters and followed up with citations for and annotations on a number of items I had not seen.

The on-line catalogs of several libraries, most especially those of the Library of Congress (Washington) and the British Library (London), allowed for checking and completion of citations in a way not possible just a few years ago. So did the telecommunications collection of David L. Solomon donated to George Washington University's Gelman Library by his widow just as this project was nearing completion. So, of course, did the IEE's database, INSPEC, which provides a search window to some 3,000 journals worldwide since 1969. And readers/collectors should not overlook the growing number of Internet secondhand book sites which offer a wealth of (sometimes fleeting!) bibliographic information. Of special value in helping to complete this volume (and add to my own collection at the same time) were both *http://www.bibliofind.com* and *http://www.abebooks.com*.

Once again, I have had the pleasure if working with my family to bring this project to fruition. My elder daughter Jennifer, operating as Spot Color in Chantilly, Virginia, has converted my computer files (how George Shiers would have loved their convenience!) into well-designed camera-ready copy, something she has undertaken for several of my previous books. Younger daughter Robin provided useful moments of humor and insight as respite from bibliographic pursuits. And—as with a widening shelf of similar efforts—my wife Ellen has been patient, understanding—and sometimes even *enthusiastic*—about the continued collection building which underlies this effort. Her continued love and support makes this sort of work possible. It is truly a joy to make such projects family affairs.

The George Washington University   CHS
Washington, DC

NOTES

1. See, for example Finn (1-007) and Shiers (1-017), pp. 114–142. In the present volume, see Section 3-A.
2. See Shiers (1-017), pp. 158–159.
3. See Shiers (1-017), pp. 174–183.
4. See for example, the excellent bibliographic guides by Cortada (1-004, 1-005), which annotate and index thousands of sources. See also Brian Randell, "An Annotated Bibliography on the Origins of Digital Computers," *ANNALS OF THE HISTORY OF COMPUTING* 1:101–208 (1979) which includes some 800 entries.
5. See Shiers (1-017), pp. 184–193, and Finn (1-007), pp. 153–161.
6. See Shiers (1-017), pp. 228–238; and Finn (1-007), pp. 288–295. For a selection of useful material In the present volume see Section 11-F.
7. See Finn (1-007) for many foreign language citations.

Chapter 1

General Reference Works

Included here are the more general resources useful for beginning research into the history of telecommunications technology. Such standard works as general purpose encyclopedias, readers guides, and handbooks have been excluded in the interest of space, though they should not be overlooked. The materials listed here are limited to those largely focused on the topic of this volume.

Section A provides citations to useful bibliographies, some of them listed in detail elsewhere in the book. Section B lists historically useful dictionaries and encyclopedias. Section C offers directories of archives, libraries and museums. Section D lists the more useful chronologies in this field. Section E surveys statistical resources and section F notes the relative handful of telecommunications atlases. See also 4-A and 5-A for more specific reference guides concerning companies and individuals, respectively.

A. Bibliographies

(Some subject-specific bibliographies appear in relevant sections later in this guide as cross-referenced here. Note as well the valuable bibliographic information found in specialized book dealer catalogues, such as those noted in the acknowledgments. See also chapter 2-A.)

See Benton, 8-014.

See Bourton, 12-132.

1-001 ** Bracken, James K., and Christopher H. Sterling. **TELECOMMUNICATIONS RESEARCH RESOURCES: AN ANNOTATED GUIDE.** Mahwah, NJ: Lawrence Erlbaum "Telecommunications," 1995, 173 pp. A companion to 1-021, see especially chapters 2 (history) and 3 (technology) for organizations and other resources not included in the present book. Includes nearly 1,200 indexed and topically-arranged citations.

1-002 ** Brightbill, George D. **COMMUNICATIONS AND THE UNITED STATES CONGRESS: A SELECTIVELY ANNOTATED BIBLIOGRAPHY OF COMMITTEE HEARINGS, 1870–1976.** Washington: Broadcast Education Association, 1978, 178 pp. Very useful indexed guide to House and Senate hearings, many of them focused on telegraphy, telephony, technical standards, radio, television, and related technical topics. See also Harris and DeBlois, 6-111.

See Chinn, 13-061a.

See Coggeshall in AMERICAN TELEGRAPHY, 6-087.

1-003 * Cooper, Isabella M., comp. **BIBLIOGRAPHY ON EDUCATIONAL BROADCASTING.** Chicago: University of Chicago Press, 1942 (reprinted by Arno Press "History of Broadcasting," 1971), 576 pp. Far more comprehensive than title suggests, this covers all of prewar radio, annotating some 1,800 studies, reports, books, and articles in a subject-divided volume with good indexes. A good deal of the material included focuses on technology and is now of historical interest.

1-004 *** Cortada, James W., comp. **A BIBLIOGRAPHIC GUIDE TO THE HISTORY OF COMPUTING, COMPUTERS, AND THE INFORMATION PROCESSING INDUSTRY.** Westport, CT: Greenwood "Bibliographies and Indexes in Science and Technology, No. 6," 1990, 644 pp. Also **SECOND BIBLIOGRAPHIC GUIDE...** "Bibliographies and Indexes in Science and Technology, No. 9," 1996, 416 pp. Massive annotated guide to thousands of sources, subject divided, and well-indexed. See also next entry.

1-005 *_____. **A BIBLIOGRAPHIC GUIDE TO THE HISTORY OF COMPUTER APPLICATIONS, 1950–1990.** Westport, CT: Greenwood "Bibliographies and Indexes in Science and Technology, No. 10," 1996, 279 pp. Divided into sections (to and after 1965), this includes nearly 1,650 annotated references to cryptology, marketing, and utilities, among other related fields. (The depth and inclusiveness of Cortada's bibliographic guides made inclusion of computer history in the present book superfluous.)

1-006 Fackelman, Mary P., and Kimberly A. Krekel. **INTERNATIONAL TELECOMMUNICATIONS BIBLIOGRAPHY.** Washington: Government Printing Office (Office of Telecommunications Special Publication 76-7), 1976, 176 pp. One entire sections deals with technology, but relevant references appear throughout. Annotated with author and subject indexes. Useful now primarily as an historical resource, though not designed with that purpose in mind.

1-007 ** Finn, Bernard S. **THE HISTORY OF ELECTRICAL TECHNOLOGY: AN ANNOTATED BIBLIOGRAPHY.** New York: Garland "Bibliographies of the History of Science and Technology No. 18," 1991, 342 pp. Prepared by a noted Smithsonian curator and authority. Part II of about 100 pp. focuses on communications and includes some foreign language citations not found in the present book. Illustrations, index.

General Reference Works

See Gluckman, 8-002.

See Hale, 8-040.

See Harris, 6-247.

See Higgens, 4-194.

1-008 ** Higgins, Thomas J. **A CLASSIFIED BIBLIOGRAPHY OF PUBLICATIONS ON THE HISTORY AND DEVELOPMENT OF ELECTRICAL ENGINEERING AND ELECTROPHYSICS.** Madison: University of Wisconsin Engineering Experiment Station Reprint No. 198, 1952, 26 pp. Reprinted from 1950–52 articles in *THE BULLETIN OF BIBLIOGRAPHY*, this includes some 1,200 books and articles classified under 23 headings. No annotations, but full publishing indicia are provided. Among the topics covered are electron tubes, electrical communication, land telegraphy, submarine telegraphy, wire telephony, radio communication, television (very brief), and electrical corporations and societies.

1-009 Not used.

See Higgins, 5-009.

See IRE Committee, 8-019.

See Jorysz, 10-061.

1-010 * Kittross, John M., comp. **A BIBLIOGRAPHY OF THESES AND DISSERTATIONS IN BROADCASTING: 1920–1973.** Washington: Broadcast Education Association, 1978, about 200 pp. Only guide of its kind—and many of the thousands of studies referenced concern technology in radio, television, and allied services. Indexed by key word and topic.

See Krauter, 9-149 through 9-150c.

See Krull, 11-075.

See MacArthur, 11-032.

See MacCafferty, 13-028.

1-011 Maynard, Katherine, ed. **A BIBLIOGRAPHY OF BIBLIOGRAPHIES IN ELECTRICAL ENGINEERING, 1918–1929.** Providence, RI: Special Libraries Assn., 1931, 156 pp. Some 2,250 entries to articles with bibliographic citations, this is especially useful for French and German material.

1-011a Moore, Charles K., and Kenneth J. Spencer. **ELECTRONICS—A BIBLIOGRAPHICAL GUIDE.** London: McDonald / New York: Macmillan, 1961; and London: McDonald / New York: Plenum, 1965, two vols.:
   VOL. 1: COVERING 1945–59. (1961), 411 pp. More than 3,000 entries.
   VOL. 2: COVERING 1959–64. (1965), 369 pp. 2,880 entries.

Entries in both volumes are arranged by topic with brief annotations and both author and subject indexes. Includes numerous references of historical value.

1-012 * Mottelay, Paul F. **BIBLIOGRAPHICAL HISTORY OF ELECTRICITY AND MAGNETISM.** London: Charles Griffin, 1922 (reprinted by Arno Press "History, Philosophy, and Sociology of Science," 1975; and by Martino Publications, 1997), 673 pp. A comprehensive record with details of publications in all languages from the earliest times up to 1821. "The present work is the definitive edition of my 'Chronological history of magnetism, electricity and the telegraph,' which had tentative publication (1891–1892) serially in... *ENGINEERING*...of London, *THE ELECTRICAL WORLD* of New York..." [and others]. Preface. Entries contain particulars of inventions, discoveries, experiments, demonstrations, and theories, with names and dates. Most are referenced. Given cut-off date, useful primarily for background. Photos, appendices, references, index.

See Nasrallah, 4-007.

See National Bureau of Standards, 8-023.

1-013 Powell, Jill H. **SELECTIVE GUIDE TO LITERATURE ON TELECOMMUNICATIONS.** Washington: American Society for Engineering Education "Engineering Literature Guides, Number 15," 1993, 15 pp. Useful annotated guide to bibliographies, indexes and abstracts, online resources, dictionaries and encyclopedias, handbooks, journals, conference proceedings, directories, and technical standards.

See Ratzlaff and Anderson, 5-220.

1-014 * Rettenmeyer, Francis X. "Radio-Electronic Bibliography," *RADIO* (May 1942–August 1945), 19 parts. A comprehensive listing of some 8,000 citations from world periodicals with some textbooks, with all titles being given in English. Generally one segment per issue: (1) aviation radio; (2) frequency modulation; (3) crystallography; (4) tubes; (5) amplification, detection, oscillators, and coils; (6) filters, sound, and loudspeakers; (7) remote control; (8) antennas and radiation; (9) propagation; (10) measurements; (11) direction finding; (12) television; (13) cathode-ray oscillographs; (14) interference and static; (15) household receivers; (16) transmitters; (17) aircraft radio; (18) velocity modulation; and (19) frequency standards (by the National Bureau of Standards).

1-015 * Ronalds, [Sir] Francis, comp. **CATALOGUE OF BOOKS AND PAPERS RELATING TO ELECTRICITY, MAGNETISM, THE ELECTRIC TELEGRAPH, ETC., INCLUDING THE RONALDS LIBRARY.** London: E. & F. N. Spon, 1880 (reprinted by Martino Publications, 1994), 564 pp. More than 13,000 entries of books, pamphlets, monographs, and other material collected by Ronalds or noted by him from other sources. Arranged alphabetically by author with biographical information on many authors. Great emphasis on foreign language works. (The library was donated to the Society of Telegraph Engineers, predecessor of the present Institution of Electrical Engineers, in London.)

See Saffady, 13-127.

General Reference Works

1-016 * Sexton, Donal J. Jr. **SIGNALS INTELLIGENCE IN WORLD WAR II.** Westport, CT: Greenwood "Bibliographies of Battles and Leaders," 1996, 163 pp. The best guide to the growing literature about ULTRA, MAGIC and other such coded telecommunications systems. More than 800 annotated citations. Index. (Note: readers are directed to Dr Sexton's work as this topic is otherwise not included in the present bibliography.)

1-017 *** Shiers, George, assisted by May Shiers. **BIBLIOGRAPHY OF THE HISTORY OF ELECTRONICS.** Metuchen, NJ: Scarecrow Press, 1972, 323 pp. The original version of the present volume, this invaluable guide was primarily devoted to communications but included some broader material (on radar, computers, and electronics generally) than is found here. Some 1,800 annotated entries. Author and subject indexes.

1-018 *** _____. **EARLY TELEVISION: A BIBLIOGRAPHIC GUIDE TO 1940.** New York: Garland "Reference Library of Social Science 582," 1997, 616 pp. Result of some two decades of work by the author and others, this is the definitive guide to writings about prewar television and closely related technologies such as facsimile—nearly 9,000 items. After 1925 each year is covered in a chapter with the author's survey of the period, tables of patents, etc. Many of the entries are annotated and some are excerpted. Includes some closely related material on telegraphy, wireless, facsimile. Tables, index.

1-019 Solomon, Richard Jay. "What Happened after Bell Spilled the Acid? Telecommunications History: A View through the Literature," *TELECOMMUNICATIONS POLICY,* 3:146–157 (June 1978). Narrative survey of major writings focusing on works still in print in the late 1970s. Notes.

See Spencer, 10-074.

1-020 * Sterling, Christopher H., ed. *COMMUNICATION BOOKNOTES QUARTERLY.* Mahwah, NJ: Lawrence Erlbaum, 1998 (Vol. 29)–date, quarterly. Continuing *COMMUNICATIONS BOOKNOTES* and earlier titles (1969–97), *CBQ* offers concise reviews across a wide variety of telecommunications topics, usually including some historical material in each issue. Good way to stay current.

1-021 * _____, James K. Bracken, and Susan M. Hill, eds. **MASS COMMUNICATIONS RESEARCH RESOURCES: AN ANNOTATED GUIDE.** Mahwah, NJ: Lawrence Erlbaum, 1998, 208 pp. Companion to 1-001, see especially chapters 2 (history) and 3 (technology) for additional organizations and sources. Includes some 1,500 citations including some to Internet resources.

See Sudalnik and Kuhl, 13-129.

1-022 Taggart, Dorothy T. **A GUIDE TO SOURCES IN EDUCATIONAL MEDIA AND TECHNOLOGY.** Metuchen, NJ: Scarecrow Press, 1975, 156 pp. More than 300 items are annotated. Index.

See Unger, 14-055.

1-023 ** Weaver, William D. **CATALOGUE OF THE WHEELER GIFT OF BOOKS, PAMPHLETS AND PERIODICALS IN THE LIBRARY OF THE AMERICAN INSTITUTE OF ELECTRICAL ENGINEERS**. New York: AIEE, 1909, 2 vols., 504 and 475 pp. (reprinted in one volume by Martino Publications, 1998). This extensive library built up over 40 years by English electrical engineer J. Latimer Clark (1822–98), was presented to the AIEE by Dr. S.S. Wheeler, and lists some 7,000 items to the end of the 19th century. First volume includes nearly 2,500 entries from 1473 to 1906. Second volume includes topical sections on telegraph; electric light, telephone, and manufacturing companies; patents and litigation; legislation, electrical expositions, and periodicals. Especially valuable in tracing early telegraph, submarine telegraph, and telephone publications. Author index.

See Wilson, 10-077, and 10-078.

B. Dictionaries and Encyclopedias
(Few of these were designed to be historical but they now have historical value for their view of their industries at the time. Most range over a variety of technologies and services despite sometimes specific titles. Items with largely current value are not included.)

1-024 Camm, Frederick J. **THE WIRELESS CONSTRUCTOR'S ENCYCLOPEDIA: A COMPLETE GUIDE, IN ALPHABETICAL ORDER, TO THE CONSTRUCTION, OPERATION, REPAIR, AND OVERHAUL OF ALL TYPES OF WIRELESS RECEIVERS AND COMPONENTS, AND CONTAINING A SPECIAL SECTION ON MODERN TELEVISION**. London: Newnes, 1932, 1935 [4th ed.], 392 pp. Construction, operation, repair, and overhaul of all receivers and radio equipment of the time. Includes 20 pages on scanning television. Diagrams.

See Cope, 4-001.

1-025 **THE CYCLOPEDIA OF ELECTRICAL ENGINEERING**. Philadelphia: Gebbie, 1891. Includes a "history of the discovery and application of electricity, with its practice and achievement from the earliest period to the present time."

1-026 * Froehlich, Fritz, and Allen Kent, eds. **THE FROEHLICH/KENT ENCYCLOPEDIA OF TELECOMMUNICATIONS**. New York: Marcel Dekker, 1991–99. 18 vols. Though largely technical in nature, this includes many historical references and entries including key figures, companies, events.

1-027 Gardner, Robert, and Dennis Shortelle. **FROM TALKING DRUMS TO THE INTERNET: AN ENCYCLOPEDIA OF COMMUNICATIONS TECHNOLOGY**. Santa Barbara, CA: ABC-Clio, 1997, 355 pp. Aimed at a general or school audience, this takes an historical approach. Photos, diagrams, references, bibliography, index

1-028 Gernsback, Sidney. **S. GERNSBACK'S RADIO ENCYCLOPEDIA**. New York: Experimenter Publishing, 1927 (reprinted by Vintage Radio, 1974), 175 pp.; 1931 [2nd ed.], 352 pp. Includes historical and even sociopolitical references, but primarily devoted to state-of-the-art radio technology as of that time. Diagrams.

General Reference Works

1-029 Houston, Edwin J. **A DICTIONARY OF ELECTRICAL WORDS, TERMS, AND PHRASES.** New York: W. J. Johnston, 1889, 640 pp.; 1892 [2nd ed.], 562 pp.; 1894 [3rd ed.], 667 pp.; 1898 [4th ed.], 945 pp. and reissued with appendix, 990 pp. A standard late-19th-century guide. Engravings, diagrams.

1-030 ** Kempner, Stanley. **TELEVISION ENCYCLOPEDIA.** New York: Fairchild, 1948, 415 pp. Historically oriented volume (including closely related technologies) in four parts: a chronology of 40 pp.; pioneer and then-contemporary people in television (90 pp.); television's technical vocabulary (260 pp.); and a brief discussion of the urban market for television. Photos, diagrams, bibliography of 250 entries.

1-031 * Lewis, E.J.G. **TELEVISION TECHNICAL TERMS AND DEFINITIONS.** London: Pitman, 1936, 95 pp. May have been the first such volume. Includes both mechanical and electronic methods. Includes some historical material. Diagrams. See also Witts 1-036.

1-032 ** Lodge, Sir Oliver, ed. **HARMSWORTH'S WIRELESS ENCYCLOPEDIA: WRITTEN FOR AMATEURS AND EXPERIENCED OPERATORS BY EXPERTS IN ALL BRANCHES OF WIRELESS SCIENCE.** London: Harmsworth, 1923, 3 vols., 2,272 pp. Originally issued in fortnightly parts, this is perhaps the most comprehensive guide to what was known at the time. Some 5,000 diagrams, photos, some fold-outs.

1-033 Manly, Harold P. **DRAKE'S RADIO CYCLOPEDIA.** Chicago: Frederick J. Drake, 1927, 869 pp.; 1931 [4th ed.], 1,042 pp.; many later editions. Thousands of entries, some quite lengthy. Covers radio, audio systems, photocells, television. Diagrams. Later retitled DRAKE'S ELECTRICAL AND RADIO DICTIONARY.

1-034 Molloy, Edward, and Ralph Stranger, eds. **COMPLETE WIRELESS: A PRACTICAL AND AUTHORITATIVE WORK FOR EVERYONE INTERESTED IN THE WIRELESS INDUSTRY.** London: George Newnes, nd but 1932, 4 vols., 1,544 pp. Originally issued in 32 weekly parts. Diagrams, photos.

1-035 Sloan, T. O'Connor. **THE STANDARD ELECTRICAL DICTIONARY.** New York: Munn, 1892, 624 pp.; 1897 [2nd ed.], 682 pp. Engravings. Useful for its very early viewpoint.

1-036 Witts, Alfred T. **TELEVISION CYCLOPAEDIA.** New York: D. Van Nostrand, 1937, 151 pp. Includes both mechanical and electronic systems as then understood. Diagrams. See also Lewis, 1-031.

C. Directories of Archives, Libraries, and Museums

(Again, this section is focused only on those topics included in this bibliography. There are many more general guides available in any good library.)

1-037 *** Bedi, Joyce E., Ronald R. Kline, and Craig Semsel. **SOURCES IN ELECTRICAL HISTORY.** New York: Center for the History of Electrical Engineering, 1989–95, 3 vols. as follows:

VOL. 1: ARCHIVES AND MANUSCRIPT COLLECTIONS IN U.S. REPOSITORIES. 1989, 234 pp. Intended to update Hounshell's 1973 guide (1-038). Describes 1,008 collections in 158 repositories, mainly university archives and state historical societies. Entries arranged by collection titles, with notes on size, contents, finding aids, and data for repositories and access. Repository and subject indexes. Identifies some 26 collections including materials for Morse; 9 for the Western Union Russian telegraph extension. An important resource for historical research, available free on-line; http://www.ieee.org/history_center/research_guides_menu.html

VOL. 2: ORAL HISTORY COLLECTIONS IN U.S. REPOSITORIES. 1992. Summarizes the contents of some 1,000 taped interviews.

VOL. 3: CORPORATE RECORDS AND ARCHIVES. 1995. Includes 130 companies.

1-038 Hounshell, David A. **MANUSCRIPTS IN U.S. DEPOSITORIES RELATING TO THE HISTORY OF ELECTRICAL SCIENCE AND TECHNOLOGY.** Washington: Smithsonian Institution Press, 1973, 116 pp. Detailed descriptions of about 250 collections, arranged under nearly 100 academic, professional, and state historical archives and libraries. Largely superceded by Bedi, et al (1-037). Name, title, and subject index.

1-039 Klaue, Wolfgang. **WORLD DIRECTORY OF MOVING IMAGE AND SOUND ARCHIVES.** Munich: K.G. Saur "Film, Television, Sound Archive Series, Vol. 5," 1993, 192 pp. Covers 577 centers in 100 countries. Concerned primarily with the *products* of recording rather than the technology that produced the products. Index.

1-040 Mehr, Linda Harris, comp. **MOTION PICTURES, TELEVISION, AND RADIO: A UNION CATALOGUE OF MANUSCRIPT AND SPECIAL COLLECTIONS IN THE WESTERN UNITED STATES.** Boston: G.K. Hall, 1977, 201 pp. Dated but still useful information on special collections, including television and radio, in academic and public libraries, historical societies, and other research centers. Index of occupations and general index.

1-041 * Yeudall, Bert, ed. **DIRECTORY OF TELEPHONE MUSEUMS.** Edmonton, Canada: Telephone Historical Centre, 1994 [7th ed.], 22 pp. Briefly annotated listing of museums worldwide including contact people and telephone numbers.

D. Chronologies
(Many other citations in this bibliography include chronologies which are usually noted in their descriptions.)

1-042 "Broadcasting at 50: Can It Adapt?" *BROADCASTING* (November 2, 1970), 18:17:74–154. Focuses primarily on business and programming aspects, but includes some technology, especially for earlier years, and with television.

1-043 Bunch, Bryan, and Alexander Hellemans, eds. **THE TIMETABLES OF TECHNOLOGY: A CHRONOLOGY OF THE MOST IMPORTANT PEO-**

PLE AND EVENTS IN THE HISTORY OF TECHNOLOGY. New York: Simon & Schuster, 1993, 490 pp. Essentially a well-annotated chronological table ranging from the stone age to modern times, divided into general periods (the electric age begins with 1879 on p. 278), always including a column on communication which in more recent years expands to columns on communications and another on electronics and computers. Many boxed features on key individuals and inventions. Index. See also Harpur, 1-053.

1-044 A CHRONOLOGICAL HISTORY OF ELECTRICAL DEVELOPMENT FROM 600 B.C. New York: National Electrical Manufacturers Association, 1946, 150 pp. Brief listings through 1944 focusing mainly on electric power. Appendix lists member companies with original name, formation year, name of founder. Index.

1-045 CHRONOLOGY: 25 YEARS OF RADIO. Washington: National Association of Broadcasters, 1945, 11 pp. The highlights, mainly business and programming.

1-046 "The Chronology of Television," *TELEVISION AND SHORT-WAVE WORLD* (July 1935), pp. 391–393; (August 1935), pp. 453–454. Especially useful for its detail of (largely mechanical) system developments in the late 1920s and early 1930s in Britain.

1-047 *** Davis, Henry B.O. **ELECTRICAL AND ELECTRONIC TECHNOLOGIES: A CHRONOLOGY OF EVENTS AND INVENTORS.** Metuchen, NJ: Scarecrow Press, 1981–85, 3 vols. as follows. Each has bibliography and index.

> **TO 1900** (1981, 221 pp.) Divided into chapters through the 15th century, the 16th century, the 17th century, the 18th century, and the 19th century, and by year within each. Entries begin with a highlighted name and vary in length from a line to a substantial paragraph.
>
> **FROM 1900 TO 1940** (1983, 220 pp.) Divided into chapters by decade. Entries begin with a highlighted name or event. Separate indexes for vacuum tubes and radio stations.
>
> **FROM 1940 TO 1980** (1985, 321 pp.) Same as previous volume except that entries for each year are now further separated by main parallel subject headings: amateur, broadcasting, committees, components, computers, etc.

1-048 *** Dummer, G[eoffrey] W[illiam] A[rnold]. **ELECTRONIC INVENTIONS AND DISCOVERIES: ELECTRONICS FROM ITS EARLIEST BEGINNINGS TO THE PRESENT DAY.** Oxford, England: Pergamon Press, 1977, 158 pp.; 1983 [3rd ed.], 233 pp.; Bristol, England, and Philadelphia: Institute of Physics Publications, 1997 [4th ed.], 284 pp. First published under a variant title, this valuable guide offers a summary of "first dates in electronic developments over a very wide field." A dozen brief initial chapters offer summary text and "history on a page" charts. There is a list of all included inventions by main topic, and then the main body of the chronology in date order. Each invention is summarized, a source is given plus one or more further references. Invaluable. Diagrams, tables, charts, index.

1-049 * Dunlap, Orrin E. **RADIO AND TELEVISION ALMANAC.** New York: Harper, 1951, 211 pp. Chronology (including month and day when available) of events focused on American broadcasting. Includes some technology, especially in earlier years of each medium. No documentation. Appendices list leadership of major organizations including FCC.

1-050 *** EVENTS IN TELECOMMUNICATIONS HISTORY. Warren, NJ: AT&T Archives, 1992, 225 pp. While emphasizing events within AT&T, this is the most comprehensive chronology of telephone communications. First published in 1958 it has been regularly revised and expanded. Bibliography, index. See also Hanscom, 1-052.

1-051 ** EVENTS IN TELECOMMUNICATIONS HISTORY. London: British Telecom Archives and Historical Information Centre, 1993 [2nd ed.], 85 pp. Very useful narrative format of a paragraph on each important event, from 1700s to early 1990s, focusing, of course, on British developments and BT specifically.

1-052 ** Hanscom, C. Dean. **DATES IN AMERICAN TELEPHONE TECHNOLOGY.** New York: Bell Telephone Laboratories, 1961, 148 pp. Though labeled "preliminary edition," no later version appeared. A classified listing of products, events, and inventions by name and date with source references. A few non-Bell and foreign items are included. Bibliography. See also EVENTS IN TELECOMMUNICATIONS HISTORY, 1-050.

1-053 Harpur, Patrick, ed. **THE TIMETABLE OF TECHNOLOGY.** London: Michael Joseph, 1982, 240 pp. Focuses on the 20th century and includes a column on communications and electronics. Photos, diagrams, bibliography, index. See also Bunch and Hellemans, 1-043.

1-054 Hudson, Robert V. **MASS MEDIA: A CHRONOLOGICAL ENCYCLOPEDIA OF TELEVISION, RADIO, MOTION PICTURES, MAGAZINES, NEWSPAPERS, AND BOOKS IN THE UNITED STATES.** New York: Garland, 1987, 435 pp. This media chronology from 1638 to 1985 can be approached by year, as arranged, or by its detailed subject index. Useful to provide context to development of radio and television.

1-055 * Malek, M.C. "Chronology of Development of Wireless Communications and Electronics," *IETE TECHNICAL REVIEW,* 3:479–522 (1986).

1-056 McNicol, Donald, comp. "A Chronological History of Electrical Communication—Telegraph, Telephone and Radio," *RADIO ENGINEERING,* Vols. 12–14 (Jan. 1932–Aug. 1934), one page per month covering inventions, discoveries, innovations, patents, and events to 1910. The 32 parts include 1,276 numbered entries. Emphasis on American personalities and affairs.

1-057 Morgan, P.F.A. "Highlights in the History of Telecommunications," *TELECOMMUNICATION JOURNAL,* 53:138–149 (March 1986). Illustrated chronology.

1-058 "Telephone Calendar," *TELEPHONY* (July 5, 1976), pp. 226–262. Brief paragraph entries from 1869 to 1976.

1-059 * U.S. Department of Commerce, Bureau of Navigation, Radio Service. **IMPORTANT DATES IN RADIOTELEGRAPHY.** Washington: Government

General Reference Works 11

Printing Office, 1916, 25 pp. Brief chronologies in four sections: peaks in the waves of wireless progress, some recent developments, radio inspection service, and wireless as a safeguard to life at sea. Useful for its early details. Available on-line; see 15-057.

See U.S. FCC, 4-049.

E. Statistics
(Many of these are regularly updated and some of the data is becoming available on-line as well.)

1-060 * COMMUNICATIONS OUTLOOK. Paris: Organization of Economic Cooperation and Development, 1991–date, biennial. Tables, charts, and text compare and contrast various measures of telecommunications growth in the world's developed nations. In addition to other subjects each report covers equipment trade, network modernization, quality of service, mobile communications, convergence between broadcasting and telecommunications, and frequency allocations.

1-061 ** CONSUMER ELECTRONICS INDUSTRY IN REVIEW. Arlington, VA: Electronic Industries Association (Alliance after 1997), 1968–date, annual. Brief booklet of data drawn from EIA's larger industry report (1-061), this offers more text and chronologies on historical trends in various electronic media. Details units sold and factory value for wide variety of consumer electronics over previous five years or so. Title has varied.

1-062 *** ELECTRONIC MARKET DATA BOOK. Arlington, VA: Electronic Industries Association (Alliance after 1997), 1954–date, annual (title has varied; initially called the ELECTRONIC INDUSTRIES YEARBOOK). Charts and graphs trace (usually for the most recent decade) trends in manufacture and sale of equipment in telecommunications and consumer electronics markets. A standard source, with far more detail in recent years.

1-063 ** European Audiovisual Observatory. STATISTICAL YEARBOOK: FILM, TELEVISION, VIDEO AND NEW MEDIA IN EUROPE. Strasbourg, France: EAO, 1994–date, annual. Increasingly valuable compendium of tabular and chart data on all European countries, including information on consumer electronic equipment availability, home video and multimedia equipment and software, television channels available and the like. Arranged by individual country and region and carefully annotated.

1-064 ** International Telecommunication Union. YEARBOOK OF COMMON CARRIER TELECOMMUNICATION STATISTICS. Geneva: ITU, 1971–date, annual. Published in French, English, and Spanish, primarily tables, and related references, for each country of the world for a decade prior to date of publication. Covers telephone, telegram, telex and data transmission services, including point-to-point radio. Includes measures of traffic, equipment installed, and investment. Retitled simply YEARBOOK OF STATISTICS with 1999 [25th ed.] edition. Available on-line (for a price) in recent years.

1-065 * Siemens AG. INTERNATIONAL TELEPHONE STATISTICS. Munich, Germany: Siemens AG, annual. Published in German, English, French, and Spanish with charts, tables, maps. Added teletext, videotex and mobile cellular services with 1988 edition.

1-066 ** Sterling, Christopher H. **ELECTRONIC MEDIA: A GUIDE TO TRENDS IN BROADCASTING AND NEWER TECHNOLOGIES, 1920–1983**. New York: Praeger Special Studies, 1984, 337 pp. See especially chapter 1 on growth of electronic media covering broadcasting, cable, and newer services. Tables, notes, bibliography.

1-067 * **TELEPHONE STATISTICS**. Washington: United States Telephone Association, annual. Usually issued in 2 vols, this source includes summary data for the local telephone service industry (Vol. 1) and detailed operating and financial statistics for each firm (Vol. 2). Tables, notes, charts.

1-068 UNESCO. **STATISTICS ON RADIO AND TELEVISION**. Paris: UNESCO, 1963–87, 3 vols. This three-part series offers comparative information on transmitters, receivers plus programming and investment in countries and major regions. Tables, charts, notes.

> **VOL. 1: 1950–1960.** "Statistical Reports and Studies," 1963, 87 pp.
>
> **VOL. 2: 1960–1976.** "Statistical Reports and Studies No. 23," 1979, 124 pp.
>
> **VOL. 3: LATEST STATISTICS ON RADIO AND TELEVISION BROADCASTING**. "Statistical Reports and Studies No. 29," 1987, 130 pp.

1-069 *** U.S. Department of Commerce, Bureau of the Census. **TELEPHONES AND TELEGRAPHS**. Washington: Government Printing Office, 1906–1939, as follows.

> **SPECIAL REPORTS: TELEPHONES AND TELEGRAPHS, 1902.** (1906, 172 pp.) A preliminary digest appeared as *Bulletin 17* in early 1905.
>
> **SPECIAL REPORTS: TELEPHONES, 1907.** (1910, 128 pp.); and **TELEGRAPH SYSTEMS, 1907.** (1910)
>
> **TELEPHONES AND TELEGRAPHS AND MUNICIPAL ELECTRIC FIRE-ALARM AND POLICE-PATROL SIGNALING SYSTEMS, 1912.** (1915, 208 pp.)
>
> **CENSUS OF ELECTRICAL INDUSTRIES, 1917: TELEGRAPHS** (1919, 59 pp); and **TELEPHONES.** (1920, 52 pp.)
>
> **CENSUS OF ELECTRICAL INDUSTRIES, 1922: TELEGRAPHS** (1924, 29 pp.); and **TELEPHONES.** (1924)
>
> **CENSUS OF ELECTRICAL INDUSTRIES, 1927: TELEGRAPHS** (1930); and **TELEPHONES.** (1930, 52 pp.)
>
> **CENSUS OF ELECTRICAL INDUSTRIES, 1932: TELEPHONES AND TELEGRAPHS.** (1934, 49 pp.)
>
> **CENSUS OF ELECTRICAL INDUSTRIES, 1937: TELEPHONES AND TELEGRAPHS.** (1939, 63 pp.)

Issued every five years (and reporting information actually gathered from 1902 to 1937), these reports provide invaluable historical indicators of technical and economic development. The first two reports (for data in 1902 and 1907 include considerable descriptive material (and photographs, largely of equipment) as well. Partially resumed and continued by next entry. Photos (reports for 1902 and 1907 only), tables, charts, notes.

1-070 *_____. ANNUAL SURVEY OF COMMUNICATION SERVICES. Washington: Government Printing Office "Current Business Reports," 1992–date, annual. Includes telephone and related services plus broadcasting, and cable/pay services providing charts and tables on financial and operating results. Reports data from 1990 to date. For earlier years see previous entry.

1-071 U.S. Department of Commerce, International Trade Commission. **U.S. INDUSTRIAL & TRADE OUTLOOK.** Washington: Government Printing Office, 1961–94, 1998–date, annual. Offered only on-line 1995–97, this annual series (which added "TRADE" to its title only in 1998) provides trends and projections for 350 manufacturing and service industries. Includes both manufacturing and services aspects of the communications industry. Photos, tables, charts.

1-072 U.S. Department of the Treasury, Bureau of Statistics. **SUBMARINE AND LAND TELEGRAPH SYSTEMS OF THE WORLD.** Washington: Government Printing Office, 1899, 22 pp.; 1902, 58 pp. Diagrams.

1-073 ** U.S. Federal Communications Commission, Common Carrier Bureau. **STATISTICS OF COMMUNICATIONS COMMON CARRIERS.** Washington: Government Printing Office, 1937–date, annual. Usually includes data for the year reported, but beginning in 1990 includes "historical tables" as well, taking trends back to 1950. Covers financial and operating information for telephone and satellite carriers. (Note: the series began with TELEGRAPH AND CABLE COMPANIES REPORTING TO THE INTERSTATE COMMERCE COMMISSION, published at least as early as 1920 and continuing until the formation of the FCC in 1934.) Available on-line at FCC web page (15-013).

1-074 _____. TRENDS IN THE INTERNATIONAL COMMUNICATIONS INDUSTRY, 1975–1989. Washington: FCC, 1990, 43 pp. Numerous measures of change and expansion in text, tables, and charts based on official information. Revised on an irregular basis. Available on-line at FCC web page (15-013).

1-075 _____. TRENDS IN TELEPHONE SERVICES. Washington: FCC, approximately annual. Tables and text define facilities and financial changes in the domestic local exchange and interexchange industry, based on data reported by the carriers. Available on-line at FCC web page (15-013).

1-076 *** THE WORLD'S TELEPHONES. New York (and elsewhere): AT&T (various corporate research offices until 1968; Long Lines 1969–81; AT&T Communications to 1986; International Marketing to 1992), 1912–93, annual (not published for 1915–18 or 1942–45; and issued biennially for last four reports: 1985–86 to 1991–92), total of 81 issues published. Title varied: TELEPHONE STATISTICS OF THE WORLD through 1955. Invaluable record of telephone expansion from before World War I into the 1990s around the world. Especially useful for comparing growth patterns outside of the U.S. Usually reports data as of January 1st of year(s) indicated. Begun on 35th anniversary of invention of the telephone. Foreign information is based on government records and foreign carrier-supplied information. Early issues compared messages by mail, telegraph, and telephone. Includes information on number of telephone sets, miles of line, other equipment measures, services, and investment. Tables, charts, maps, notes.

1-077 U.S. INFORMATION TECHNOLOGY INDUSTRY TRADE ANALYSIS, 1960–1991. Washington: Computer and Business Equipment Manufacturers Association, 1992, 99 pp. Data tracing the dramatic rise in imports of all types of consumer and industry telecommunications equipment over this three-decade period.

F. Atlases

1-078 CABLE & STATION COVERAGE ATLAS. Washington: TV Digest, later Warren Publications, 1968–date, annual. State maps with A and B contour maps showing coverage of VHF and UHF television broadcast stations, cable TV system locations, and related directories. Tables, maps.

1-079 MAP OF COAST STATIONS OPEN TO PUBLIC CORRESPONDENCE. Geneva: International Telecommunication Union, 1965 [9th ed.], 1976 [10th ed.], 144 pp; 1984 [11th ed.], 65 pp.; 1994 [12th ed.]. Actually a small atlas, this "map" includes 17 double-page two-color maps locating coastal stations, plus related tables. Covers Morse radiotelegraphy, direct-printing radiotelegraphy, and radiotelephony. Trilingual (French, English, Spanish).

1-080 STANDARD BROADCAST ALLOCATION MAPS: 540 KC TO 1600 KC. New York: Mutual Broadcasting System, 1954, ca 150 pp. National map for each AM channel showing daytime and nighttime coverage patterns for AM stations in Canada, Cuba, Mexico, and the U.S.

1-081 * U.S. Department of Commerce, Office of Telecommunications. GEOGRAPHICAL AREAS SERVICED BY BELL AND INDEPENDENT TELEPHONE COMPANIES IN THE UNITED STATES. Washington: Government Printing Office (OT Report 73-1), 1973, 117 pp. State-by-state maps with delineation of which firms served which areas. Tables, maps.

1-082 _____. FM BROADCAST COVERAGE OF THE COTERMINOUS UNITED STATES. Washington: Government Printing Office (OT Report 76-93), 1977, 40 pp. Largely a discussion of how the several maps presented were derived. Useful to show the medium's expanding geographical reach.

Chapter 2

Serial Publications

By far the largest portion of historical literature on telecommunications technology has appeared in a host of both general and specialized periodicals over the years. This material varies tremendously in historical value, much of what is now of historical interest having been written as current description at the time.

This chapter begins with (A) guides to finding the right periodicals and (B) some of the more important technical and scientific indexes useful in locating specific material in that wealth of publications, some of which include books and reports as well as periodical citations. All periodicals cited in this book then appear either in (C) general and science periodicals, or (D) electrical and telecommunications periodicals. (Listings in Parts C and D do not have entry numbers as they support entries in other chapters.)

A. Guides to Periodicals

(These are helpful in identifying useful periodicals along with information on their history and title changes.)

2-001 * Ardis, Susan B., edited by Jean A. Poland. **A GUIDE TO THE LITERATURE OF ELECTRICAL AND ELECTRONIC ENGINEERING**. Littleton, CO: Libraries Unlimited "Reference Sources in Science and Technology Series," 1987. 190 pp. Annotated guide divided into sections by type of literature (handbooks, reference works, dictionaries, periodicals, etc.). Though dated, still useful.

2-002 ** Bolton, Henry C. **A CATALOGUE OF SCIENTIFIC AND TECHNICAL PERIODICALS, 1665–1895, TOGETHER WITH CHRONOLOGICAL TABLES AND A LIBRARY CHECK-LIST**. Washington: Smithsonian Institution, 1885; 1897 [2nd ed.] (reprinted by Johnson Reprint, 1965), 1,247 pp.

More than 8,600 entries of serials published in several languages with foreign titles in the original. Includes 95 electricity titles, 85 on inventions, and 29 on telephone and telegraph. Invaluable for early titles.

2-003 Brown, Michael. "Radio Magazines and the Development of Broadcasting: *Radio Broadcast* and *Radio News*, 1922–1930," *JOURNAL OF RADIO STUDIES*, 5:68–81 (Winter 1998). A content analysis of these two monthlies. Tables, charts, references.

2-004 England, Rosemary. UNION LIST OF PERIODICALS ON ELECTRONICS AND RELATED SUBJECTS. London: Aslib Electronics Group, 1961, 62 pp. About 1,250 entries.

2-005 Loughney, Katharine. FILM, TELEVISION, AND VIDEO PERIODICALS: A COMPREHENSIVE ANNOTATED LIST. New York: Garland, 1991, 431 pp. A quite useful alphabetical listing which includes many technical journals and provides some background on each of them.

2-006 Lyons, Floyd. "Publication Date: Early U.S. Radio Magazines," *THE OLD TIMER'S BULLETIN*, 16:2:24–25; 16:3:33 (1975). Useful list tracing title changes and first and last issue dates.

2-007 Scudder, Samuel H. CATALOGUE OF SCIENTIFIC SERIALS OF ALL COUNTRIES, INCLUDING THE TRANSACTIONS OF LEARNED SOCIETIES IN THE NATURAL, PHYSICAL, AND MATHEMATICAL SCIENCES, 1633–1876. Cambridge, MA: Library of Harvard University, 1879 (reprinted by Kraus, 1965), 358 pp. To a large extent replaced by Bolton, 2-002.

2-008 Slide, Anthony, ed. INTERNATIONAL FILM, RADIO, AND TELEVISION JOURNALS. Westport, CT: Greenwood, 1985, 428 pp. Useful extensively annotated guide to 150 periodicals, some of them technical in nature. Includes listing by subject area and country of publication.

2-009 ** Sova, Harry W., and Patricia L. Sova. COMMUNICATION SERIALS, 1992–93 EDITION: "AN INTERNATIONAL GUIDE TO PERIODICALS IN COMMUNICATION, POPULAR CULTURE, AND THE PERFORMING ARTS." Virginia Beach, VA: Sovacom, 1992, 1,040 pp. Invaluable guide, unfortunately never revised, to thousands of periodicals, many of them technical journals relevant here. Especially useful for tracing title changes.

2-010 UNION LIST OF PERIODICALS: SCIENCE–TECHNOLOGY–ECONOMICS. Palo Alto, CA: Special Libraries Assn., San Francisco Bay Region Chapter, 1966, 235 pp.; Stanford, 1971 [2nd ed.], 447 pp. Library holdings in the Bay Area.

2-011 ** WORLD LIST OF SCIENTIFIC PERIODICALS PUBLISHED IN THE YEARS 1900–19xx. London: Oxford University Press, 1925–27, 2 vols. (covers to 1921); 1934 [2nd ed.], 779 pp. (covers to 1933); Butterworths Scientific Publications/New York: Academic Press, 1952 [3rd ed.], 1,058 pp. (covers to 1950); Washington: Butterworths, 1963–65 [4th ed.], 3 vols., (extends list to 1960). Final version includes more than 50,000 titles, and shows which libraries hold

Serial Publications

which titles. Extended by SUPPLEMENT: NEW PERIODICAL TITLES 1960–1968 (Butterworths, 1970), and by the *WORLD LIST OF SCIENTIFIC PERIODICALS* (quarterly with annual cumulations), 1964–1971.

B. Indexes to Periodical Literature

(There are a host of such guides with new on-line resources developing at all times. Included here are only those most focused on the subject matter of this bibliography.)

2-012 *APPLIED SCIENCE AND TECHNOLOGY INDEX* (Monthly). New York: H.W. Wilson, 1913–date (available on-line from 1983). Was titled INDUSTRIAL ARTS INDEX to 1958. Subject index to selected list of engineering and trade periodicals.

See Cortada, 1-004, 1-005.

2-013 CUMULATIVE INDEX TO ENTIRE IEEE GROUP, 1951–1971. Tokyo: Nichigai Associates, 1973, two vols. One volume for author, one for subjects to all publications of the IEEE and its predecessors. For continuation, see INDEX TO IEEE PUBLICATIONS, 2-020.

2-014 ** CURRENT BIBLIOGRAPHY IN THE HISTORY OF TECHNOLOGY. Supplement to *TECHNOLOGY AND CULTURE*, 1964–date, annual. Became separate issue as of 1990. Classified semi-annotated subject listing (including both English and other languages) arranged by specific periods (see "Communication and Records" under "General and 20th Century" and "19th Century" sections) with author index. Available on RLIN on-line service since 1982.

2-015 *** CURRENT BIBLIOGRAPHY OF THE HISTORY OF SCIENCE AND ITS CULTURAL INFLUENCES, the January issue of *ISIS*, 1975–date, annual. Covers some 600 journals in a variety of languages. Check under "Technology" headings under both 19th- and 20th-century parts of the listing. Available on RLIN on-line service since 1982. For indexed cumulations, see:

Whitrow, Magda. **ISIS CUMULATIVE BIBLIOGRAPHY: A BIBLIOGRAPHY OF THE HISTORY OF SCIENCE FORMED FROM ISIS CRITICAL BIBLIOGRAPHIES 1–90, 1913–1965.** London: Mansell, 1971–84, 6 vols. issued in 7 parts. Vols. 1–2 indexed by personalities; 3 covers subjects; 4–5 civilizations and periods; 6 forms an overall index.

Neu, John, ed. **ISIS CUMULATIVE BIBLIOGRAPHY, 1966–1975...**London: Mansell, 1980–85, 2 vols. First supplement.

Neu, John, ed. **ISIS CUMULATIVE BIBLIOGRAPHY, 1976–1985...**Boston: G.K. Hall, 1989, 2 vols. Second supplement.

Neu, John, ed. **ISIS CUMULATIVE BIBLIOGRAPHY, 1986–1995...** Canton, MA: Science History Publications, 1997, 4 vols. Third Supplement. Vols. 1–2 cover personalities and institutions, vol. 3 covers subjects, and vol. 4 is civilizations and periods.

2-016 *CURRENT CONTENTS: ENGINEERING, TECHNOLOGY, & APPLIED SCIENCES* (Weekly). Philadelphia: Institute for Scientific Information, 1970–date. Also available on CD-ROM and on-line. Reprints tables of contents of a host of technical and academic journals.

2-017 ** *ELECTRICAL AND ELECTRONICS ABSTRACTS* (Monthly). London: Institution of Electrical Engineers, 1898–date, with cumulations, and annual author and subject index. Available on-line. Formerly ELECTRICAL ENGINEERING ABSTRACTS (1941–66). Over a third of the entries for recent years are concerned with electronics, telecommunications, and related topics. Available on-line as INSPEC since 1969.

2-018 *ELECTRONICS AND COMMUNICATIONS ABSTRACTS JOURNAL* (Bimonthly). Riverdale, MD: Cambridge Scientific Abstracts, 1967–date with annual cumulation, and on-line versions. Was titled *ELECTRONIC ABSTRACTS JOURNAL*. Covers some 5,000 items a year.

2-019 *** *ENGINEERING INDEX* (Monthly). New York: Engineering Information Inc., 1884–date with annual multi-volume cumulations, and on-line as COMPENDEX (1970–date). Most important guide to telecommunications literature, most of it not designed to be historical. Abstracts and indexes some 4,500 journals of a scholarly, technical, and professional nature. As one indication of the growing literature, the annual cumulation grew to entail five volumes by 1977 and ten by 1995.

2-020 * INDEX TO IEEE PUBLICATIONS. New York: IEEE, 1973–date, annual; 2 vols. since 1987 (one each by author and subject); became 3 vols. with 1994 edition (2 subject volumes). Look under "History" and specific technologies. (Was titled INDEX TO IEEE PERIODICALS, 1971–72.)

See Moore and Spencer, 1-011a.

See Rettenmeyer, 1-014.

See Shiers, 1-018.

2-021 *SCIENCE CITATION INDEX* (Bimonthly). Philadelphia: Institute for Scientific Information, 1961–date with annual cumulations. Author, cited reference and keyword indexes for some 4,500 journals.

C. Periodicals—General

(Includes nearly all nontelecommunications serial publications mentioned in the body of this bibliography—save for a few general circulation titles such as Fortune. *These include general science, social science, history, and related titles. They are arranged alphabetically with prior and subsequent titles, if known, as well as indexes where known. Place of publication and publisher is listed as of the present time if the periodical is still published. Older titles are cross-referenced. Journals still publishing are preceded by a #.)*

AMERICAN JOURNAL OF PHYSICS (Monthly). College Park, MD: American Institute of Physics, 1933–date.

Serial Publications 19

AMERICAN JOURNAL OF SCIENCE (Quarterly, later monthly). New Haven, CT: 1818–?. Formerly *AMERICAN JOURNAL OF SCIENCE AND ARTS (SILLIMAN'S JOURNAL...)*. Appeared in several series. Became the *PROCEEDINGS OF THE BOSTON SOCIETY OF NATURAL HISTORY.*

\# *AMERICAN JOURNAL OF SOCIOLOGY* (Bimonthly). University of Chicago Press, 1985–date.

\# *ANNALS OF THE AMERICAN ACADEMY OF POLITICAL AND SOCIAL SCIENCE* (Quarterly). Philadelphia: the Academy (later, Thousand Oaks, CA: Sage), 1890–date.

\# *ANNALS OF SCIENCE* (Bimonthly). London: Taylor & Francis, 1936–date. "The history of science and technology since the Thirteenth Century." Cumulative Index to Vols. 26–43 (1970–86), 1988, 78 pp.

\# *BRITISH JOURNAL FOR THE HISTORY OF SCIENCE* (Quarterly). Cambridge University Press and British Society for the History of Science, 1962–date.

BRITISH JOURNAL OF APPLIED PHYSICS (Monthly, later biweekly). London: Institute of Physics, 1950–68. Superceded by *JOURNAL OF PHYSICS*, Section D.

\# *BUSINESS AND ECONOMIC HISTORY* (Semiannual). Williamsburg, VA: College of William and Mary, Department of Economics, 1975–date.

\# *BUSINESS HISTORY REVIEW* (Quarterly). Boston: Harvard Business School Press, 1926–date.

\# *CALIFORNIA MANAGEMENT REVIEW* (Quarterly). Berkeley: University of California, Haas School of Business, 1958–date.

CASSIER'S MAGAZINE (Monthly). New York: Cassiers, 1891–1913.

\# *COMPARATIVE STUDIES IN SOCIETY AND HISTORY* (Quarterly). Cambridge, England: Cambridge University Press, 1959–date.

DISCOVERY (Monthly). London, 1920–66. Absorbed by *SCIENCE JOURNAL* (1965–71), which later merged into *NEW SCIENTIST.*

\# *ENDEAVOUR* (Quarterly). London: Imperial Chemical Industries (later Pergamon), 1942–date. "A Review of the Progress of Science and Technology in the Service of Mankind."

\# *THE ENGINEER* (Weekly). London: Morgan Grampian, 1856–date.

\# *ENGINEERING* (Weekly; monthly since 1995). London and other cities: various publishers, 1866–date.

ENGINEERING MANAGEMENT INTERNATIONAL. See *JOURNAL OF ENGINEERING AND TECHNICAL MANAGEMENT.*

ENGINEERING SCIENCE AND EDUCATION JOURNAL (Bimonthly). Stevenge, England: Institute of Electrical Engineers, 1992–date.

ESSAYS IN ECONOMIC AND BUSINESS HISTORY (Annual). Columbus: Ohio State University Department of History, 1978–date.

FORTNIGHTLY REVIEW (Biweekly). London: 1865–1934. Became *THE FORTNIGHTLY* (1934–54). Merged into *CONTEMPORARY REVIEW*.

HISTORIA SCIENTIARUM (three per year). Tokyo: History of Science Society of Japan, 1962–date. Formerly *JAPANESE STUDIES IN THE HISTORY OF SCIENCE* (to 1980).

HISTORICAL JOURNAL OF FILM, RADIO, AND TELEVISION (three per year). Abingdon, England: Carfax Publishing for the International Association for Media and History, 1981–date.

HISTORY & TECHNOLOGY: AN INTERNATIONAL JOURNAL (Quarterly). Basle, Switzerland: Harwood Academic Publishers, 1983–date.

HISTORY OF TECHNOLOGY (Annual). London: Mansell Publishing, 1976–date.

INDUSTRIAL AND CORPORATE CHANGE (Quarterly). Oxford, England: Oxford University Press, 1991–date.

INTERNATIONAL HISTORY REVIEW (Quarterly). Burnaby, BC: Simon Fraser University, 1979–date.

ISIS (Quarterly). Chicago: University of Chicago Press and the History of Science Society, 1913–date. "An International Review Devoted to the History of Science and Its Cultural Influences."

JAPANESE STUDIES IN THE HISTORY OF SCIENCE. See *HISTORIA SCIENTIARUM*.

JOURNAL OF APPLIED PHYSICS (Semimonthly). College Park, MD: American Institute of Physics, 1931–date. Formerly *PHYSICS*.

JOURNAL OF THE ACOUSTICAL SOCIETY OF AMERICA (Monthly). New York: American Institute of Physics for the Acoustical Society, 1929–date.

JOURNAL OF ECONOMIC HISTORY (Quarterly). Cambridge, England: Cambridge University Press and the Economic History Association, 1941–date.

JOURNAL OF ENGINEERING AND TECHNOLOGY MANAGEMENT (Quarterly). Amsterdam: Elsevier Science, 1981–date. Was *ENGINEERING MANAGEMENT INTERNATIONAL* to 1989.

JOURNAL OF ENGINEERING EDUCATION (Quarterly). Washington: American Society for Engineering Education, 1910–date.

JOURNAL OF HISTORICAL GEOGRAPHY (Quarterly). London: Academic Press, 1975–date.

JOURNAL OF PHYSICS (Semimonthly). Bristol, England: English Institute of Physics. See *PROCEEDINGS OF THE PHYSICS SOCIETY.*

JOURNAL OF SOCIAL HISTORY (Quarterly). Pittsburgh, PA: Carnegie-Mellon University Press, 1967–date.

JOURNAL OF THE FRANKLIN INSTITUTE (Monthly). Philadelphia: Franklin Institute (later Elmsford, NY: Pergamon Press), 1826–1988. Began as *THE FRANKLIN JOURNAL AND AMERICAN MECHANIC'S MAGAZINE* (to 1827). Issued in several different series to 1851.

JOURNAL OF THE RÖNTGEN SOCIETY (Quarterly). London: 1904–23. Merged into *ARCHIVES OF CLINICAL SKIAGRAPHY.*

JOURNAL OF THE ROYAL SOCIETY OF ARTS (Quarterly). London: Royal Society of Arts, Manufactures and Commerce, 1852–date.

JOURNAL OF URBAN HISTORY (Quarterly to 1977, then bimonthly). Thousand Oaks, CA: Sage, 1974–date.

MECHANIC'S MAGAZINE (Weekly). London: 1823–73. Numerous subtitle changes. Became *IRON.*

MILITARY AUTOMATION.

NATIONAL ACADEMY OF SCIENCE BIOGRAPHICAL MEMOIRS (Irregular). Washington: National Academy Press, 1953–date.

NATURE (Weekly). London: Macmillan Magazines, 1869–date.

NEW SCIENTIST (Weekly). London: IPC Magazines, 1956–date.

NEWCOMEN—see *TRANSACTIONS OF THE....*

NOTES AND RESEARCHES OF THE ROYAL SOCIETY. London.

PHILOSOPHICAL MAGAZINE (Monthly). London: Taylor & Francis, 1798–date. Various titles and numerous series, including *LONDON, EDINBURGH AND DUBLIN PHILOSOPHICAL MAGAZINE AND JOURNAL OF SCIENCE* (1840–1948). Series "B" has included electronics since 1978.

PHILOSOPHICAL TRANSACTIONS (Monthly). London: The Royal Society, 1665–date. Several subseries; important here is Series A: Mathematical and Physical Papers (1887–1990); Physical Sciences and Engineering (1990–95); Mathematical, Physical, and Engineering Sciences (1996–date).

PHYSICAL REVIEW (Monthly or bimonthly). College Park, MD: American Physical Society, 1913–date. There are now multiple different subparts, designated

by letters, of this publication. For details and tables of contents, see *http://publish.aps.org/*

PHYSICS. See *JOURNAL OF APPLIED PHYSICS*.

PHYSICS TODAY (Monthly). College Park, MD: American Institute of Physics, 1948–date.

PHYSIS: REVISTA INTERNAZIONALE DI STORIA DELLA SCIENZA (three per year). Florence, Italy: Casa Editrice Leo S. Olshki, 1959–date (multilingual).

PROCEEDINGS OF THE AMERICAN PHILOSOPHICAL SOCIETY (Quarterly). Philadelphia: American Philosophical Society, 1838–date.

PROCEEDINGS, PHYSICAL SOCIETY. London: Institute of Physics, 1874–1968. Issued in two parts from 1949. Became *JOURNAL OF PHYSICS* in 1968.

PROCEEDINGS OF THE ROYAL INSTITUTION (Biennial 1931–69; annual from 1970). London: Royal Institution of Great Britain, 1931–date. Formerly *NOTICES OF THE PROCEEDINGS AT THE MEETINGS OF MEMBERS OF THE ROYAL SOCIETY* (1851–1928, triennial).

PROCEEDINGS OF THE ROYAL SOCIETY (Various; monthly by 1989). London: Taylor & Francis, later Harrison & Son, 1854–1995. Split into Series A, Mathematical and Physical Sciences, and Series B, Biological Sciences in 1905. Numerous published indexes from 1913 on. Title has varied.

REPORTS ON PROGRESS IN PHYSICS (Monthly). Bristol, England: Institute of Physics, 1934–date.

RESEARCH IN ECONOMIC HISTORY (Irregular). Greenwich, CT: JAI Press, 1976–date (Vol. 16 appeared in 1996.)

RÉSEAUX: THE FRENCH JOURNAL OF COMMUNICATION (Semiannual). Luton, England: John Libbey Media/University of Luton, 1993–date.

REVIEWS OF MODERN PHYSICS (Quarterly). College Park, MD: American Physical Society, 1929–date.

SCIENCE AND INVENTION (Monthly). New York: Gernsback, 1920–31.

SCIENCE MONTHLY (Monthly). London, 1883–?

SCIENTIFIC AMERICAN (Weekly, monthly as of November 1921). New York, Scientific American Inc., 1845–date.

SCIENTIFIC AMERICAN SUPPLEMENT (Weekly). New York: Scientific American, 1876–1919. Retitled as *SCIENTIFIC AMERICAN MONTHLY*, 1920–21 (monthly). Absorbed by *SCIENTIFIC AMERICAN*.

THE SCIENTIFIC MONTHLY (Monthly). Washington: American Asociation for the Advancement of Science, 1915–57. Merged with *SCIENCE.*

SOCIAL SCIENCE HISTORY (Quarterly). Durham, NC: Duke University Press, 1976–date.

TECHNOLOGY AND CULTURE (Quarterly). Chicago: University of Chicago Press for the Society for the History of Technology, 1960–date.

TECHNOLOGY IN SOCIETY (Quarterly). Oxford: Pergamon, 1979–date.

TECHNOLOGY REVIEW (Eight per year). Boston: MIT, 1899–date. "MIT's national magazine of technology and policy."

TRAINING IN BUSINESS AND INDUSTRY (Monthly). New York: Gellert Publications (later Minneapolis: Lakewood Publications), 1963–74 when it became *TRAINING.*

TRANSACTIONS OF THE NEWCOMEN SOCIETY (Annual). London: Newcomen Society for the Study of the History of Engineering and Technology, 1922–date.

VICTORIAN STUDIES (Quarterly). Bloomington: Indiana University Press, 1957–date.

WORLD POWER (Monthly). London, 1924–37.

D. Periodicals—Electrical and Telecommunications

(Includes all telecommunications serial publications mentioned in the body of this bibliography—plus a few additional titles likely to be of value in historical research. They are arranged alphabetically with prior and subsequent titles, if known, as well as indexes where known. Place of publication and publisher is listed as of the present time if the periodical is still published. Older titles are cross-referenced. Journals still publishing are preceded by a #.)

AES: JOURNAL OF THE AUDIO ENGINEERING SOCIETY. See *JOURNAL OF THE AUDIO ENGINEERING SOCIETY.*

ALL-WAVE RADIO (Monthly). New York: Manson Publications, 1935–38. "Official organ of the Radio Signal Survey League." Covered Ultra-shortwave, shortwave, broadcast, and long-wave transmissions. Merged into *RADIO NEWS.*

AT&T TECHNOLOGY (Bimonthly). New York: AT&T, 1986–date. "Products, Systems, Services." See *BELL LABORATORIES RECORD.*

THE A.W.A. REVIEW (annual), see 9-092.

AUDIO (Monthly). New York and other cities, 1917–date. "The World of Sound." Formerly *PACIFIC RADIO NEWS* (1917–21); *RADIO* (1921–47); and *AUDIO ENGINEERING* (1947–54).

AUDIO ENGINEERING. See *AUDIO.*

BELL LABORATORIES RECORD (Monthly to 1980; then 10 per year and finally [1986] bimonthly). New York: AT&T, 1925–86. Continued as *AT&T TECHNOLOGY.*

BELL LABS TECHNICAL JOURNAL. See *BELL SYSTEM TECHNICAL JOURNAL.*

BELL SYSTEM TECHNICAL JOURNAL (Bimonthly). New York: AT&T, 1922–83. CUMULATIVE INDEX 1922–83 (1983, 399 pp.) is an integrated author, topic, title index. Continued as *AT&T TECHNICAL JOURNAL* (1984–96), and as a quarterly, *BELL LABS TECHNICAL JOURNAL* (Murray Hill, NJ: Lucent Technologies, 1996–date).

BELL TELEPHONE MAGAZINE (Monthly). New York: AT&T, 1922–83. Original title was *BELL TELEPHONE QUARTERLY* (1922–41).

BELL TELEPHONE QUARTERLY. See *BELL TELEPHONE MAGAZINE.*

BKSTS JOURNAL. See *IMAGE TECHNOLOGY.*

BOSCH TECHNISCHE BERICHTE (Bimonthly). Darmstadt, Germany: Robert Bosch GmbH, 1974–date.

BRITISH KINEMATOGRAPH... See *IMAGE TECHNOLOGY.*

BRITISH TELECOMMUNICATION ENGINEERING. See *POST OFFICE ELECTRICAL ENGINEERS JOURNAL.*

BROADCAST ENGINEERING (Monthly). Overland Park, KS: Intertec, 1959–date.

BROADCAST NEWS (Quarterly, then bimonthly). New York, later Camden, NJ: RCA, 1931–85. In 1968 title modified to *RCA BROADCAST NEWS.*

BROADCASTING (Biweekly to 1941; weekly thereafter). Washington (later New York): 1931–date. Called *BROADCASTING & TELECASTING* in 1950s. Became *BROADCASTING & CABLE* in 1995. See 50th anniversary issue, 1-042.

BT TECHNICAL JOURNAL (Quarterly). London: Thomson Science, 1983–date. Also referred to as *BRITISH TELECOMMUNICATIONS TECHNICAL JOURNAL.*

CHANNELS (Bimonthly, later biweekly). New York: Media Commentary Council, later Act III Publishing, 1981–90. See also the annual *CHANNELS FIELD GUIDE,* 13-041.

COMMUNICATIONS (Monthly). New York, 1937–49. Incorporated *RADIO ENGINEERING* (1924–37); and *COMMUNICATION AND BROADCAST ENGINEERING.*

COMMUNICATION AND ELECTRONICS (Monthly). New York: American Institute of Electrical Engineers, 1952–63.

Serial Publications 25

COMMUNICATION ENGINEERING (Monthly). Great Barrington, MA: Sleeper Publishing, 1948–54. Began as *FM*. See *FM AND TELEVISION*.

COMMUNITY ANTENNA TELEVISION JOURNAL (Monthly). Oklahoma City, OK: Television Publications Inc., years not known. "Official journal of the Community Television Association."

db: THE SOUND ENGINEERING MAGAZINE (Bimonthly). Commack, NY: DB, 1967–date.

EBU REVIEW (Quarterly). Geneva: European Broadcasting Union, 1950–92. Several former titles; continued by *EBU TECHNICAL REVIEW*. For special issue on "A Century of Wireless," see 9-089.

THE ELECTRIC CLUB JOURNAL. Pittsburgh: Electric Club, 1904–5. Became *ELECTRIC JOURNAL* and continued until 1939.

ELECTRIC WORLD (Monthly). New York: McGraw-Hill, 1874–1919. Founded as *THE OPERATOR AND ELECTRICAL WORLD: A JOURNAL FOR TELEGRAPHISTS, TELEPHONISTS, ELECTRICIANS AND ELECTRICAL ENGINEERS* (weekly). Title changed in 1883. Became *RADIO NEWS*.

ELECTRICAL COMMUNICATION (Quarterly). New York (later Harlow, England): ITT, 1922–93. Became *ALCATEL TELECOMMUNICATIONS REVIEW* (Paris).

THE ELECTRICAL ENGINEER (Weekly). New York, 1882–99. For first year called *THE ELECTRICIAN*. Later became *ELECTRICAL WORLD AND ELECTRICAL ENGINEER*.

ELECTRICAL ENGINEERING (Monthly). New York: American Institute of Electrical Engineers, 1884–1963. Formerly *TRANSACTIONS OF THE AIEE* (1884–1905); *PROCEEDINGS OF THE AIEE* (1905–19); and *JOURNAL OF THE AIEE* (1920–30). Became *IEEE SPECTRUM*.

ELECTRICAL EXPERIMENTER (Monthly). New York: Gernsback, 1913–20. Became *SCIENCE AND INVENTION*.

THE ELECTRICAL JOURNAL (Weekly). London: David Adams, later Benn Bros., 1952–62. See *THE ELECTRICIAN*.

ELECTRICAL REVIEW (Weekly, later biweekly). Sutton, England: Reed Business, 1872–date. Formerly *THE TELEGRAPHIC JOURNAL AND ELECTRICAL REVIEW* (1872–91). Assumed present title in 1892.

ELECTRICAL TIMES (Monthly). London: Reed Business, 1891–date. Began as *LIGHTNING* (to 1902).

THE ELECTRICIAN (Weekly, later monthly). London, 1861 (not published 1865–78)–1952. Incorporated into *ELECTRICAL REVIEW*. Became *THE ELECTRICAL JOURNAL* in 1952.

ELECTRONIC AGE (Quarterly). New York: RCA, 1941–71 Formerly *RADIO AGE* (1941–57).

ELECTRONIC AND RADIO ENGINEER. See *EXPERIMENTAL WIRELESS.*

ELECTRONIC ENGINEERING (Monthly). London: Morgan-Gampian Technical Press, 1941–date. Formerly *TELEVISION: THE WORLD'S FIRST TELEVISION JOURNAL* (1928–35, edited by Alfred Dinsdale); *TELEVISION AND SHORT-WAVE WORLD* (1935–39); *ELECTRONICS AND TELEVISION AND SHORT-WAVE WORLD* (1939–41). For 60th anniversary issue, see 3-004.

ELECTRONICS (Monthly; biweekly as of 1977). New York: McGraw-Hill (later San Jose, CA: Penton), 1930–date. For 50th anniversary issue, see 3-005.

ELECTRONICS & COMMUNICATIONS ENGINEERING JOURNAL (Bimonthly). London: Institution of Electrical Engineers, 1989–date. For background, see *RADIO AND ELECTRONIC ENGINEER.*

ELECTRONICS AND TELEVISION AND SHORT-WAVE WORLD. See *ELECTRONIC ENGINEERING.*

ELECTRONICS AND POWER (Monthly). London: IEE, 1964–87. Became part of *IEE PROCEEDINGS.*

ELECTRONICS WORLD (Monthly). London: Reed Business, 1911–date. Began as *WIRELESS WORLD* (1922–83) and became *ELECTRONICS WORLD AND WIRELESS WORLD* (1983–89).

EXPERIMENTAL WIRELESS (Monthly). London: Experimental Wireless, later Iliffe & Sons, 1923–31. Became *WIRELESS ENGINEER AND EXPERIMENTAL WIRELESS* (1931–35); *WIRELESS ENGINEER* (1936–57); *ELECTRONIC AND RADIO ENGINEER* (1957–59); *ELECTRONIC TECHNOLOGY* (1960–62); and *INDUSTRIAL ELECTRONICS* (1962–69). Absorbed by *ELECTRONICS WEEKLY.*

FM AND TELEVISION (Monthly). New York: 1940–48. Considerable variation in title. Simply *FM* (to 1942); then *FM RADIO-ELECTRONIC ENGINEERING & DESIGN* (1942–44); title shown (1944–48). Became *COMMUNICATION ENGINEERING.*

FREQUENCY MODULATION (Monthly). New York: 1946–47. Title varied: *JOURNAL OF FREQUENCY MODULATION* for first four issues, later *FREQUENCY MODULATION BUSINESS.*

GENERAL ELECTRIC REVIEW (Monthly; bimonthly after 1952). Schenectady, NY: General Electric, 1903–58.

HI-FI TAPE RECORDING (Bimonthly; monthly in 1956). New York: A-TR Publications, 1953–71. Many title changes; began as *TAPE AND FILM RECORDING* but quickly became *MAGNETIC FILM AND TAPE RECORDING* (1954–56); then present title (1956–60), and finally simply *TAPE RECORDING.* Absorbed by *TAPE RECORDING MAGAZINE* in 1971.

HIGH FIDELITY (Monthly). New York: ABC Consumer Magazines, 1958–89. Absorbed by *STEREO REVIEW*.

IEEE ANTENNAS AND PROPAGATION MAGAZINE (Bimonthly). New York: Institute of Electrical and Electronics Engineers, 1952–date.

IEEE COMMUNICATIONS MAGAZINE (Monthly). New York: Institute of Electrical and Electronic Engineers, 1953–date. Was *IEEE COMMUNICATIONS SOCIETY MAGAZINE*. For special historical issue, see 3-047.

IEEE JOURNAL OF SOLID STATE CIRCUITS (Bimonthly). New York: Institute of Electrical and Electronics Engineers, 1966–date.

IEEE SPECTRUM (Monthly). New York: Institute of Electrical and Electronics Engineers, 1964–date. For prior history, see under *ELECTRICAL ENGINEERING*.

IEEE TECHNOLOGY AND SOCIETY MAGAZINE (Quarterly). New York: Institute of Electrical and Electronics Engineers, 1982–date.

IEEE TRANSACTIONS ON AEROSPACE AND ELECTRONIC SYSTEMS (Bimonthly). New York: Institute of Electrical and Electronics Engineers, 1965–date. Merged several earlier titles.

IEEE TRANSACTIONS ON COMMUNICATIONS (Bimonthly to 1971, then monthly). New York: Institute of Electrical and Electronics Engineers, 1953–date. Formerly *IRE TRANSACTIONS ON COMMUNICATIONS SYSTEMS* (to 1964); *IEEE TRANSACTIONS ON COMMUNICATIONS AND TECHNOLOGY* (1964–71).

IEEE TRANSACTIONS ON CONSUMER ELECTRONICS (Bimonthly to 1974; quarterly to 1979; then five per year). New York: Institute of Electrical and Electronic Engineers, 1952–date. Was *IRE TRANSACTIONS ON BROADCAST AND TELEVISION RECEIVERS* (to 1962). For special historical issue, see 3-063.

IEEE TRANSACTIONS ON MICROWAVE THEORY & TECHNOLOGY (16 per year). New York: Institute of Electrical and Electronic Engineers, 1953–date.

IEEE TRANSACTIONS ON VEHICULAR TECHNOLOGY (Quarterly). New York: Institute of Electrical and Electronic Engineers, 1952–date.

IETE TECHNICAL REVIEW (Bimonthly). New Delhi: Institute of Electrical and Telecommunications Engineering, 1984–date.

IMAGE TECHNOLOGY [JOURNAL OF THE BRITISH KINEMATOGRAPH, SOUND AND TELEVISION SOCIETY] (Monthly). London: BKSTS, 1919–date. Several title changes: *BRITISH KINEMATOGRAPH SOCIETY PROCEEDINGS* (to 1947); *BRITISH KINEMATOGRAPHY* (to 1965); *BRITISH KINEMATOGRAPHY, SOUND AND TELEVISION* (1966–73); and *BKSTS JOURNAL* (1974–86); present title.

IRE TRANSACTIONS ON COMMUNICATIONS SYSTEMS. See: *IEE TRANSACTIONS ON COMMUNICATIONS*.

IRE TRANSACTIONS ON BROADCAST AND TELEVISION RECEIVERS. See *IEEE TRANSACTIONS ON CONSUMER ELECTRONICS.*

JOURNAL OF THE AIEE. See *ELECTRICAL ENGINEERING.*

JOURNAL OF THE AUDIO ENGINEERING SOCIETY (Quarterly to 1968; then ten per year). New York: the Society, 1953–date. Became *AES: JOURNAL OF THE AUDIO ENGINEERING SOCIETY.*

JOURNAL OF THE BRITISH IRE. See *PROCEEDINGS OF THE IERE.*

*JOURNAL OF THE BRITISH KINEMATOGRAPHY...*See *IMAGE TECHNOLOGY.*

JOURNAL OF BROADCASTING & ELECTRONIC MEDIA (Quarterly). Washington: Broadcast Education Association, 1956–date. Formerly *JOURNAL OF BROADCASTING* (to 1985). AUTHOR & TOPIC INDEX 1956–81 (Vols. 1–25), Spring 1982, 107 pp. (more recent indexes are available on disc; inquire of publisher).

JOURNAL OF FREQUENCY MODULATION. See *FREQUENCY MODULATION.*

JOURNAL OF THE IEE (Monthly). London: Institution of Electrical Engineers, 1889–date. Issued general indexes approximately every ten years. Several title changes; began as periodical of the Society of Telegraph Engineers (see below). Divided in 1941: Part 1 (General); Part 2 (Power Engineering); Part 3 (Radio and Communication Engineering). New Series (1955–date) divided publication into a broader series of topics designated by letters A though J.

JOURNAL OF RADIO STUDIES (four annual issues 1992–97; biennial 1998–date). Washington: Broadcast Education Association, 1998–date.

JOURNAL OF THE SMPTE. See *SMPTE JOURNAL.*

JOURNAL OF THE TELEGRAPH (Semimonthly; monthly in 1882). New York: James D. Reid (first 4 vols.); Western Union Telegraph Co., 1867–1914. Absorbed *TELEGRAPHER* in 1877.

JOURNAL OF THE TELEVISION SOCIETY. See *TELEVISION.*

MAGNETIC FILM AND TAPE RECORDING. See *HI-FI TAPE RECORDING.*

MARCONIGRAPH (Monthly). London: Marconi, 1911–13. American edition became *WIRELESS AGE* published in New York; British edition became *WIRELESS WORLD.*

MICROWAVE JOURNAL (Monthly). Norwood, MA: Horizon House, 1958–date.

THE OLD TIMER'S BULLETIN (Quarterly). Various towns in NY: Antique Wireless Association, 1959–date. "Published for the Collector, Historian and Old-Time Radio Operator."

PACIFIC RADIO NEWS. See *AUDIO.*

 # *PHILIPS TECHNICAL REVIEW* (Quarterly, then three per year). Hilversum, Netherlands: Philips Communications Systems, 1956–85; 1990–date. Superceded *COMMUNICATION NEWS* (1940–56); title was briefly changed to *PHILIPS TELECOMMUNICATION AND DATA SYSTEMS REVIEW* (1985–90).

POPULAR RADIO (Monthly). New York: Popular Radio Inc., 1922–28. Absorbed *WIRELESS AGE* in 1925; continued by *POPULAR RADIO AND TELEVISION.*

POST OFFICE ELECTRICAL ENGINEERS JOURNAL (Quarterly). London: Institution of Post Office Electrical Engineers, 1908–82. Became *BRITISH TELECOMMUNICATION ENGINEERING.* For special commemorative issues, see 7-141.

PROCEEDINGS OF THE AIEE (Monthly). New York: American Institute of Electrical Engineers, 1884–1963. Continued as *PROCEEDINGS OF THE IEEE.*

 # *PROCEEDINGS OF THE IEE* (Monthly). London: Institution of Electrical Engineers, 1889–date. Divided in 1955 to Part A (Power Engineering); Part B (Electronics and Communications); Part C (Monographs). See also 3-011.

 # *PROCEEDINGS OF THE IEEE* (Monthly). New York: Institute of Electrical and Electronics Engineers, 1963–date. Earlier see: *PROCEEDINGS OF THE IRE.* For special historical issue see 3-012.

PROCEEDINGS OF THE IERE (Irregular). London: Institution of Electronic and Radio Engineers, 1926–39. Became *JOURNAL OF THE BRITISH INSTITUTION OF RADIO ENGINEERS* (1939–62), and finally (in 1988) *ELECTRONICS & COMMUNICATIONS ENGINEERING JOURNAL.*

PROCEEDINGS OF THE IRE (Monthly). New York: Institute of Radio Engineers, 1913–63. Indexes published for 1909–42 (Vols. 1–30); and 1943–47 (Vol. 36:2, June 1948). See also CUMULATIVE INDEX OF IRE PUBLICATIONS 1948–53, and continuation for 1954–58. For 50th anniversary issue, see 3-036. For an historical analysis, see 4-087 and 4-088. Continued as *PROCEEDINGS OF THE IEEE.*

 # *PROCEEDINGS, RADIO CLUB OF AMERICA* (Quarterly, then irregular). New York: Radio Club of America, 1920–date. For 50th and 75th anniversary issues, see 4-092 and 4-093.

PROCEEDINGS OF THE SOCIETY OF INFORMATION DISPLAY (Quarterly). San Jose, CA: The Society, 1976–91. Became *JOURNAL OF THE SOCIETY FOR INFORMATION DISPLAY.*

 # *QST* (Monthly). Various cities: American Radio Relay League, 1915–date.

RADIO. See *AUDIO.*

RADIO AGE. See *ELECTRONIC AGE.*

RADIO BROADCAST (Monthly). Garden City, NY: Doubleday, 1922–30. Merged with *RADIO-DIGEST.*

RADIO AND ELECTRONIC ENGINEER (Monthly). London: British Institution of Radio Engineers; later the Institution of Electronic and Radio Engineers, 1963–84. Became *JOURNAL OF THE IERE* (1985–88); then *ELECTRONICS & COMMUNICATIONS ENGINEERING JOURNAL.*

RADIO & TELEVISION NEWS. See *RADIO NEWS.*

RADIO-CRAFT (Monthly). New York: Gernsback Publications, 1929–41. See *RADIO-ELECTRONICS.*

RADIO DIGEST (Monthly). Chicago: Radio Digest Publishing, 1922–33. Absorbed *RADIO BROADCAST* in 1930; continued by *RADIO FAN-FAIR* (which ended in 1933 as well).

RADIO-ELECTRONICS (Monthly). New York: Gernsback Publications, 1929–1948. Began as *RADIO-CRAFT* (1929–41); then a variety of closely related titles.

RADIO ENGINEERING. See *COMMUNICATIONS.*

RADIO NEWS (Monthly). Chicago: Ziff-Davis: 1919–59. Originally *RADIO AMATEUR NEWS* (1919–20); Many minor title changes 1932–38; *RADIO & TELEVISION NEWS* (1948–59). Became *ELECTRONICS WORLD* in 1959.

RADIO REVIEW (Monthly). London: Wireless Press, 1919–22. Absorbed by *WIRELESS WORLD.*

RADIO WORLD (Weekly, then monthly). New York: Hennessy Radio Publications, 1922–39. "The first and only national radio weekly."

RCA REVIEW (Quarterly). New York, later Princeton, NJ: RCA, 1936–85.

RECORDED SOUND (Quarterly to 1980, then semiannual). London: British Institute of Recorded Sound, 1961–84, 86 numbers published (issues numbered 1–44 from 1961–71). Superceded *BRITISH INSTITUTE OF RECORDED SOUND BULLETIN* (1956–60, quarterly, 18 issues). Sometimes referred to as the *JOURNAL* of the British Institute.

ROYAL TELEVISION SOCIETY JOURNAL. See *TELEVISION: JOURNAL OF THE...*

SMPTE JOURNAL (Monthly). White Plains, NY: Society of Motion Picture and Television Engineers, 1976–date. Formerly *TRANSACTIONS OF THE SMPE* (quarterly, 1916–30); *JOURNAL OF THE SMPE* (monthly, 1930–50); *JOURNAL OF THE SMPTE* (adding "Television", 1950–76). Has issued 13 cumulative subject-author indexes; latest covers 1991–95, issued as part two of the January 1996 (Vol. 105:1) issue. See also 4-095.

SOCIETY OF TELEGRAPH ENGINEERS JOURNAL (Monthly). London: Society of Telegraph Engineers, 1872–1880. Began as *PROCEEDINGS* and changed to this title in 1872. Became *SOCIETY OF TELEGRAPH ENGINEERS AND OF ELECTRICIANS JOURNAL,* 1881–1888. Became *JOURNAL OF THE IEE.*

TAPE AND FILM RECORDING. See *HI-FI TAPE RECORDING.*

TAPE RECORDING. See *HI-FI TAPE RECORDING*

TELECOMMUNICATION JOURNAL (Monthly). Geneva: International Telecommunication Union, 1934–93. Formerly *JOURNAL TELEGRAPHIQUE* (1869–1933). Replaced by *NEWSLETTER OF THE INTERNATIONAL TELECOMMUNICATION UNION.*

TELECOMMUNICATIONS POLICY (Quarterly, then bimonthly, increasing to 10 issues per year). Oxford, England: Elsevier Science, 1976–date.

TELEGRAPH AND TELEPHONE AGE (Semimonthly). New York: 1883–1953. Formerly *TELEGRAPH AGE* (to 1909) described as "Devoted to land-line telegraphs, submarine cable interests, radio-telegraphy and allied industries." Became *WIRE AND RADIO COMMUNICATIONS* (1953–59) and then absorbed by *COMMUNICATION NEWS.*

THE TELEGRAPHER (Monthly, then weekly). New York: McGraw, 1864–1877. Originally published by National Telegraphic Union (to 1870). Merged into *JOURNAL OF THE TELEGRAPH.*

TELEPHONE ENGINEER & MANAGEMENT (Semimonthly). Cleveland, OH: Harcourt Brace, 1909–94. Continued as *AMERICA'S NETWORK* (Cleveland: Advanstar Publishing). For 75th anniversary issue, see 7-075.

TELEPHONY (Weekly). Chicago, Intertec, 1901–date.

TELEVISION (Monthly). London: Reed Business, 1934–date.

TELEVISION. See *ELECTRONIC ENGINEERING.*

TELEVISION: JOURNAL OF THE [ROYAL] TELEVISION SOCIETY (Quarterly, then 8 per year). London: [Royal] Television Society, 1934–date. Formerly *ROYAL TELEVISION SOCIETY JOURNAL* (1961–75) and *TELEVISION SOCIETY JOURNAL.*

TELEVISION AND SHORT-WAVE WORLD. See *ELECTRONIC ENGINEERING.*

TELEVISION QUARTERLY (Quarterly). New York: National Academy of Television Arts and Sciences, 1962–date.

TELEVISION SOCIETY JOURNAL. See *TELEVISION: JOURNAL OF THE...*

TRANSACTIONS OF THE AIEE (Monthly). New York: American Institute for Electrical Engineering, 1922–64. Published indexes for vols. 41–57, 1922–38 (1939), and vols. 58–68, 1939–49 (1950). See *ELECTRICAL ENGINEERING*.

TRANSACTIONS OF THE IRE (Monthly). New York: IRE, 1951–64. Became *TRANSACTIONS OF THE IEEE*.

TRANSACTIONS OF THE SMPE. See *SMPTE JOURNAL*.

WESTERN UNION TECHNICAL REVIEW (Quarterly). New York: Western Union.

WIRELESS AGE (Monthly). New York: Wireless Press, 1913–25. Merged with *POPULAR RADIO* in 1925.

WIRELESS ENGINEER. See *EXPERIMENTAL WIRELESS*.

WIRELESS WORLD (Monthly). London: Marconi Press Agency, 1913–83. Formerly *MARCONIGRAPH*. Continued by *ELECTRONICS & WIRELESS WORLD*.

Chapter 3

General Surveys

These entries range from very broad or even popular approaches to others with a more technical or specific tone, but all cover topics more broadly (usually including both wired *and* wireless systems) than works listed in the following subject-specific chapters. Material included here is *highly* selective; only those works with content directly related to the overall history of telecommunications technology are included.

The chapter begins with (A) broad contextual surveys of electricity and electrical engineering development. This is followed by (B) citations on the process of invention and patenting (for more specifics including individual inventor biographies, see chapter 5). Section C includes broad histories of telecommunication (see 4-C for surveys of telecommunication *industries*), while D surveys histories of electronic media. Section E reviews development of foreign telecommunications systems, divided into general and British Commonwealth subsections. Section F includes citations to international (cross-border) telecommunications with references to international broadcasting and technical aspects of propaganda. The final two sections survey the history of telecommunications with (G) military and (H) maritime applications, both further divided into general and foreign, and U.S. material. For related Internet resources, see 15-B.

A. Electricity and Electrical Engineering
(This is but a selection from a much larger literature which stresses items containing useful telecommunications historical material.)

3-001 ** Antébi, Elizabeth. **THE ELECTRONIC EPOCH**. New York: Van Nostrand Reinhold, 1983, 256 pp. Historical survey of electronics and telecommunications in the 19th and 20th centuries. Includes contributions by many authorities. Photos, charts, bibliography, index.

3-002 Beauchamp, K.G. **EXHIBITING ELECTRICITY**. Stevnage, England: IEE "History of Technology Series 21," 1997, 338 pp. The first book-length history of major national and international electrical exhibitions from 1851 to the present, including many references to telecommunications devices and services. Photos, chapter references, index.

3-003 * Brittain, James E., ed. **TURNING POINTS IN AMERICAN ELECTRICAL HISTORY**. New York: IEEE Press "Selected Reprint Series," 1976, 399 pp. Sixty-four papers are reproduced covering events over 225 years; 18 papers specifically concern electrical communication systems. Photos, diagrams, notes, index.

3-004 "Diamond Jubilee Edition, 1928–1988," *ELECTRONIC ENGINEERING* (March 1988), pp. 3-190. Special commemorative issue of the British journal that began as *TELEVISION* in 1928, including historical surveys of the cathode ray tube, evolution of television technology, and general industry trends.

3-005 * *ELECTRONICS*, Editors of. **AN AGE OF INNOVATION: THE WORLD OF ELECTRONICS, 1930–2000**. New York: McGraw-Hill, 1981, 274 pp. Book format reissue of the 50th anniversary issue published the year before. It provides a useful survey of people and events over a half century—and some predictions. Photos, index.

3-006 "The Electronics Challenge: An Historical Perspective," special issue of *HISTORY AND TECHNOLOGY,* 11:113–157 (1994). Eight articles on the development of the transistor, telecommunications, and computers in the U.S., France, Britain, and Norway. References.

See Fleming, 5-124.

3-007 Grivet, Pierre. "60 Years of Electronics," *ADVANCES IN ELECTRONICS AND ELECTRON PHYSICS,* 50:89–174 (1980). Emphasis is more on technology than history, but a useful survey. Photos, diagrams, references.

3-008 Hawks, Ellison. **THE BOOK OF ELECTRICAL WONDERS**. New York: Dial Press, 1936, 316 pp. A popular account, including chapters on the telephone, telephone exchange, telegraphing without wires, wireless telephony, electrical transmission of pictures, and television. Photos, diagrams, index.

3-008a * Houston, Edwin J. **ELECTRICITY IN EVERY-DAY LIFE**. New York: P.F. Collier, 1904, 3 vols., 584, 566, and 609 pp. A standard treatment. Vol. 3 includes nearly 400 pages on the telegraph, telephone, phonograph, and related inventions and services. Photos, engravings, diagrams, index.

3-009 * Hughes, Thomas P. **NETWORKS OF POWER: ELECTRIFICATION IN WESTERN SOCIETY, 1880–1930**. Baltimore: Johns Hopkins University Press, 1983, 474 pp. Erudite history of the expansion of electricity and its applications, primarily in North America and Britain. Photos, diagrams, notes, index.

See Mclaurin, 9-108.

See McMahon, 4-085.

General Surveys 35

3-009a Morton, David. "The Electric Century," *PROCEEDINGS OF THE IEEE,* January 1999–date. Just as this bibliography was going to press, this new series was announced, one article to appear each month in this journal, devoted to some aspect of electricity in the 20th century.

See Ryder, 4-086.

3-010 Seitz, Frederick. "The Tangled Prelude to the Age of Silicon Electronics," *PROCEEDINGS OF THE AMERICAN PHILOSOPHICAL SOCIETY,* 140:289–337 (1996). The electronics industry in the first half of the 20th century.

3-011 * "Special Issue: The History of Electrical Engineering," *PROCEEDINGS OF THE IEE,* Part A, 132:489–601 (December 1985); 136:265–340 (November 1989); and 140:173–248 (May 1993). Each of these includes six to eight historical articles, one or two of which deal with telecommunications topics and are listed elsewhere in this book. Illustrations, diagrams, maps, bibliography.

3-012 * "Special Issue on Two Centuries in Retrospect," *PROCEEDINGS OF THE IEEE,* 64:1265–1456 (September 1976). Special issue of 23 historical papers, eight of which concentrate on telecommunications concerns. Photos, references.

3-013 Tannenbaum, Morris. "The Historical Evolution of U.S. Telecommunications," *TECHNOLOGY IN SOCIETY,* 15:263–272 (1993). A very concise survey.

3-014 * Walmsley, Robert M. **ELECTRICITY IN THE SERVICE OF MAN**. London: Cassell, 1890, 891 pp.; 1896 [2nd ed.], 976 pp.; 1911–19 [3rd ed.], 3 vols., 840, 692, and 724 pp. Based on a German work first published in 1885 (by Alfred R. von Urbanitzky), this is "a popular and practical treatise on the applications of electricity in modern life." Each edition includes hundreds of pages on the telegraph and telephone and related inventions. Engravings, index.

B. The Process of Invention and Patents

(Again, this is a very selective survey generally including only those resources with considerable focus on telecommunications. See also biography [chapter 5], and the several technology-specific chapters.)

3-015 Abbot, Charles Greeley. **GREAT INVENTIONS**. New York: Scribner's "Smithsonian Scientific Series, Vol. 12," 1932, 383 pp. A traditional "great men" approach which is heavily focused on electricity and electronics, and includes two chapters specifically on wired and wireless communication. Abbot was secretary of the Smithsonian at the time. Photos, some text references, diagrams, index.

3-016 Auger, C. Peter, ed. **INFORMATION SOURCES IN PATENTS**. London: Bowker-Saur "Guides to Information Sources," 1992, 187 pp. Covers the U.S., Britain, and other industrial countries with a variety of finding aides and search techniques including online research. Glossary, index.

3-017 Baker, R. **NEW & IMPROVED: INVENTORS AND INVENTIONS THAT HAVE CHANGED THE MODERN WORLD**. London: British Museum Publications, 1976, 168 pp. Reproducing key pages from British patents, this is a survey of invention from the 19th and 20th centuries, arranged by topic and including many within the realm of this bibliography. Photos, diagrams, bibliography, chronology indexed by inventor, and topic.

3-018 * Beniger, James R. **THE CONTROL REVOLUTION: TECHNOLOGICAL AND ECONOMIC ORIGINS OF THE INFORMATION SOCIETY.** Cambridge, MA: Harvard University Press, 1986, 493 pp. A sociological approach to determining "causes" of the modern era (including industrialization, processing speed, and how changing technology changed means of business and political control) as well as its initial impacts, especially in business. References, index.

See Bunch and Hellemans, 1-043.

3-019 * Burlingame, Roger. **ENGINES OF DEMOCRACY: INVENTIONS AND SOCIETY IN MATURE AMERICA.** New York: Scribner's, 1940, 606 pp. Quite useful contextual history of invention and society for the period after 1865, including chapters on the telephone, business communication, telegraph, print and film media, and radio. Includes some material on English and European contributions. Continues the author's earlier MARCH OF THE IRON MEN which took the story to 1865. Photos, diagrams, notes, bibliography, chronology, index.

See Dummer, 1-048.

See Harpur, 1-053.

3-020 ** Jewkes, John, David Sawyers, and Richard Stillerman. **THE SOURCES OF INVENTION.** London: Macmillan / New York: St. Martin's Press, 1958, 428 pp. Classic survey of inventors, inventions, corporate research in part one followed by 50 brief case studies, nine referring to electronics (magnetic recording, radar, radio, television, the transistor). Notes, index.

See Maclaurin, 9-108.

3-021 * U.S. Department of Commerce, Patent and Trademark Office, Office of Technology Assessment and Forecast. **PATENT PROFILES.** Washington: Government Printing Office, 1979–84, occasional. Relevant volumes trace foreign- and domestic-owned patents in:

> **MICROELECTRONICS** (1981), covers integrated circuit structures, and systems using a computer central processing unit.
>
> **MICROELECTRONICS—II** (1983, 179 pp.) covers 1963–82 in digital logic circuits, semiconductor memories, and speech analysis and synthesis.
>
> **TELECOMMUNICATIONS** (1984, 260 pp.) covers 1963–83 in telephony, light wave communication, multiplex communication, analog carrier wave systems, digital and pulse communication, television and facsimile.

3-022 * Van Deusen, Edmond L. "The Inventor in Eclipse," *FORTUNE,* 50:132–135, 197–202 (December 1954). Discusses the decline of the individual inventor.

General Surveys 37

Valuable two-page color chronological chart of the forebears to radio and television development.

3-023 Walker, Richard D. **PATENTS AS SCIENTIFIC AND TECHNICAL LITERATURE.** Metuchen, NJ: Scarecrow Press, 1995, 533 pp. Six chapters on interpretation of patent documents, world patent procedures, and bibliographic resources. Appendix, bibliography, index.

See Winston, 3-057 and 3-058.

C. Telecommunications

(Survey histories of industries and companies are found in chapter 4; telegraph or telephone in chapters 6 and 7, respectively, wireless in chapter 9, television in chapter 12, and newer services in chapters 13–15. Material included below covers more than one telecommunications service.)

3-024 Asmann, Edwin N. **THE TELEGRAPH AND THE TELEPHONE: THEIR DEVELOPMENT AND ROLE IN THE ECONOMICS HISTORY OF THE UNITED STATES—THE FIRST CENTURY, 1844–1944.** Lake Forest, IL: Lake Forest College, 1980. 350 pp. The first 100 pages are devoted to the telegraph, the remainder to the varied history of both means of transmission, and their roles in transport, agriculture, industry, and domestic commerce. Illustrations, tables, appendices (chiefly documents), bibliography.

3-025 Beck, Arnold H.W. **WORDS AND WAVES.** New York: McGraw-Hill, 1967, 255 pp. Chapters 3–5 cover the history of telegraph, telephone, and radio. Useful technical explanations for layperson. Photos, diagrams, bibliography, index.

3-026 Black, Robert M. **A HISTORY OF ELECTRIC WIRES AND CABLES.** Stevenage, England: Peter Peregrinus with the Science Museum "History of Technology 4," 1983, 204 pp. Covers most aspects into the 1950s, but includes some discussion of optical fiber cables as well. Photos, references, index.

See Bray, 5-018.

See Brock, 4-053.

3-027 Clarke, Arthur C. **HOW THE WORLD WAS ONE: BEYOND THE GLOBAL VILLAGE.** New York: Bantam Books, 1992, 300 pp. To a considerable degree this is an amalgamation of the author's earlier history of the Atlantic cable and communications satellites (6-222), with considerably more detail on the former.

3-028 Colpitts, Edwin H., and O.B. Blackwell. "Carrier Current Telephony and Telegraphy," *TRANSACTIONS OF THE AIEE*, 40:205–300 (February 1921); reprinted in *PROCEEDINGS OF THE IEEE*, 52:340–359 (April 1964). Historical treatment from Gray's multiple harmonic telegraph to commercial systems of the World War I period. Bibliography, notes, photos.

3-029 Crawley, Chetwode. **FROM TELEGRAPHY TO TELEVISION: THE STORY OF ELECTRICAL COMMUNICATIONS.** London: Frederick Warne,

1931, 212 pp. Based largely on a series of articles in *TELEVISION*, this is a popular narrative. Photos, diagrams, index.

3-030 * Czitrom, Daniel J. **MEDIA AND THE AMERICAN MIND FROM MORSE TO McLUHAN**. Chapel Hill: University of North Carolina Press, 1982, 254 pp. Insightful chapters on telegraph and radio (and another on film) plus several more reviewing theoretical approaches to American media development. Notes, bibliography, index.

3-031 Denman, R.P.G. **THE HISTORY OF ELECTRICAL COMMUNICATION**. London: HMSO, 1926, 55 pp. Catalogue of the line telegraphy and telephony collections in the Science Museum. Photos.

3-032 Dunlap, Orrin E. **COMMUNICATIONS IN SPACE: FROM MARCONI TO MAN ON THE MOON**. New York: Harper & Row, 1962, 175 pp.; 1964 [rev. ed.], 260 pp.; 1970 [3rd ed.], 338 pp. A broad survey by a longtime communications journalist with an historical viewpoint, limited to American contributions and focusing on space communications. Photos, notes, index. (Title varied.)

3-033 Enochs, Hugh. **ELECTRONICS RESEARCH IN THE SPACE AGE**. Palo Alto, CA: Palo Alto Chamber of Commerce, 1962 [3rd ed.], 32 pp. Story of electronics research and development in the Palo Alto–Stanford University area. Photos, chronology.

3-034 * Flichy, Patrice. **DYNAMICS OF MODERN COMMUNICATION: THE SHAPING AND IMPACT OF NEW COMMUNICATION TECHNOLOGIES**. London: Sage "Media, Culture and Society Series," 1995, 181 pp. Translated from the French original. Divided into three parts: 1790–1870 (semaphore telegraphs, networks and electricity, and the electric telegraph); 1870–1930 (recordings, the telephone, and radio); and 1930–90 (more telephony, television, computers, and private communications). Chapter notes, index.

3-035 Friedlander, Amy. **NATURAL MONOPOLY AND UNIVERSAL SERVICE: TELEPHONES AND TELEGRAPHS IN THE U.S. COMMUNICATIONS INFRASTRUCTURE, 1837–1940**. Reston, VA: Corporation for National Research Initiatives, 1995, 87 pp. Useful as a concise survey of a large literature, culling out major organizational and policy trends based on technological change and patent control. Photos, notes, selected bibliography.

3-036 *** Goldsmith, Alfred N., ed. "The Fiftieth Anniversary Issue, 1912–1962," *PROCEEDINGS OF THE IRE*, 50:529–1448 (May 1962). Invaluable collection of 113 papers divided into 28 topics, most providing concise historical surveys. The first 100 pages offer a series of brief glimpses of the future—the year 2012. The histories include discussion of antennas and propagation, audio, broadcast and television receivers, broadcasting, communication systems, components, electron devices, microwaves, military electronics, radio frequency interference, and many other topics beyond the realm of the present volume. Photos, diagrams, references.

3-037 * Grant, August E., and Jennifer Harman Meadows, eds. **COMMUNICATIONS TECHNOLOGY UPDATE**. Woburn, MA: Focal Press, 1998 [6th ed.], 372 pp. First published in 1992, this is a very useful "handbook" of contributed

General Surveys

chapters reviewing recent developments in electronic media, computers, consumer electronics, telephone services, and satellite technologies. Updated regularly on-line: http://www.tfi.com/ctu. Tables, diagrams, notes, glossary (no index).

3-038 ** Harlow, Alvin F. **OLD WIRES AND NEW WAVES: THE HISTORY OF THE TELEGRAPH, TELEPHONE, AND WIRELESS**. New York: Appleton-Century, 1936 (reprinted by Arno Press "Telecommunications," 1971), 548 pp. A classic and widely cited volume, this remains one of the better broad surveys that, while not documented, offers an integrated view of wired and wireless services up to World War II. Photos, bibliography, index.

3-039 Houston, Edwin J. **THE ELECTRIC TRANSMISSION OF INTELLIGENCE**. New York: W.J. Johnston, 1893, 330 pp. Electric telegraph and telegraph cables, annunciators, alarms, time systems, the telephone, and other applications of electricity. Engravings.

3-039a* Jones, Frank A.K.L., ed. **SEMAPHORES TO SHORT WAVES: PROCEEDINGS OF A CONFERENCE ON THE TECHNOLOGY AND IMPACT OF EARLY TELECOMMUNICATIONS**. London: Arts Council of England, 1998. Sponsored by the British Society for the History of Science, the Newcomen Society, and the Royal Society of Arts, this volume includes papers on all types of telecommunication services. References.

3-040 ** King, W. James. **THE TELEGRAPH AND THE TELEPHONE**. Washington: Smithsonian Institution "United States National Museum Bulletin 228; Contributions from the Museum of History and Technology, Paper 29," 1962, 60 pp. Standard paper—a technical review from the Chappe optical telegraph to the electrical telegraph and telephone equipment and systems of the 1880s. Footnotes, photos, diagrams.

3-041 * LeBow, Irwin. **INFORMATION HIGHWAYS & BYWAYS: FROM THE TELEGRAPH TO THE 21st CENTURY**. New York: IEEE Press, 1995, 307 pp. Assuming little technical background, the author provides an historical survey of expanding technological options; roughly half devoted to postwar developments. Photos, notes, index.

3-042 ** Lubar, Steven. **INFOCULTURE: THE SMITHSONIAN BOOK OF INFORMATION AGE INVENTIONS**. Boston: Houghton Mifflin, 1993, 408 pp. A useful combination of history and current survey of all of telecommunications, inspired by an exhibit at the Smithsonian, and written by one of its curators. It is divided into sections on communications (wired and wireless), entertainment (recorded sound, films, and broadcasting), and information (computers and software). Photos, diagrams, further reading, notes, index.

3-043 ** Marland, Edward A. **EARLY ELECTRICAL COMMUNICATION**. London: Abelard-Schumann, 1964, 220 pp. Survey of British and American telecommunications during the 19th century, focusing on the land and submarine telegraph, and the early telephone, with emphasis upon inventions and apparatus. Notes, photos, references, index.

3-044 * Marvin, Carolyn. **WHEN OLD TECHNOLOGIES WERE NEW: THINKING ABOUT COMMUNICATIONS IN THE LATE 19th CENTURY.** New York: Oxford University Press, 1988, 269 pp. Popular conceptions of the electric light, telegraph, and telephone in the last quarter of the century. Focus is on relationships of technological progress and social and cultural life. Photos, notes, index.

3-045 McGillem, Clare D., and William P. McLauchlan. **HERMES BOUND: THE POLICY AND TECHNOLOGY OF TELECOMMUNICATIONS.** West Lafayette, IN: Purdue University Office of Publications "Series in Science, Technology, and Human Values," 1978, 284 pp. Considerable history backs up the discussion on experimental development of telecommunication, broadcasting, two-way systems, and impact on society.

3-046 Morgan, Jane. **ELECTRONICS IN THE WEST: THE FIRST FIFTY YEARS.** Palo Alto, CA: National Press, 1967, 194 pp. Story of electronics research and industry in the San Francisco Bay region, designed for younger readers. Emphasis on men and inventions.

3-047 ** "100 Years of Communications Progress," *IEEE COMMUNICATIONS MAGAZINE,* 22:5–126 (May 1984). A centennial celebration (of the formation of AIEE, a predecessor to IEEE). Includes 12 major papers on technology (deep space and communication satellites, optical fibre communication, computer communication and switching); international perspectives (the ITU, and developments in Europe, Japan, and Korea); and several personal reminiscences by pioneers. Photos, charts, references.

3-048 * Oslin, George P. **THE STORY OF TELECOMMUNICATIONS.** Macon, GA: Macon University Press, 1992, 507 pp. A long-time Western Union public relations official writes of American telecommunication—and especially telegraph (the first 210 pp.)—history. More anecdotal than the implied full-length history of the industries and their underlying technologies, it remains useful for the insights of its author, who was in his nineties when the book appeared. Photos, diagrams, chapter notes, bibliography, index.

3-049 Pierce, John R., and A. Michael Noll. **SIGNALS: THE SCIENCE OF TELECOMMUNICATIONS.** New York: W.H. Freeman "Scientific American Library 32," 1990, 247 pp. Combines historical review and description of current methods including modulation and encoding, multiplexing, electricity and light waves, the physics of communication systems, switching and signaling, networks for voice and data, and policy questions. Photos, diagrams, further readings, index.

3-050 Pool, Ithiel de Sola. **TECHNOLOGIES OF FREEDOM: ON FREE SPEECH IN AN ELECTRONIC AGE.** Cambridge, MA: Harvard University Press, 1983, 399 pp. Thoughtful analysis of the many impacts of changing technology on basic and accepted social and cultural norms. Focuses on media, but includes discussion of First Amendment rights of telecommunications common carriers as well. Notes, index.

3-051 _____, edited by Eli Noam. **TECHNOLOGIES WITHOUT BOUNDARIES: ON TELECOMMUNICATIONS IN A GLOBAL AGE.** Cambridge, MA: Harvard University Press, 1990, 283 pp. Sequel to item immediately above, this

General Surveys

broadens the focus to include newer means of transmission, and especially the role of satellites and computers. Notes, index.

3-052 Randell, Wilfrid L. **MESSENGERS FOR MANKIND.** London: Hutchinson & Co. "Conquest of Space and Time," 1939, 240 pp. Extensively illustrated account of British communications including chapters on telegraphy, telephone, wireless, cables and cable laying, transmission of pictures (facsimile), and broadcasting. Useful especially for its prewar viewpoint. Photos, diagrams, tables, maps, index.

3-053 * Rowland, Wade. **SPIRIT OF THE WEB: THE AGE OF INFORMATION FROM TELEGRAPH TO INTERNET.** Toronto: Somerville House, 1997, 420 pp. Broadscale historical survey by a respected Canadian science and technology author. Offers 31 chapters divided into analog (telegraph, telephone, and wireless broadcasting) and digital (computer and internet) eras, with considerable focus on who controls information. Notes, index.

3-054 Squire, George O. **TELLING THE WORLD.** Baltimore: Williams & Wilkins Co. "Century of Progress Series," 1933, 163 pp. Readable story of electricity, telegraph, telephone, radio, and the Army's signal corps (by its then chief). References.

3-055 Still, Alfred. **COMMUNICATION THROUGH THE AGES, FROM SIGN LANGUAGE TO TELEVISION.** New York: Murray Hill Books, 1946, 201 pp. A popular story with some philosophical features. Chronology, photos and diagrams, notes, index.

3-056 Williams, Archibald. **TELEGRAPHY AND TELEPHONY.** London: Thomas Nelson "Highroads of Modern Knowledge," 1928, 340 pp. Title is somewhat misleading as the majority is devoted to historical developments of wired and wireless systems. Coverage includes birth of the telegraph, trans-Australian telegraph, telegraph lines in Central Africa, early submarine cables, invention of the telephone, early stages of wireless, wireless and navigation, and television. Photos, diagrams, maps, index.

3-057 ** Winston, Brian. **MISUNDERSTANDING MEDIA.** Cambridge, MA: Harvard University Press, 1986, 419 pp. Mistitled in that it focuses on the innovation process in telecommunications, with chapters detailing inventors and inventions in television, the computer, the integrated circuit, the communications satellite, and the telephone. Winston proposes and utilizes a new model of the process of innovation. Diagrams, notes, index. See next entry.

3-058 *** _____. **MEDIA TECHNOLOGY AND SOCIETY: A HISTORY FROM THE TELEGRAPH TO THE INTERNET.** London and New York: Routledge, 1998, 374 pp. Update and revision of previous entry placing history as the central theme with the model in a supportive and organizing role. Covers most telecommunications and electronic media services and again applies the Winston model. Diagrams, notes, references, index.

3-059 Woodbury, David O. **COMMUNICATION: STORIES OF MAN'S ACHIEVEMENTS.** New York: Dodd, Mead, 1931, 280 pp. Popular history of telegraph, telephone, and radio. Photos, bibliography (arranged by chapter), index.

D. Electronic Media
(See also chapters 9-C/D, 12-C/D, and 13.)

3-060 ** Besen, Stanley M., and Leland L. Johnson. **COMPATIBILITY STANDARDS, COMPETITION, AND INNOVATION IN THE BROADCASTING INDUSTRY.** Santa Monica, CA: RAND Corporation, 1986, 120 pp. Very useful historical study of radio and television technical standards and how their derivation and preservation has impacted the pace of development in the field. Tables, notes.

3-061 * **BROADCASTING TECHNOLOGY—PAST, PRESENT & FUTURE.** Stevenage, England: Institution of Electrical Engineers, 1973, 104 pp. Nine seminar papers survey development of the domestic receiver, studio–transmitter links, the TV camera tube, satellite and cable systems, and several more general overviews. Photos, diagrams, index.

3-062 Curtis, Philip J. **THE FALL OF THE U.S. CONSUMER ELECTRONICS INDUSTRY: AN AMERICAN TRADE TRAGEDY.** Westport, CT: Quorum Books, 1994, 339 pp. A longtime counsel with Zenith outlines the rise and fall of the American industry as seen through the many Zenith legal battles over patent control and trade incursions, chiefly by the Japanese, but often, he argues, with collusion of American firms. Notes, appendices, bibliography.

3-063 ** "A History of Consumer Electronics: Commemorating a Century of Electrical Progress, 1884–1984," *IEEE TRANSACTIONS ON CONSUMER ELECTRONICS,* CE30:11–211 (May 1984). Divided into two parts—13 invited papers, and reprints of key historical material from a 1945 meeting, the May 1952 40th anniversary IRE issue, the January 1954 color issue (see 12-134), and the May 1962 IRE 50th anniversary issue (see 3-036) all focused primarily on radio and television broadcasting.

3-064 ** Inglis, Andrew F. **BEHIND THE TUBE: A HISTORY OF BROADCASTING TECHNOLOGY AND BUSINESS.** Stoneham, MA: Focal Press, 1990, 527 pp. While focusing largely on RCA (where the author worked), this is a useful survey of American developments in AM and FM radio, black-and-white and color television, audio and video recording methods, cable, and satellite television. It is strongest in the 150 pages devoted to television. Photos, tables, diagrams, notes, glossary, index.

3-065 * Sterling, Christopher H., and John M. Kittross. **STAY TUNED: A CONCISE HISTORY OF AMERICAN BROADCASTING.** Belmont, CA: Wadsworth, 1978, 562 pp.; 1990 [2nd ed.], 705 pp. Initial sections of each chapter focus on technical developments and changes over a specific period of years. Photos, diagrams, tables, glossary, bibliography, index.

3-066 Zavada, Roland J. "'Managing' the Moving Image—From an Engineering Point of View," *SMPTE JOURNAL,* 101:148–166 (March 1992). Includes film, television, and video recording methods. Diagrams, bibliography.

General Surveys 43

E. Foreign Telecommunications Systems

(Includes surveys of telecommunication services within one or more specific countries or regions. See also chapters 6-C through 6-E and 7-E for foreign and international telegraph and telephone links, and 9-D for foreign radio.)

General

3-067 * Brown, Frank J. THE CABLE AND WIRELESS COMMUNICATIONS OF THE WORLD. London: Pitman, 1927, 148 pp.; 1930 [2nd ed.], 153 pp. About two-thirds devoted to cables with wireless assessed as an international competitive carrier. Index, diagrams, foldout map.

3-068 Cerver, Francisco Asensio, ed. COMMUNICATION TOWERS: NEW ARCHITECTURE 5. Barcelona: Ediciones Atrium, nd but ca 1993, 255 pp. As the introduction begins, "the dividing line between architecture and engineering, always subtle, seems to become blurred when we are discussing telecommunications towers." This book details 26 of the dramatic structures, most of them in Europe or Japan. Only book of its kind. Diagrams, photos. Spanish and English.

3-069 * Fransman, Martin. JAPAN'S COMPUTER AND COMMUNICATIONS INDUSTRY: THE EVOLUTION OF INDUSTRIAL GIANTS AND GLOBAL COMPETITIVENESS. Oxford, England: Oxford University Press, 1995, 540 pp. Based on hundreds of interviews and documents, this includes chapter-length analyses of the switching industry, optical fiber industry, the evolution of NEC, the role and future of NTT and its chief competitor. Tables, charts, appendices, notes, bibliography, index.

3-070 Stumpers, F.L. "The History, Development and Future of Telecommunications in Europe," *IEEE COMMUNICATIONS MAGAZINE*, 22:84–95 (May 1984). Broad survey.

3-071 ** UNESCO. PRESS, FILM, RADIO. Paris: UNESCO, 1947–51, 7 vols. (Reprinted in 3 vols. by Arno Press, "International Propaganda and Communications," 1972). This "survey on the structure, work and equipment of the news agencies, press, cinema and radio in 157 countries and territories" is valuable for its historical record. In each volume, all the information on radio is in one place making for ready reference. Tables, maps, notes, cumulative index.

3-072 ** _____. WORLD COMMUNICATIONS: PRESS, RADIO, FILM, TELEVISION. Paris and New York: UNESCO, 1950 (Publication 700), 220 pp.; 1951 [2nd ed., Publication 942], 223 pp.; 1956 [3rd ed.], 262 pp.; 1964 [4th ed.], 380 pp.; and 1975 [5th ed.], 533 pp. Invaluable comparative data for growing number of countries. Each edition includes general discussion, then country-by-country details, with each edition including more countries (the final volume includes data on 200). Also includes information on press and film and ancillary fields. Television not included in first edition (for more on television, see UNESCO, 12-107). Maps, tables, charts, bibliography. To some extent continued by next entry.

3-073 * _____. WORLD COMMUNICATION REPORT: THE MEDIA AND THE CHALLENGE OF THE NEW TECHNOLOGIES. Paris:

UNESCO, 1989, 551 pp.; 1998 [2nd ed.], 298 pp. Rather than being arranged by country as in previous entry, this takes a worldwide approach, reviewing trends and impacts. Photos, tables, charts, notes, bibliography, glossary.

3-074 ** WORLD TELECOMMUNICATIONS. New York: Arthur D. Little, 1970, 4 vols.; 1980 [2nd ed.], 4 vols., 1,029 pp. Massive client survey of development and current status of telecommunication facilities and services around the world. Each edition reviews the prior decade and makes projections for the decade to come—the series thus traces trends over three decades. Maps, tables, charts. Both editions are arranged in similar fashion (details are for 1980 ed.):

> VOL. 1: OVERVIEW AND TECHNOLOGICAL TRENDS. (216 pp.) Includes overview, international facilities, technological trends, major manufacturers, bibliography, glossary.
>
> VOL. 2: THE AMERICAS AND OCEANIA. (279 pp.) Is a country-by-country review of North and South America, and the Pacific region.
>
> VOL. 3: EUROPE. (237 pp.)
>
> VOL. 4: AFRICA AND ASIA. (297 pp.)

British Commonwealth

(Many further histories concerning a specific technology [e.g., telegraph] appear in the relevant chapter [e.g., 6].)

3-075 BRITAIN AND COMMONWEALTH TELECOMMUNICATIONS. London: Central Office of Information, 1963, 36 pp. Historical record of telegraph, telephone, and cable systems, plus radio and television. Tables, reading list.

3-076 Coates, Austin. QUICK TIDINGS OF HONG KONG. Hong Kong: Oxford University Press, 1990, 198 pp. Informal history of telecommunications to, in, and near the British colony, from telegraphy through modern day telecommunications. Photos, maps, index.

3-077 * Collins, Robert. A VOICE FROM AFAR: THE HISTORY OF TELECOMMUNICATIONS IN CANADA. Toronto: McGraw-Hill Ryerson, 1977, 304 pp. Covers both telegraphy and telephony into the age of satellites. Photos, chronology, bibliography, index.

3-078 Kuhn, M. "Telecommunications in Canada: A Century of Symbiotic Development," *IEEE COMMUNICATIONS MAGAZINE,* 22:104–114 (May 1984). Broad survey of various systems and companies. Illustrations, diagrams, maps, bibliography.

3-079 * Livingston, Kevin T. THE WIRED NATION CONTINENT: THE COMMUNICATION REVOLUTION AND FEDERATING AUSTRALIA. Melbourne: Oxford University Press, 1996, 232 pp. Discusses the development and impact of the domestic telegraph, Pacific cables, and the role of Australia within the

British Commonwealth. Appendix on rates, maps, notes, bibliography, index. See also Moyal, 3-081 and 4-250.

3-079a McCarthy, Michael, et al. **THE VOICE OF GENERATIONS: A HISTORY OF COMMUNICATIONS IN NEWFOUNDLAND.** St. John's: Robinson Blackmore, 1994, 428 pp. Issued for the 75th anniversary of the Newfoundland Telephone Company, this covers services from telegraphy and telephony through wireless and data communications. Photos, maps, bibliography, appendices, index.

3-080 McPhail, Thomas L., and David C. Coll, eds. **CANADIAN CONTRIBUTIONS TO TELECOMMUNICATIONS: AN OVERVIEW OF SIGNIFICANT ACTIVITIES.** Calgary: University of Calgary Graduate Program in Communication Studies, 1986, 245 pp. Part one includes nine papers on the overall development and role of telecommunications; part two includes six technical papers which are not historical. Photos, diagrams, references.

3-081 Moyal, Ann. "The History of Telecommunications in Australia: Aspects of the Technological Experience, 1854–1930," in Nathan Reingold and Marc Rothenberg, eds. **SCIENTIFIC COLONIALISM: A CROSS-CULTURAL COMPARISON.** Washington: Smithsonian Institution Press, 1987, pp. 35–54. See also 4-250.

3-082 Smith, Robert C. **RHODESIA: A POSTAL HISTORY—ITS STAMPS, POSTS AND TELEGRAPHS.** Salisbury: Kingstone Ltd., 1967, 453 pp. Chronological survey of posts and electrical telecommunications including chapters on telegraphs, the Africa Trans-Continental Telegraph Line, telephones, overseas communications, and broadcasting. Photos, bibliography.

3-083 * Wilson, A.C. **WIRE AND WIRELESS: A HISTORY OF TELECOMMUNICATIONS IN NEW ZEALAND, 1860** [cover says **1890**]**–1987**. Palmerston North: Dunmore Press, 1994, 236 pp. The international context, domestic factors, impact of specific politicians, and the role of technology in developing electronic communications to and within New Zealand up to the formation of Telecom New Zealand. Photos, notes, bibliography, index.

F. International Telecommunications

(Includes studies of links between and among nations, divided into sections on general and technical aspects of propaganda.)

General

3-084 * **THE CABLE AND WIRELESS COMMUNICATIONS OF THE WORLD: SOME LECTURES AND PAPERS ON THE SUBJECT, 1924–1939.** Cambridge, England: Cable and Wireless, 1939, 282 pp. The issues as seen in 18 presentations by directors and officers of the company. Each is dated, though original titles of published papers are not provided. Photos, notes, color folding map.

3-085 * Cherry, Colin. **WORLD COMMUNICATION: THREAT OR PROMISE? A SOCIO-TECHNICAL APPROACH.** New York: Wiley-Interscience, 1971, 229 pp; 1978 [rev. ed.], 229 pp. History and impact of technological developments on international relations. Diagrams, tables, bibliography, index.

3-086 Clark, Keith. **INTERNATIONAL COMMUNICATIONS: THE AMERICAN ATTITUDE.** New York: Columbia University Press, 1931 (reprinted by AMS Press, 1968), 261 pp. Relates the development of—and American attitudes toward—the Universal Postal Union, the International Telegraph Union, submarine cables, and the International Radio Union. Primarily a political analysis. Notes, tables, index.

3-087 Eward, Ronald S. **PRESENT AND PROJECTED BUSINESS UTILIZATION OF INTERNATIONAL TELECOMMUNICATIONS: 1985.** Washington: Government Printing Office (NTIA Contractor Report 85–35), 1985, 88 pp. Reviews past and likely future demand for leased circuits, public services, and user concerns about capacity. Tables, charts, references.

3-088 ** Fortner, Robert S. **INTERNATIONAL COMMUNICATION: HISTORY, CONFLICT, AND CONTROL OF THE GLOBAL METROPOLIS.** Belmont, CA: Wadsworth, 1993, 390 pp. Chapter 3 reviews technology, chapters 4–9 provide a detailed history from 1835. Photos, diagrams, appendices, glossary, references, index.

3-089 ** Headrick, Daniel R. **THE INVISIBLE WEAPON: TELECOMMUNICATIONS AND INTERNATIONAL POLITICS, 1851–1945.** New York: Oxford University Press, 1991, 289 pp. Explores the impact of changing technology: submarine cables, land telegraphy, wireless, shortwave services, and wartime communications intelligence. Basically an economic and political study, this unique work includes regular reference to the baseline technologies involved. Notes, bibliographic essay, index.

3-090 _____. "Shortwave Radio and Its Impact on International Telecommunications," *HISTORY AND TECHNOLOGY,* 11:21–32 (1994). See also author's related book immediately above.

3-090a ** Hugill, Peter J. **GLOBAL COMMUNICATIONS SINCE 1844: GEOPOLITICS AND TECHNOLOGY.** Baltimore: Johns Hopkins University Press, 1999, 277 pp. Fascinating study of geography, history, and technology tracing the international control of telecommunications from the inception of the telegraph through the telephone and wireless up to 1945 with an epilogue on events since. Includes chapters on military uses of telecommunications and closely related technologies such as radar. Maps, charts, glossary, references, index.

3-091 INTERNATIONAL FACILITIES STUDY. Washington: Office of Telecommunications Policy, Executive Office of the President, May 1971, 130 pp. A staff report which surveys American international carriers and the varied technologies used. Tables, notes.

3-092 Li, George S. **ANALYSIS OF SUBMARINE CABLE AND COMMUNICATION SATELLITE SYSTEMS RELIABILITIES.** Washington: Government Printing Office (Office of Telecommunications Contractor Report 77-5), 1977, 26 pp. Brief text and series of tables with comparative data for 1970–75 period showing system outages or interruptions.

General Surveys

3-093 * Mance, Sir H. Osborne. INTERNATIONAL TELECOMMUNICATIONS. London: Oxford University Press "International Transport and Communications," 1944, 90 pp. Explores international policy and technical problems concerning telegraphy, telephony, and wireless. Notes, note on sources, index.

See Michaelis, 4-016.

3-094 Schreiner, George Abel. CABLES AND WIRELESS: THE POSITION OF THE TELELECTRIC MEANS OF COMMUNICATION IN INTERNATIONAL RELATIONS AND WORLD PUBLIC ORDER. Boston: Stratford Co., 1924, 269 pp. Largely policy but with some study of technical factors. Notes, tables, appendices, two foldout maps, index.

3-095 * STUDY OF INTERNATIONAL TELECOMMUNICATIONS POLICIES, TECHNOLOGY, AND ECONOMICS and SUPPLEMENT and APPENDICES. Menlo Park, CA: Stanford Research Institute, 1966, 3 vols. 207, 318, and ca. 750 pp. Prepared for a federal intragovernmental committee, this reviews in some detail the history, technology, and finances of American international carriers. The supplement offers three largely economic studies comparing use of cables and satellites. The six appendices include study of the operating environment and competition among international carriers, demand projections, existing and planned facilities, and projections of likely future technology. Notes, tables, charts, maps (many foldouts)

3-096 * Tribolet, Leslie Bennett. THE INTERNATIONAL ASPECTS OF ELECTRICAL COMMUNICATIONS IN THE PACIFIC AREA. Baltimore: Johns Hopkins Press "Studies in Historical and Political Science," 1929 (reprinted by Arno Press "International Propaganda and Communications," 1972), 282 pp. Arranged by region: South America, China, the Pacific, Japan, and the Philippines. Focuses primarily on cable and wireless links. Notes, bibliography, index.

3-097 * Ungerer, Herbert. TELECOMMUNICATIONS IN EUROPE: FREE CHOICE FOR THE USER IN EUROPE'S 1992 MARKET. Brussels: Commission of the European Communities "European Perspectives Series," 1988, 259 pp. While largely a policy review, sections II and III provide a detailed assessment of technological changes, especially development of convergence in voice and data markets. Tables, charts, glossary.

3-098 Williams, Francis. TRANSMITTING WORLD NEWS: A STUDY OF TELECOMMUNICATIONS AND THE PRESS. Paris: UNESCO, 1953 (reprinted by Arno Press "International Propaganda and Communications," 1972), 95 pp. Discusses leased wire services, facsimile and telephoto services, point-to-point radio channels, cable systems, etc. Maps, tables, references, index.

Technical Aspects of Propaganda

(The literature on propaganda is huge—noted here are only citations to the telecommunications aspects of propaganda.)

3-099 Browne, Donald R. **INTERNATIONAL RADIO BROADCASTING: THE LIMITS OF THE LIMITLESS MEDIUM**. New York: Praeger Special Studies, 1982, 369 pp. A largely social and political history of the many major services, with some reference to technology. Notes, appendices, bibliography, index.

3-100 **INTERNATIONAL AUDIO BROADCASTING FOR THE TWENTY-FIRST CENTURY**. Washington: National Academy Press, 1989, 67 pp. Assesses short-wave radio services and potential future digital and direct broadcast satellite options. Maps, charts, references.

3-101 Lean, Tangye. **VOICES IN THE DARKNESS: THE EUROPEAN RADIO WAR**. London: Secker and Warburg, 1943, 243 pp. Useful wartime view of propaganda radio efforts with more on content and politics than transmission. Photos, map, appendices, index.

3-102 McGregor, James P. **JAMMING OF WESTERN RADIO BROADCASTS TO THE SOVIET UNION AND EASTERN EUROPE**. Washington: U.S. Information Agency, Office of Research, Research Report R-4-83, 1983, 25 pp. Clear description of how jamming is accomplished and its impact. Charts, tables, notes.

3-103 * Nelson, Michael. **WAR OF THE BLACK HEAVENS: THE BATTLES OF WESTERN BROADCASTING IN THE COLD WAR**. Syracuse, NY: Syracuse University Press, 1997, 277 pp. Excellent history—with considerable discussion of transmitters and jamming—of the successes and failures of international propaganda radio. Photos, chronology, notes, bibliography, index.

3-104 Rigby, C.A. **THE WAR ON THE SHORT WAVES**. London: Lloyd Cole, 1944 (reprinted by Arno Press "International Propaganda and Communications," 1972), 68 pp. While aimed at amateur listeners, this offers a valuable wartime record of the technical side of propaganda broadcasting. Photos, diagrams, tables.

3-105 * Ward, John E., Ithiel De Sola Pool, and Richard J. Solomon. **A STUDY OF FUTURE DIRECTIONS FOR THE VOICE OF AMERICA IN THE CHANGING WORLD OF INTERNATIONAL BROADCASTING**. Cambridge, MA: MIT Laboratory for Information and Decision Systems, Final Report LIDS-FR-1296, April 25, 1983, 140 pp. Focuses on short wave signal delivery and spectrum utilization issues, drawbacks of shortwave, and looks ahead to digital services and satellites. Charts, references.

3-106 *** Wood, James. **HISTORY OF INTERNATIONAL BROADCASTING**. London: Peter Peregrinus and Science Museum "History of Technology Series 19," 1992, 258 pp. Only technical history available, this focuses on transmission, the author's specialty. Photos, bibliography, appendices, index.

3-107 *** _____. **HISTORY OF INTERNATIONAL BROADCASTING, VOLUME 2**. London: IEE "History of Technology Series 23," 1998, 270 pp. Con-

G. Military Telecommunications

General and Foreign

3-108 Adams, M.R. THROUGH TO 1970: ROYAL SIGNAL CORPS GOLDEN JUBILEE. London: Royal Signals Institution, 1970, 122 pp. Handsome illustrated album surveying story of British army signals and the technologies applied. Photos, maps, diagrams, bibliography. See also, Nalder, 3-123, 3-124.

3-109 Baker, Duncan C., and Brian A. Austin. "Wireless Telegraphy circa 1898–99: The Untold South African Story," *IEEE ANTENNAS AND PROPAGATION MAGAZINE*, 37:48–58 (December 1995). Some of the earliest use of wireless for military communications, during the Boer War. Photos, diagrams.

3-110 Bangay, Raymond D. "Wireless Telegraphy for Army Purposes," *THE ARMY REVIEW*, 3:23–41 (July–October 1912). Reviews current technology and applications of wireless to land military operations. Table, plates.

3-111 Bright, Charles, Sir. TELEGRAPHY, AERONAUTICS AND WAR. London: Constable, 1918, 407 pp. Collects a number of his papers and addresses.

3-112 de Arcangelis, Mario. ELECTRONIC WARFARE: FROM THE BATTLE OF TSUSHIMA TO THE FALKLANDS AND LEBANON CONFLICTS. Poole, England: Blandford Press, 1985, 320 pp. Translation from the Italian original. Historical survey of the development and applications of the "invisible war" technologies from the Russo-Japanese war (1904–05) to developments in space. Diagrams, maps, photos, notes, bibliography, index.

3-113 * Devereux, Tony. MESSENGER GODS OF BATTLE: RADIO, RADAR, SONAR, THE STORY OF ELECTRONICS IN WAR. London: Brassey's, 1991, 322 pp. Historical review of the role of electronics in military history, including the development and applications of radio, radar, sonar, and space electronic warfare. Diagrams, photos, maps, notes, bibliography, index.

3-114 Fitch, Richard D. "The Telegraph in World War I," *RADIO-ELECTRONICS*, 58:2:88–91 (February 1987).

3-115 Fox, Barry. "What Telecommunications did in the War," *NEW SCIENTIST* (October 30, 1986), pp. 35–38. World War II. Illustrations.

3-116 Fraser, David. A MODERN CAMPAIGN; OR, WAR AND WIRELESS TELEGRAPHY IN THE FAR EAST. London: Methuen, 1905, 356 pp. Author was a correspondent with the London *Times* for the early part of the Russo-Japanese war. Photos, maps.

3-117 Harris, L.H. **SIGNAL VENTURE.** Aldershot, England: Gale & Polden, 1951, 278 pp. A personal narrative on the role of telegraph, telephone, and radio with British forces in World War I and World War II. Photos, map, index.

3-118 Hippisley, Richard L. **HISTORY OF THE TELEGRAPH OPERATIONS DURING THE WAR IN SOUTH AFRICA, 1899–1902.** London: HMSO, 1903, 85 pp. With 100 diagrams.

3-119 Lansford, Willis R. "Aircraft Communication in World War I," *RADIO NEWS,* 29:50–54, 208–216 (June 1943). Historical survey from 1910 to October 1917 covering the earliest air-to-ground communications. Photos.

3-120 Meulstee, Louis. **WIRELESS FOR THE WARRIOR: A TECHNICAL HISTORY OF RADIO COMMUNICATION EQUIPMENT IN THE BRITISH ARMY**

> **VOL. 1: WIRELESS SETS 1–88.** Broadstone, England: G.C. Arnold, 1995, 200 pp. Detailed technical history of the devices used, beginning with those of the early 20th century. Photos, tables.

> **VOL. 2: STANDARD SETS OF WORLD WAR II.** 1998, 722 pp. Includes some Canadian and Australian equipment. Photos, drawings, tables, glossary.

> **VOL. 3: RECEPTION SETS.** in progress (probably 2001) will include selected equipment into the 1960s.

3-121 * Moir, John S., ed. **HISTORY OF THE ROYAL CANADIAN CORPS OF SIGNALS 1903–1961.** Ottawa: Royal Canadian Corps of Signals, 1962, 366 pp. Carries the story through the Korean War. Photos, maps (folding and endpaper), index.

3-122 Moreau, Louise. "The Military Communications Explosion, 1914–1918," **THE A.W.A. REVIEW,** 6:135–154 (1991). Photos, references.

3-123 ** Nalder, R.F.H. **THE HISTORY OF BRITISH ARMY SIGNALS IN THE SECOND WORLD WAR: GENERAL SURVEY.** London: Royal Signals Institution, 1953, 377 pp. See next entry. Charts, appendices, index.

3-124 *** _____. **THE ROYAL CORPS OF SIGNALS: A HISTORY OF ITS ANTECEDENTS AND DEVELOPMENT (CIRCA 1800–1955).** London: Royal Signals Institution, 1958, 672 pp. With previous entry, best histories of the British army's developing use of signaling technology. Photos, maps, appendices. Index.

3-125 Rohwer, Jürgen. "The Wireless World at War: 1939–1945," *INTERNATIONAL HISTORY REVIEW,* 16:536–548 (August 1994). Review essay of six books concerning intelligence, radar, and codebreaking during World War II.

3-126 *** Scheips, Paul J., ed. **MILITARY SIGNAL COMMUNICATIONS.** New York: Arno Press "Historical Studies in Telecommunications," 1980, 2 vols., ca

General Surveys

750 pp. Covers the history of American and British army use of the telegraph, telephone, and other means of electrical communications. Vol. 1 includes 27 papers, many covering the Civil War and telegraph eras. Vol. 2 includes another 29 covering signaling techniques with semaphore, telegraph, telephone, and radio service. Photos, diagrams, maps.

3-127 Turner, H[enry] F. **NOTES ON MILITARY TELEGRAPH INSTRUMENTS WITH DIAGRAMS OF CONNECTIONS.** London: HMSO, 1884, 1885 [rev. ed.] 25 pp. "The [33] diagrams form a special feature of the booklet" (Weaver, 1-023, I: 2377).

3-128 Warner, Philip. **THE VITAL LINK: THE STORY OF ROYAL SIGNALS, 1945–1985.** London: Leo Cooper, 1989, 353 pp. Endpaper maps, index.

3-129 * Woods, David. **A HISTORY OF TACTICAL COMMUNICATION TECHNIQUES.** Orlando, FL: Martin-Marietta Corp, 1965 (reprinted by Arno Press "History of Telecommunications," 1977), 300 pp. Only history of its type, cutting across the years from messengers and signal flags up to various electrical means. Photos, diagrams, bibliography.

U.S.

3-130 *ARMY TIMES*, Editors of. **A HISTORY OF THE UNITED STATES SIGNAL CORPS.** New York: Putnam, 1961, 192 pp. Popular history in 12 chapters from the Civil War to pioneering satellite operations. Photos, bibliography, index. See also 3-137.

3-131 Bates, David Homer. **LINCOLN IN THE TELEGRAPH OFFICE: RECOLLECTIONS OF THE UNITED STATES MILITARY TELEGRAPH CORPS DURING THE CIVIL WAR.** New York: The Century Co., 1907 (reprinted in "Appleton Dollar Library," 1939; by Univ. of Nebraska "Bison Books," 1995; and Old Books Pub., 1996), 431 pp. The view from the top, 1861–65 showing how the president could better control his generals thanks to telegraphy. Photos, appendix, index.

3-132 BATTLE TALK. New York: Western Electric, 1943, 38 pp. Wartime applications of telephone equipment. Photos and diagrams.

3-133 * Bergen, John D. **MILITARY COMMUNICATIONS: A TEST FOR TECHNOLOGY—THE U.S. ARMY IN VIETNAM.** Washington: Government Printing Office, Center for Military History "United States Army in Vietnam," 1986, 515 pp. Highly detailed survey in 20 chapters of changing communication technology from the 1950s into the mid-1970s, with one chapter on North Vietnamese developments. Maps, photos, notes, index. See also Rienzi, 3-144.

3-134 * Brown, J. Willard. **THE SIGNAL CORPS, U.S.A. IN THE WAR OF THE REBELLION.** Boston: U.S. Veteran Signal Corps Association, 1896 (reprinted by Arno Press "Telecommunications," 1974, minus roster of Signal Corps. members; and again by Walter Mathers, 1996), 711 pp. Very detailed survey of the technologies, men, and campaigns of the war, with some mention of Confederate developments. Suffers from lack of index. Photos, engravings, maps. See also, Plum, 3-140.

3-135 Harlow, Alvin F. **BRASS POUNDERS: YOUNG TELEGRAPHERS OF THE CIVIL WAR.** Denver: Sage Books, 1962, 159 pp. Written for young people, this concentrates on telegraph operators, many of whom were boys. Photos.

3-136 * Lavine, A. Lincoln. **CIRCUITS OF VICTORY.** Garden City, NY: Doubleday Doran, 1921, 634 pp. Anecdotal history of American telephone facilities on the Western Front, 1917–19. Includes some discussion of both telegraph and wireless applications as well. Photos, maps.

3-137 Marshall, Max L., ed. **THE STORY OF THE U.S. ARMY SIGNAL CORPS.** New York: Franklin Watts, 1965, 305 pp. Divided into two parts—half historical articles and half dealing with the Signal Corps as of the time of publication. Photos, appendices, index. See also 3-130.

3-138 MILITARY TELEVISION: THE FIRST PUBLIC DEMONSTRATION OF COMBAT TELEVISION. New York: Radio Corporation of America, 1954, 16 pp. Promotional booklet. Photos.

3-139 O'Brien, John Emmet. **TELEGRAPHING IN BATTLE: REMINISCENCES OF THE CIVIL WAR.** Scranton, PA: Raeder Press, 1910, 312 pp. A professional autobiography; the author and his brother served in Civil War telegraph units. Photos.

3-140 * Plum, William R. **THE MILITARY TELEGRAPH DURING THE CIVIL WAR IN THE UNITED STATES.** Chicago: Jansen, McClurg & Co., 1882, 2 vols. (reprinted in single vol. by Arno Press "Telecommunications," 1974), 377 + 390 pp. Massively detailed history of the U.S. Military Telegraph Corps of the Union Army as well as the telegraph in the Confederacy in all theaters of war. Engravings, documents, maps, index. See also Brown, 3-134.

3-141 * Price, Alfred. **THE HISTORY OF U.S. ELECTRONIC WARFARE.** Alexandria, VA: Association of Old Crows, 1984, 1989 (2 vols.). Dealing primarily with radar and jamming efforts, this set (sponsored by present and retired military personnel concerned with electronic warfare) also discusses other kinds of electronic warfare and its countermeasures. Photos, notes, diagrams, index.

> **VOL. 1: THE YEARS OF INNOVATION—BEGINNINGS TO 1946.** (1984, 312 pp.) Devoted largely to World War II, and discusses Allied and enemy equipment.
>
> **VOL. 2: THE RENAISSANCE YEARS, 1946 TO 1964.** (1989, 390 pp.) Deals with the postwar period to the eve of Vietnam.
>
> **VOL. 3: 1964–PRESENT.** (in preparation) Will focus on Vietnam War and after.

3-142 *** Raines, Rebecca Robbins. **GETTING THE MESSAGE THROUGH: A BRANCH HISTORY OF THE U.S. ARMY SIGNAL CORPS.** Washington: Government Printing Office "Army Historical Series," 1996,

General Surveys

464 pp. First comprehensive history of the Signal Corps including trends in technology of instruments used. Photos, maps, appendices, (excellent) bibliographical note, glossary, index.

3-143 ** **REPORT OF THE CHIEF SIGNAL OFFICER TO THE SECRETARY OF WAR, 1919.** Washington: Government Printing Office, 1919 (reprinted by Arno Press "Telecommunications," 1974), 547 pp. Invaluable and very detailed study of the use of telecommunications in World War I by the U.S. Army. Suffers only from lack of an index. Tables, photos, maps. (Note: such annual reports were published from 1860 to 1920 when they were discontinued in an economy move. Taken as a series, the reports allow one to trace the Army's growing use of telegraphy, telephony, and wireless.)

3-144 Rienzi, Thomas M. **COMMUNICATIONS-ELECTRONICS, 1962–1970.** Washington: Government Printing Office "Department of the Army Vietnam Studies," 1972, 184 pp. Useful contemporary view of the role of signals communications in the Vietnam War. Photos, maps, index. See also Bergen, 3-133.

3-145 * Scriven, George P. **THE TRANSMISSION OF MILITARY INFORMATION.** Governors Island, NY: Journal of the Military Service Institution, 1908, 153 pp. Articles reprinted from the *Journal* on the Signal Corps and the mobile army and wireless, electrical and visual communication, and relation of Signal Corps to the coast artillery. Useful for contemporary view as wireless was becoming more widely used. Diagrams, photos.

See Sexton, 1-016.

3-146 "Signal Corps Centennial Issue," *TRANSACTIONS OF THE INSTITUTE OF RADIO ENGINEERS,* MIL-4: 396–607 (October 1960). Thirty-five papers on high frequency communications, radar, and other developments in military electronics, including historical surveys. Photos, references.

3-147 ** **THE SIGNAL CORPS.** Washington: Government Printing Office "U.S. Army in World War II: The Technical Services," 1956–66, 3 vols. as follows. The series offers a well-documented and detailed assessment of both technology development and applications, as well as Signal Corps operations during the war. Photos, tables, maps, index.

> **VOL. 1:** Terrett, Dulany. **THE EMERGENCY.** (1956, 383 pp.) Deals with the period up to Pearl Harbor attack.

> **VOL. 2:** Thompson, George Raynor, et al. **THE TEST.** (1957, 621 pp.) Covers the initial American role in the war, December 1941 to July 1943.

> **VOL. 3:** Thompson, George Raynor, and Dixie R. Harris. **THE OUTCOME** (1966, 720 pp.) Completes the story— from mid-1943 through 1945.

3-148 Snyder, Thomas S., gen. ed. **THE AIR FORCE COMMUNICATIONS COMMAND: PROVIDING THE REINS OF COMMAND 1938-1981.** Scott Air Force Base, IL, I: AFCC Office of History, Scott Air Force Base, 1981, 231 pp.

Well-illustrated history of the organization and technology of aircraft communications in the military. Photos, indexes.

H. Maritime and Naval Telecommunications

General and Foreign

3-149 Baarslag, Karl. **SOS TO THE RESCUE**. New York: Oxford University Press, 1935, 310 pp. (London: Methuen, 1937, 243 pp.). Popular narrative of wireless in sea disasters written by a sea-going radio operator. Photos, index.

3-150 Booth, John, and Sean Coughlan. **TITANIC: SIGNALS OF DISASTER**. Westbury, England: White Star Publications, 1993, 184 pp. Reproduces most of the Marconi traffic to and from the liner which sank on an infamous maiden voyage in April 1912. Photos, index.

3-151 Copplestone, Bennet [pseud. for Frederick H. Kitchin]. **TALES OF S.O.S. AND T.T.T.** Edinburgh: William Blackwood / Boston: Houghton Mifflin, 1927, 235 pp. The role of wireless at sea told though a number of specific disasters. No documentation or index.

3-152 Dubreuil, Stephan. **COME QUICK, DANGER—A HISTORY OF MARINE RADIO IN CANADA**. Ottawa: Canadian Government Publishing, 1998, 283 pp. Largely based on first-person accounts, this details the Canadian Coast Guard's efforts to save ships and crews off Canada's three coasts. Photos, notes, bibliography.

3-153 Erskine-Murray, James. "Wireless Communication over Sea," *TRANSACTIONS OF THE INSTITUTION OF ENGINEERS AND SHIPBUILDERS,* 51:162–192 (1908). Survey of the contemporary state of maritime wireless technology. Includes discussion between author and other prominent engineers.

3-154 Faulkner, Brian. **WATCHERS OF THE WAVES**. Broadstone, England: G.C. Arnold, 1996, 117 pp. A history of maritime coast radio stations in Britain from 1896 to 1995. Diagrams, maps, photos.

3-155 * Goodwin, W.D. **ONE HUNDRED YEARS OF MARITIME RADIO**. Glasgow: Brown Son & Ferguson, 1995, 259 pp. A thorough history of the maritime applications of radio with a heavy emphasis on British shipping. Photos, bibliography, index.

3-156 *** Hancock, Harry E. **WIRELESS AT SEA: THE FIRST FIFTY YEARS**. Chelmsford, England: Marconi International Marine Communication Co., 1950 (reprinted by Arno Press "Telecommunications," 1974), 263 pp. Focusing on the company, this provides a valuable historical survey of wireless on board merchant ships. Photos, endpaper maps, index.

3-157 * Hezlet, Sir Arthur. **ELECTRONICS AND SEA POWER**. New York: Stein & Day, 1975, 318 pp. The role of wireless and other electronics (including radar and code breaking) in naval warfare from the late 19th century into the 1960s. Photos, notes, bibliography, index.

3-158 * Kent, Capt. Barrie. **SIGNAL: A HISTORY OF SIGNALING IN THE ROYAL NAVY.** Hampshire, England: Hayden House, 1993, 371 pp. From signal flags to wireless and on to modern methods. Photos.

3-159 * Pocock, R.F., and G.R.M. Garratt. **THE ORIGINS OF MARITIME RADIO: THE STORY OF THE INTRODUCTION OF WIRELESS TELEGRAPHY IN THE ROYAL NAVY BETWEEN 1896 AND 1900.** London: HMSO "Science Museum Survey," 1972, 60 pp. Based on official files. Photos, appendices, references.

3-160 WIRELESS AND SHIPPING: A SYMPOSIUM OF SPECIAL ARTICLES WRITTEN BY EXPERTS AND REPRINTED FROM *THE SHIPPING WORLD*. London: Marconi International Marine, 1932, 87 pp. Just that. Tables, photos.

3-161 *** Woods, David L., ed. **SIGNALING AND COMMUNICATING AT SEA.** New York: Arno Press "Historical Studies in Telecommunications," 1980, 2 vols., about 800 pp. Extensive anthology of contemporary papers and documents. Vol. 1 includes 31 items, most devoted to visual signaling. Vol. 2 continues with 27 more which are largely devoted to wireless, including research. Photos, diagrams, tables, references.

U.S.

3-162 Douglas, Susan. "Technological Innovation and Organizational Change: The Navy's Adoption of Radio, 1899–1919," in Merritt Roe Smith, ed. **MILITARY ENTERPRISE AND TECHNOLOGICAL CHANGE.** Cambridge, MA: MIT Press, 1985, pp. 117–173. Focuses on the role of Stanford C. Hooper.

3-163 ** Gebhard, Louis A. **EVOLUTION OF NAVAL RADIO-ELECTRONICS AND CONTRIBUTIONS OF THE NAVAL RESEARCH LABORATORY.** Washington: Government Printing Office (Naval Research Laboratory Report 8300), 1979, 448 pp. An important study of the development of improving wireless technology. Includes discussion of radar and electronic countermeasures as well as radio communication. Photos, chapter references, index. See also Taylor, 3-166.

3-164 *** Howeth, L[inwood] S., Capt. **HISTORY OF COMMUNICATIONS-ELECTRONICS IN THE UNITED STATES NAVY.** Washington: Government Printing Office, 1963, 657 pp. Well-documented history focusing on wireless and radio applications. The 42 chapters appear in sections on the decade of development (to World War I), the golden age (1914 to the early 1920s), and the age of electronics (through World War II). Numerous appendices, photos, notes, bibliography, index.

See Hudgins, 9-014

See Robison, 9-035

3-165 * Schroeder, Peter B. **CONTACT AT SEA: A HISTORY OF MARITIME RADIO COMMUNICATIONS**. Ridgewood, NJ: The Gregg Press, 1967, 150 pp. Chronological narrative of the development of wireless and radio for maritime use, with an emphasis on safety of life at sea and international conventions concerning same. Photos, appendices, bibliography, notes, index.

3-166 * Taylor, A. Hoyt. **RADIO REMINISCENCES: A HALF CENTURY**. Washington: U.S. Naval Research Laboratory, 1949 (republished 1960), 248 pp. The author's experiences from the turn of the century to World War II, with a host of details on people and technical developments. See also Gebhard, 3-163.

3-167 U.S. Navy Department. "Radio Telegraphy," in **HISTORY OF THE BUREAU OF ENGINEERING, NAVY DEPARTMENT: DEVELOPMENT DURING THE PERIOD OF THE WAR**. Washington: Government Printing Office, 1922, pp. 90–132. Explores the role of wireless during World War I. Photos. Reprinted in Woods, 3-161.

Chapter 4

Institutional and Company Histories

Most telecommunications technical development has taken place within the private sector—companies, research centers, trade and professional associations. Section A begins with reference resources such as directories of company histories, chronologies, and archives, and other sources of company information. See 1-C for some related reference material.

Historical treatments of international and U.S. federal government bodies concerned with telecommunication (except the military which appear at the end of chapter 3) are found in section B. Remarkably little historical material exists on these agencies whose decisions are often central to industry developments. Save for an occasional thesis or dissertation, little seems to have been written, for example, about the historical role of the state department in telecommunications. No overall history of the FCC exists—or has been attempted to date.

Section C cites the relative handful of broad telecommunication industries surveys available. Review the cross-referenced material from chapters 1 and 3 for additional approaches. (Most histories, of course, are either of specific companies, for which see section E of this chapter, or of technical developments.)

The histories of professional societies noted in section D mirror activities and events that constitute progress of the respective science or technology. Such a record is also a kind of collective biography. These histories offer information concerning professional and engineering standards, the growth of technical education, professional publications, and such special events as conventions and exhibitions.

Specific company histories, surveyed in sections E (American companies) and F (foreign firms and administrations), record leading events in research and industrial development, progress in production and services, and tell of many of the people who contributed to these advances. Included here are a few company or industry leader biographies, appearing with other entries about their firms (e.g., Vail with AT&T, Young with GE, Sarnoff with RCA, Siemens with his firm). Biographies of inventors and other technical people appear in chapter 5. For related Internet resources, see 15-C.

A. Reference Resources

See Bedi, 1-037.

4-001 Cope, S.T., comp. **GLOSSARY OF ABBREVIATIONS FOR NAMES OF TECHNICAL, SCIENTIFIC, INDUSTRIAL AND PROFESSIONAL ORGANIZATIONS, WITH PARTICULAR REFERENCE TO THE TELECOMMUNICATIONS INDUSTRY.** Chelmsford, England: Marconi's Wireless Telegraph Company, 1957, [2nd ed.], 38 pp.

4-002 Ingham, John N. **BIOGRAPHICAL DICTIONARY OF AMERICAN BUSINESS LEADERS.** Westport, CT: Greenwood, 1983, 4 vols. Some 1,100 key figures are included, with indexing by industry.

4-003 *** **INTERNATIONAL DIRECTORY OF COMPANY HISTORIES.** Chicago: St. James Press, 1988–date, 25 vols. Global in scope, this is exceptionally useful for informative histories of electronics, telecommunications, and information companies. First six vols. arranged alphabetically by industry; vol. 2 includes electronics companies; vol. 4 includes telecommunications carriers. From vol. 7 on, material is arranged alphabetically by company name. References include bibliographies. Index to all companies appears in each of the later volumes.

4-004 Jeremy, David J. **DICTIONARY OF BUSINESS BIOGRAPHY: A BIOGRAPHICAL DICTIONARY OF BUSINESS LEADERS ACTIVE IN BRITAIN IN THE PERIOD 1860–1980.** London: Butterworths, 1984, 5 vols. More than 1,000 British CEOs and others are profiled at some length with citations of both unpublished and published reference sources. Portraits, index.

4-005 ** Martin, Susan Boyles, ed. **NOTABLE CORPORATE CHRONOLOGIES.** Detroit: Gale Research, 1995, 2 vols., 2,038 pp. Some 1,150 companies worldwide are included, each with at least 30 historical dates (some larger firms such as AT&T go on for pages), and further reading for most. Geographical, alphabetical, and anniversary indexes.

4-006 MOODY'S INDUSTRIAL MANUAL. New York: Moody's Investor Services, 1909–date, annual. Over the years this reference allows one to trace company developments and changes for some 2,000 companies (by the early 1990s). MOODY'S PUBLIC UTILITY MANUAL (1914–date, annual) includes all public and private utilities, including telegraph and telephone carriers.

4-007 * Nasrallah, Wahib. **UNITED STATES CORPORATE HISTORIES: A BIBLIOGRAPHY, 1965–1990.** New York: Garland, 1986; 1991 [2nd ed.], 511 pp. Unannotated listing of books, articles, company publications, and dissertations, arranged by firm. Indexed by industry and executive officer. Second edition added 1985–90 material.

Institutional and Company Histories

4-008 Roess, Anne C. **PUBLIC UTILITIES: AN ANNOTATED GUIDE TO INFORMATION SOURCES.** Metuchen, NJ: Scarecrow Press, 1991, 340 pp. Section IV offers about 180 citations on the telephone industry.

4-009 * **STANDARD & POOR'S INDUSTRIAL SURVEYS.** New York: Standard & Poor/McGraw-Hill, weekly with annual cumulation. As of 1996, coverage expanded to 52 categories including broadcasting and cable, communications equipment, several computer categories, movies and home entertainment, semiconductors (2), wireless telecommunications, and wireline telecommunication. Concise analytic narrative and statistics.

B. Government Bodies

(Includes international bodies and elements of the United States federal government. For foreign carriers and administrations, see 4-F below.)

International

International Telecommunication Union (ITU)

4-010 "CCIR 50th Anniversary Special," *TELECOMMUNICATION JOURNAL,* 45:6/7:266–323, 357–404 (June–July 1978). The International Consultative Committee on Radio (CCIR) is reviewed historically with some discussion of its study groups. The focus is on technical standards. Photos, charts.

4-011 Channing, Ian, commissioning ed. **INTERNATIONAL TELECOMMUNICATION UNION: 130 YEARS, 1865–1995.** London: International Systems and Communications, 1995, 612 pp. Though half devoted to advertising by governments and international firms, this still offers useful historical survey articles, descriptions of the three sectors of the ITU, technology surveys of telecommunication, radio services, personal communications and the like. Photos, charts, bibliography.

4-012 *** Codding, George A. Jr. **THE INTERNATIONAL TELECOMMUNICATION UNION: AN EXPERIMENT IN INTERNATIONAL COOPERATION.** Leiden, Netherlands: E.J. Brill, 1952 (reprinted by Arno Press "International Propaganda and Communications," 1972), 505 pp. Definitive history of the ITU to time of publication. Notes, tables, bibliography, index.

4-013 ** Codding, George A. Jr., and Anthony M. Rutkowski. **THE INTERNATIONAL TELECOMMUNICATION UNION IN A CHANGING WORLD.** Norwood, MA: Artech House, 1982, 414 pp. Administrative study building on previous entry. While dated (the ITU's structure has changed substantially since), this remains useful as an update of previous title. Tables, annexes, bibliography, index.

4-014 * Codding, George A. Jr. "Evolution of the ITU," *TELECOMMUNICATION POLICY,* 15:271–285 (August 1991). A brief survey by an American authority—see also his books above.

4-015 Lee, Kelley. **GLOBAL TELECOMMUNICATIONS REGULATION: A POLITIAL ECONOMY PERSPECTIVE.** London: Pinter, 1996, 210 pp. Most of this is devoted to a study of the ITU's development and operations. Notes, bibliography, glossary, index.

4-016 * [Michaelis, Anthony R.] FROM SEMAPHORE TO SATELLITE: PUBLISHED ON THE OCCASION OF THE CENTENARY OF THE INTERNATIONAL TELECOMMUNICATION UNION. Geneva: ITU, 1965, 343 pp. Highly illustrated work with brief supporting text providing a short and accurate, though non-technical, account devoted mainly to international affairs. Brief bibliography. No documentation or index. Photos, diagrams, maps.

4-017 Savage, James G. THE POLITICS OF INTERNATIONAL TELECOMMUNICATIONS REGULATION. Boulder, CO: Westview Press, 1989, 240 pp. Focused on the ITU this relates the organization's history and offers a chapter on the ITU's role in spectrum use and management and another on the setting of international technical standards. Notes, index.

4-018 Tomlinson, John D. THE INTERNATIONAL CONTROL OF RADIO-COMMUNICATIONS. Ann Arbor, MI: J.W. Edwards, 1945 (reprinted by Arno Press "International Propaganda and Communications," 1972), 314 pp. Review of the spectrum management process before and after it became part of the ITU in 1932. Notes, appendices, index.

International Telecommunications Satellite Organization (INTELSAT)
(See also Comsat, 4-124 through 4-126.)

4-019 ** Alper, Joel, and Joseph N. Pelton, eds. THE INTELSAT GLOBAL SATELLITE SYSTEM. New York: American Institute of Aeronautics and Astronautics "Progress in Astronautics and Aeronautics Vol. 93," 1984, 425 pp. Fifteen contributed papers review the first two decades of INTELSAT development; others discuss satellites, earth segment, transmission techniques, INTELSAT operations, relations with the ITU, planning, research and development, etc. Tables, charts, diagrams, references, glossary.

4-020 * ANNUAL REPORT. Washington: INTELSAT, 1973–date, annual. Useful summation of the world organization's operations and changing satellite technology. Tables, photos, charts.

4-021 COMSAT GUIDE TO THE INTELSAT SATELLITE SYSTEM. Washington: COMSAT, annual. Booklet details the full series of satellites from "Early Bird" in 1965 to the present in charts, photos, and data plus a list of ground stations.

4-022 * Edelson, Burton I. "Global Satellite Communications," *SCIENTIFIC AMERICAN* (February 1977), 237:58–73. Survey of the first twelve years. Photos, charts, maps.

4-023 INTELSAT MEMOIRS. Washington: INTELSAT, 1979, 48 pp. Brief collection of essays and articles on the first 15 years. Photos.

Institutional and Company Histories 61

4-024 INTELSAT RESEARCH AND DEVELOPMENT PROGRAM. Washington: INTELSAT, various years. Occasional booklet publication reviewing current trends.

4-025 * Snow, Marcellus S. **THE INTERNATIONAL TELECOMMUNICATIONS SATELLITE ORGANIZATION (INTELSAT).** Baden-Baden, Germany: Nomos Verlagsgesellschaft "Law and Economics of International Telecommunications 2," 1987, 200 pp. Background and structure as well as relationships with member countries. References, bibliography.

4-026 TWENTY-FIFTH ANNIVERSARY HISTORICAL CHRONOLOGY. Washington: INTELSAT, 1989, poster. Includes key dates from 1945 to 1989. Photos.

United States Government
(For U.S. and foreign military services, see chapter 3-G and 3-H.)

U.S. Congress

4-027 Knauer, Leon T., et al. **TELECOMMUNICATIONS ACT HANDBOOK: A COMPLETE REFERENCE FOR BUSINESS.** Rockville, MD: Government Institutes, 1996, 620 pp. Articles about and a full transcript of the important amendments, showing clearly what was added and subtracted from the 1934 act. Tables, charts, glossary, index.

4-028 * Paglin, Max D., ed. **A LEGISLATIVE HISTORY OF THE COMMUNICATIONS ACT OF 1934.** New York: Oxford University Press, 1989, 981 pp. Pulls together the key documents (including committee hearings and floor debate) concerning the passage of the landmark legislation which created the FCC and is, though amended, still in force. Index.

U.S. Department of Agriculture

4-029 * Gray, Gary Craven. **RADIO FOR THE FIRELINE: A HISTORY OF ELECTRONIC COMMUNICATION IN THE FOREST SERVICE.** Washington: U.S. Department of Agriculture, Forest Service (FS Publication 369), March 1982, 304 pp. Useful survey of a unique application of wireless service. Photos, diagrams, charts, tables, reference notes, appendices (including one on specific radio equipment utilized).

U.S. Department of Commerce

4-030 FROM LIGHTHOUSES TO LASERBEAMS: A HISTORY OF THE U.S. DEPARTMENT OF COMMERCE, 1913–1988. Washington: GPO, 1988, 117 pp. Includes discussion of the National Telecommunications and Information Administration (and its predecessors), and the Patent Office, among other units. Photos, tables, references.

4-031 Kittross, John M., ed. **DOCUMENTS IN AMERICAN TELECOMMUNICATIONS POLICY.** New York: Arno Press "Historical Studies in Telecommunications," 1977, 2 vols. First volume includes all relevant portions of Department of

Commerce reports concerning wireless and radio, 1909–32, and recommendations of National Radio Conferences of 1922–24 called by the department.

See Rosen, 4-046.

4-032 * Snyder, Wilbert F., and Charles L. Bragaw. **ACHIEVEMENT IN RADIO: SEVENTY YEARS OF RADIO SCIENCE, TECHNOLOGY, STANDARDS, AND MEASUREMENT AT THE NATIONAL BUREAU OF STANDARDS.** Washington: Government Printing Office, 1986, 842 pp. Some 30 chapters relate the development of radio work from early in the century to the present. Photos, notes, appendices, index.

U.S. Executive Office of the President

4-033 Coase, Ronald H. "The Interdepartment Radio Advisory Committee," *JOURNAL OF LAW AND ECONOMICS*, 5: 17–47 (October 1962). Later a Nobel laureate, the author traces the history and role of IRAC in determining government frequency assignments.

4-034 President's Communications Policy Board. **TELECOMMUNICATIONS: A PROGRAM FOR PROGRESS.** Washington: GPO, 1951, 238 pp. Results of Korean War-era analysis of how telecommunications technology was changing and what government polices were needed to encourage both national security and private sector needs. Charts, notes. Reprinted in Kittross, 4-031.

4-035 President's Task Force on Communication Policy. **FINAL REPORT.** Washington: GPO, 1969, 528 pp. The "Rostow" report assesses both media and telecommunications needs on the domestic and international level as well as what government structure and policies would be useful.

4-036 U.S. Executive Office of the President, Office of Telecommunications Policy. **THE INTERDEPARTMENT RADIO ADVISORY COMMITTEE: 50 YEARS OF SERVICE.** Washington: GPO "TEL-IRAC-72-30," 1972, 32 pp. Brief survey of the main government body allocating and licensing federal spectrum users. Tables, photo.

4-037 Will, Thomas E. **TELECOMMUNICATIONS STRUCTURE AND MANAGEMENT IN THE EXECUTIVE BRANCH OF GOVERNMENT, 1900–1970.** Boulder, CO: Westview, 1978, 214 pp. Focuses on the role of telecommunications advisor in the office of the president. Charts, notes, references.

4-038 Not used.

U.S. Federal Communications Commission (FCC)
(Includes references to the Federal Radio Commission.)

4-039 Emery, Walter B. "FCC Chronology and Leadership from 1934 to 1970," in **BROADCASTING AND GOVERNMENT: RESPONSIBILITIES AND REGULATIONS.** East Lansing, MI: Michigan State University Press, 1971 [2nd

Institutional and Company Histories

ed.], pp. 455–499. Though stressing electronic media concerns, this is one of the few attempts at a history of the FCC.

4-040 "Federal Communications Commission," *FORTUNE* (May 1938), pp. 17:5:60–62, 124–135; and "Government by Commission," *FORTUNE* (May 1943), pp. 29:5:86–89, 202–209 (May 1943). Two contemporary surveys, a mere five years apart, of the FCC's developing role. Second article focuses on network regulation.

4-041 Flannery, Gerald V. **COMMISSIONERS OF THE FCC, 1927–1994**. Lanham, MD: University Press of America, 1994, 228 pp. Useful brief (averaging three pages) profiles of chairs and members of both the FRC and FCC.

4-042 Herring, James M., and Gerald C. Gross. **TELECOMMUNICATIONS: ECONOMICS AND REGULATION**. New York: McGraw-Hill, 1936, (reprinted by Arno Press "Telecommunications," 1974), 544 pp. Despite title, offers very useful historical background on wired (telegraph, submarine cable, telephone) and wireless industries, and development of FCC activities growing out of 1934 act, including technical regulation. Tables, notes, appendices, index.

4-043 Horwitz, Robert Britt. **THE IRONY OF REGULATORY REFORM: THE DEREGULATION OF AMERICAN TELECOMMUNICATIONS**. New York: Oxford Unversity Press, 1989, 31:414. Argues that far more remains regulated despite years of rhetoric and action to *de*regulate both telephone and broadcast industries. Notes, bibliography, index.

4-044 [Malia, Thomas.] **FIFTY YEARS OF *TELECOMMUNICATIONS REPORTS***. New York: Business Research Publications, 1985, 220 pp. Year-by-year story from 1934 through 1984 of telephone industry regulation as reported from pages of the trade weekly.

4-045 McChesney, Robert W. **TELECOMMUNICATIONS, MASS MEDIA, & DEMOCRACY: THE BATTLE FOR THE CONTROL OF U.S. BROADCASTING, 1928–1935**. New York: Oxford University Press, 1993, 393 pp. Strongly critical analysis of developing radio regulation from early FRC through the first year of the FCC, noting arguments against the system as developed. Notes, bibliography, index.

4-046 Rosen, Philip T. **THE MODERN STENTORS: RADIO BROADCASTERS AND THE FEDERAL GOVERNMENT, 1920–1934**. Westport, CT: Greenwood, 1980, 267 pp. How broadcasters at first supported, and then resisted, the growth of regulation through the FRC and early FCC. Photos, notes, bibliography, index.

4-047 Schmeckebier, Laurence F. **THE FEDERAL RADIO COMMISSION: ITS HISTORY, ACTIVITIES AND ORGANIZATION**. Washington: Brookings Institution "Service Monographs of the United States Government No. 65," 1932, 162 pp. Just that. Appendices, notes, index. Partially reprinted in Kittross, 4-031.

4-048 * U.S. Federal Communications Commission. **ANNUAL REPORT**. Washington: Government Printing Office, 1935–date (first 20 issues [for 1935–55] reprinted by Arno Press "History of Broadcasting," 1971). Valuable means of tracing FCC policy developments and decisions. Though less apparent in recent years, has often included extensive statistics.

4-049 _____, Office of Public Affairs. **50th ANNIVERSARY, 1934–1984: A CHRONOLOGICAL HISTORY**. Washington: FCC, 1984, 30 pp. Brief entries, some of which deal with technology.

4-050 U.S. Federal Radio Commission. **ANNUAL REPORT**. Washington: Government Printing Office, 1927–33 (reprinted in one volume by Arno Press "History of Broadcasting," 1971). Predecessor to the FCC, this body concentrated on radio broadcasting technical and economic problems. Tables, charts, maps (including foldouts).

U.S. Post Office

4-051 U.S. Postmaster General. **GOVERNMENT OWNERSHIP OF ELECTRICAL MEANS OF COMMUNICATION**. Washington: Government Printing Office, 1914 (partially reprinted in Kittross, 4-031), 150 pp. Surveys history of Post Office operation of telegraph in mid-1840s, and attempts in years thereafter to regain control as demonstrated in reprinted Postmaster General reports, 1845–1913.

C. Electronics/Telecommunications Industry Surveys

See A.W.A. REVIEW, 9-092.

4-052 Borchardt, Kurt. **STRUCTURE AND PERFORMANCE OF THE U.S. COMMUNICATIONS INDUSTRY: GOVERNMENT REGULATION AND COMPANY PLANNING**. Cambridge, MA: Harvard University Graduate School of Business Administration, 1970, 180 pp. Largely economic. Appendices, reference notes.

4-053 ** Brock, Gerald W. **THE TELECOMMUNICATIONS INDUSTRY: THE DYNAMICS OF MARKET STRUCTURE**. Cambridge, MA: Harvard University Press, 1981, 336 pp. A standard history of the industry to the eve of the AT&T divestiture with insightful comments on technology's impact on both policy and applications. Notes, index.

4-054 Bussey, Gordon. **WIRELESS—THE CRUCIAL DECADE: HISTORY OF THE BRITISH WIRELESS INDUSTRY, 1924–1934**. London: Peter Peregrinus with Science Museum "IEE History of Technology Series 13," 1990, 126 pp. Chapters on broadcasting trends, radio valves (tubes), receiver developments, British domestic wireless, and home construction and set use. Photos, tables, chapter references, bibliography, index. See also, Pocock, 4-067.

See CONSUMER ELECTRONICS INDUSTRY IN REVIEW, 1-061.

4-055 Cortada, James W. **HISTORICAL DICTIONARY OF DATA PROCESSING ORGANIZATIONS**. Westport, CT: Greenwood, 1987, 311 pp. While

Institutional and Company Histories 65

most of the firms focus on computers and software, some also played a broader telecommunications role—e.g., AT&T, Bell Labs, and RCA. References, index.

4-056 ** Douglas, Alan. **RADIO MANUFACTURERS OF THE 1920s**. Vestal, NY: Vestal Press, 1988-90 (reprinted by Chandler Press, 1998) 3 vols. Very useful narrative of companies large and small, plus illustrations of their products, many from contemporary articles and advertisements. Many of the firms included lasted far beyond the 1920s while others are virtually unknown today. Photos, diagrams, index.

VOL. 1: A-C DAYTON TO J.B. FERGUSON, INC. (1988, 225 pp.)

VOL. 2: FREED-EISEMANN TO PREISS. (1989, 266 pp.)

VOL. 3: RCA TO ZENITH. (1991, 292 pp.)

4-057 Dowling, Michael J. **STRATEGIC INVESTMENTS IN INNOVATION: THE TELECOMMUNIATIONS EQUIPMENT INDUSTRY, 1975-1986.** New York: Garland "Studies in Entrepreneurship," 1992, 201 pp. Tables, bibliography, index.

4-058 * Economic Commission for Europe. **THE TELECOMMUNICATIONS INDUSTRY: GROWTH AND STRUCTURAL CHANGE.** New York: United Nations, 1987, 292 pp. Reviews technical trends (switching, transmission, terminal equipment, computer and telecommunications equipment interface), international standard setting, and demand for services, including some country-specific information. Tables, charts, references.

See ELECTRONIC MARKET DATA BOOK, 1-062.

See Fransman, 3-069.

See Friedlander, 3-035.

4-059 ** Geddes, Keith, and Gordon Bussey. **THE SETMAKERS: A HISTORY OF THE RADIO AND TELEVISION INDUSTRY.** London: British Radio & Electronic Equipment Manufacturer's Association, 1991, 464 pp. A model of an industry history, this relates the full story of British manufacturers, with a host of illustrations including many period advertisements—much of it in color. Photos, chapter references, bibliography, index.

4-060 Glover, John G., and William B. Cornell, eds. **THE DEVELOPMENT OF AMERICAN INDUSTRIES.** New York: Prentice Hall, 1932, 930 pp.; 1941 [2nd ed.], 1,005 pp.; 1951 [3rd ed.], 1,121 pp.; and, edited by Glover and Rudolph L. Lagai, Simmons-Boardmann, 1959 [4th ed.], 835 pp. A classic anthology of articles on all types of industry, written by those active in the specific field. Content varied by edition, but included chapters on telegraph, telephone, radio, the recording industry, and later, television. Tables, figures, index.

4-061 Harris, William B. "The Electronic Business," *FORTUNE*, 55: 136-143, 216-226 (April 1957); 134-138, 286-290 (May 1957); 136-139, 292-298 (June 1957). A useful, contemporary three-part survey of companies, products, and markets at the dawn of the digital revolution, with some historical background. Photos, charts, diagrams.

4-062 Irwin, Manley R. **THE TELECOMMUNICATIONS INDUSTRY: INTEGRATION VS COMPETITION.** New York: Praeger Special Studies, 1971, 230 pp. Largely economic, but relating the industry to its baseline technologies. Tables, bibliography.

4-063 ** Mayes, Thorn L. **WIRELESS COMMUNICATION IN THE UNITED STATES: THE EARLY DEVELOPMENT OF AMERICAN RADIO OPERATING COMPANIES.** East Greenwich, RI: New England Wireless and Steam Museum, 1989, 242 pp. Compendium of a quarter-century of the author's research and articles tracing companies up to World War I. Other sections detail progress in equipment development, growth in land and ship stations, and the Alexanderson Alternator. Photos, diagrams, charts, notes, index.

4-064 Maclaren, Malcolm. **THE RISE OF THE ELECTRICAL INDUSTRY DURING THE NINETEENTH CENTURY.** Princeton, NJ: Princeton University Press, 1943, 225 pp. Helps set early electrical communications systems within an industry-wide context, though the focus here is largely on power and lighting. Photos, bibliography, index.

4-065 *MacPherson, Andrew. **INTERNATIONAL TELECOMMUNICATIONS STANDARDS ORGANIZATIONS.** Norwood, MA: Artech House, 1990, 317 pp. Only book-length treatment of the many entities, including the ITU, which play a part in recommending standards. Tables, charts, notes, index.

See Mclaurin, 9-108.

See McMeal, 7-057.

See Morgan, 3-046.

4-066 Passer, Harold C. **THE ELECTRICAL MANUFACTURERS, 1875–1900: A STUDY IN COMPETITION, ENTREPRENEURSHIP, TECHNICAL CHANGE, AND ECONOMIC GROWTH.** Cambridge, MA: Harvard University Press "Studies in Entrepreneurial History," 1953, 412 pp. Discusses arc lighting, incandescent lighting, and electric power—useful as background to the growing importance of telecommunications manufacture. Tables, diagrams, notes, bibliography, index.

See Pleasance, 7-088.

4-067 Pocock, Rowland F. **THE EARLY BRITISH RADIO INDUSTRY.** Manchester, England: Manchester University Press, 1988, 184 pp. Based on archives of the Post Office, the Admiralty, and Marconi. Notes, index. See also Bussey, 4-054.

4-068 Sanjek, Russell, and David Sanjek. **AMERICAN POPULAR MUSIC BUSINESS IN THE 20th CENTURY.** New York: Oxford University Press, 1991, 334 pp. Includes extensive discussion of technological advances in recording as well as interindustry competition among recording, radio, television, and film businesses. Bibliography, index.

4-069 Skinner, Wickham, and David C.D. Rogers. **MANUFACTURING POLICY IN THE ELECTRONICS INDUSTRY: A CASEBOOK OF MAJOR PRODUCTION PROBLEMS.** Homewood, IL: Richard D. Irwin, 1957; 1964; 1968 [3rd ed.], 289 pp. Harvard case studies useful for their portrayal of then-current methods in, among other companies examined, National Video, RCA, Honeywell, and Zenith Radio. Photos, diagrams, tables, notes, index.

4-070 Not used.

4-071 * U.S. Congress, House of Representatives. **REPORT ON COMMUNICATION COMPANIES.** 73rd Cong., 2nd Sess., House Report 1273, 1934–35, three parts issued in seven vols, 4,266 pp. Massive study of telegraph, telephone, and radio industries conducted as part of lead-up to the Communications Act of 1934. While largely financial in nature, the "Splawn" report (after special counsel Walter M. W. Splawn under whose direction it was assembled) includes useful details on services provided (especially by many smaller companies) and some of the baseline technologies utilized. Tables, charts, index.

See U.S. Department of Commerce, 1-071.

4-072 * U.S. Federal Trade Commission. **REPORT OF THE FEDERAL TRADE COMMISSION ON THE RADIO INDUSTRY...** Washington: Government Printing Office, 1924 (reprinted by Arno Press, "Telecommunications," 1974), 347 pp. Focus is on point-to-point radio services, with chapters on development of the industry, control of patents, traffic agreements, and trade practices on manufacture of radio apparatus. Invaluable for the early years of the business. Tables.

See U.S. INFORMATION TECHNOLOGY INDUSTRY TRADE ANALYSIS, 1-077.

4-073 Witte, E[berhard], ed. **GLOBAL PLAYERS IN TELECOMMUNICATIONS.** Berlin: Springer-Verlag "Publication of Münchner Kleis, Vol. 20," 1994, 157 pp. Papers from a 1994 conference include chapters assessing the changing roles of France Telecom, MCI, BellSouth, AT&T, and Cable &Wireless. Tables, charts, references.

D. Societies and Institutions
(This section includes both American and foreign–largely British–organizations.)

American Institute of Electrical Engineers (AIEE)

4-074 "Fiftieth Anniversary of the AIEE," *ELECTRICAL ENGINEERING,* 53:641–848 (May 1934). Valuable historical collection of 51 articles including all aspects of AIEE from 1884. Most are devoted to power engineering and heavy machinery. Photos.

4-075 Not used.

American Radio Relay League (ARRL) and Amateur Radio

See Berg, 9-133.

4-076 ** DeSoto, Clinton B. **TWO HUNDRED METERS AND DOWN: THE STORY OF AMATEUR RADIO**. West Hartford, CT: American Radio Relay League "Radio Amateur's Library, No. 13," 1936, 184 pp. Divided into sections on pioneers, development and recognition, and international high-frequency communication. Numerous names, dates, call numbers, and extracts in the text. No index.

4-077 _____. **CALLING CQ: ADVENTURES OF SHORT-WAVE RADIO OPERATORS**. New York: Doubleday Doran, 1941, 291 pp.

4-078 * **FIFTY YEARS OF A.R.R.L.** Newington, CT: ARRL, 1965, 151 pp. Reprints series of articles that originally ran in *QST* to celebrate the 50th anniversary. Photos, diagrams, maps.

4-079 Jahnke, Debra, and Katharine A. Fay, eds. **FROM SPARK TO SPACE: A PICTORIAL JOURNEY THROUGH 75 YEARS OF AMATEUR RADIO**. Newington, CT: American Radio Relay League, 1989, 95 pp. Just that.

4-080 Tiffany, Willard D., et al. **AMATEUR RADIO: AN INTERNATIONAL RESOURCE FOR TECHNOLOGICAL, ECONOMIC, AND SOCIOLOGICAL DEVELOPMENT**. Menlo Park, CA: Stanford Research Institute (Project No. M-5436), August 1966, 110 pp. Based on a survey, this reviews the various impacts of amateur operators. Tables, bibliography.

British Institution of Radio Engineers (BIRE)

4-081 Clifford, Graham, D., et al., eds. **A TWENTIETH CENTURY PROFESSIONAL INSTITUTION: THE STORY OF THE BRIT. I.R.E.** London: BIRE, 1960, 119 pp.; 1989 [2nd ed.], 331 pp. Overall history of what became the Institution of Electronic and Radio Engineers in 1988. The second edition reprints the first as part one, and then in part two tells the 1961–88 story. Chronology, photos, appendices.

Electronic Industries Association (EIA)

4-082 Secrest, James E. **EIA 50: ELECTRONIC INDUSTRIES ASSOCIATION: THE FIRST FIFTY YEARS**. Washington: EIA, 1974, 248 pp. Beginning with its inception as the Radio Manufacturers Association in 1924, this history details the association's growth and services and relates some background on the growth of electronics as well. Photos, chronology, index.

Institute of Electrical and Electronic Engineers (IEEE)

4-083 Gannett, Elwood K. "*Proceedings of the IEEE*: The First 75 Years," *PROCEEDINGS OF THE IEEE*, 76:1268–1279 (October 1988). Focuses especially on the journal's editors. Photos, bibliography.

Institutional and Company Histories

4-084 Kline, Ronald. "An Overview of Twenty-Five Years of Electrical and Electronic Engineering in the *Proceedings of the IEEE*, 1963–1987," **PROCEEDINGS OF THE IEEE**, 78:469–485 (March 1990). Bibliography.

4-085 * McMahon, A. Michael. **THE MAKING OF A PROFESSION: A CENTURY OF ELECTRICAL ENGINEERING IN AMERICA.** New York: IEEE Press, 1984, 304 pp. A history of the field of electrical engineering, seen through key players, as well as the IEEE and its predecessor organizations. Photos, notes, index.

4-086 Ryder, John D., and Donald G. Fink. **ENGINEERS & ELECTRONS: A CENTURY OF ELECTRICAL PROGRESS.** New York: IEEE Press, 1984, 251 pp. Issued on the centenary of the IEEE's predecessor organizations, this is a popular history of American electrical developments, including telecommunication, as seen through key people, inventions, and organizations. Photos, chapter further reading notes, index.

Institute of Radio Engineers (IRE)

4-087 * Brittain, James E. "The Evolution of Electrical and Electronic Engineering and the *Proceedings of the IRE*: 1913–1937," **PROCEEDINGS OF THE IEEE**, 77:837–856 (June 1989). Covers the first quarter century showing how the journal reflected the industry. References.

4-088 *_____. "The Evolution of Electrical and Electronic Engineering and the *Proceedings of the IRE*: 1938–1962," **PROCEEDINGS OF THE IEEE**, 85:762–797 (May 1997). Continues previous entry. Illustrations, references.

See Goldsmith, 3-036.

4-089 Whittemore, Laurens E. "The Institute of Radio Engineers—Forty-Five Years of Service," **PROCEEDINGS OF THE IRE**, 45:597–635 (May 1957). A full survey of the IRE's history and activities. Photos, graphs, tables, list of publications.

Institution of Electrical Engineers (IEE)

4-090 Appleyard, Rollo. **THE HISTORY OF THE INSTITUTION OF ELECTRICAL ENGINEERS (1871–1931).** London: IEE, 1939, 342 pp. History of electrical engineering with technical and biographical details, and numerous extracts. Photos (largely portraits), tables, appendices, index.

4-091 Reader, W.J. **A HISTORY OF THE INSTITUTION OF ELECTRICAL ENGINEERS, 1871–1971.** London: Peter Peregrinus, 1987, 327 pp. Centennial history of the foremost British association, this focuses on the second 50 years, thus supplementing Appleyard (4-090). Photos, chapter references, appendices, bibliography, index.

Institution of Post Office Electrical Engineers (IPOEE)

See *THE POST OFFICE ELECTRICAL ENGINEERS JOURNAL*, 7-141.

Radio Club of America

4-092 FIFTIETH ANNIVERSARY GOLDEN YEARBOOK. New York: Radio Club of America, 1959, 216 pp. Official history of the group with statistics, many historical papers, lists of members, awards, and activities. Photos, diagrams.

4-093 SEVENTY-FIFTH ANNIVERSARY DIAMOND JUBILEE YEARBOOK, 1909-1984. Westwood, NJ: Radio Club of America, 1984, 297 pp. As above, but with a focus on 1959-84 period and reprinting dozens of articles on key inventors and developments. Photos, diagrams.

Radio Society of Great Britain

4-094 Clarricoats, John. **WORLD AT THEIR FINGERTIPS: THE STORY OF AMATEUR RADIO IN THE UNITED KINGDOM AND A HISTORY OF THE RADIO SOCIETY OF GREAT BRITAIN.** London: Radio Society of Great Britain, 1967, 307 pp. A detailed record of people and events from the early years of radio. Photos, tables, appendices, notes.

Society of Motion Picture and Television Engineers (SMPTE)

4-095 Anniversary issues of the *SMPTE JOURNAL*, all of which include articles on the society and its publications and other activities. Specifically:

> **"50th Anniversary,"** 75 (September 1966) includes Glenn Matthews, "Historic Aspects of the SMPTE" (pp. 856-867) which concludes with a listing of nearly 250 historical articles from previous years.
>
> **"60th Anniversary Issue,"** 85:449-590 (July 1976) includes 14 historical papers on changing film and television technology, and education and the society in the early years.
>
> **"Special 75th Anniversary Issue,"** 100:484-527 (July 1991) includes surveys of publication history, the annual television conference, the technical conference, and the society's role in standardization. See also Friedman, 4-096.
>
> **"SMPTE 80th Anniversary, 1916-1996,"** 105:581-640 (October 1996) includes an historical paper from each of the society's eight decades, plus a bibliography of historical papers published in the journal.

4-096 Friedman, Jeffrey. **MILESTONES IN MOTION PICTURE AND TELEVISION TECHNOLOGY: THE SMPTE'S 75th ANNIVERSARY COLLECTION.** White Plains, NY: SMPTE, 1991, 241 pp. Reprints papers from the *SMPTE JOURNAL* and its predecessors. Photos, references.

Institutional and Company Histories

E. Specific Companies—U.S.

(A few biographies of industry and company leaders are included here rather than in chapter 5 as they deal substantially with company affairs. For regional telephone company histories see 7-C.)

American Telephone & Telegraph Co. (AT&T)

(See also Bell Telephone Laboratories and Western Electric)

See Boettinger, 7-069.

See BROADCASTING NETWORK SERVICE, 9-094.

4-097 Brooks, John. **TELEPHONE: THE FIRST HUNDRED YEARS.** New York: Harper & Row, 1976, 369 pp. Popular narrative relating development of both the instrument and AT&T's exploitation of it, but offering more on people than technology. Photos, notes, index.

4-098 Carlson, W. Bernard. "Entrepreneurship in the Early Development of the Telephone: How Did William Orton and Gardner Hubbard Conceptualize This New Technology?" *BUSINESS AND ECONOMIC HISTORY,* 23:161–192 (Winter 1994). Pioneering decisions in the late 1870s.

4-098a Coon, Horace C. **AMERICAN TEL AND TEL: THE STORY OF A GREAT MONOPOLY.** New York: Longmans, Green, 1939 (reprinted by Books for Libraries, 1971), 276 pp. General history from the invention of the telephone. Includes telegraph, radio, and sound recording developments. Bibliography, index.

4-099 ** Danielian, N[oobar] R. **A.T. AND T.: THE STORY OF INDUSTRIAL CONQUEST.** New York: Vanguard, 1939 (reprinted by Arno Press, "Telecommunications," 1974), 460 pp. Author was a central figure in a major FCC investigation (see 4-108) and offers a highly critical business history. Three chapters (5, 6, and 7) deal with technology including patents, inventions, litigation, and agreements relating to early electron tubes, circuits, wire and radio communication, broadcasting, sound motion pictures, and other products and services. Notes, tables, references, index.

See EVENTS IN TELECOMMUNICATIONS HISTORY, 1-050.

4-100 * Galambos, Louis, "Theodore N. Vail and the Role of Innovation in the Modern Bell System," *BUSINESS HISTORY REVIEW,* 66:95–126 (Spring 1992). Focuses on the period from 1907 to 1919, the second era when Vail headed AT&T.

4-101 ** Garnet, Robert W. **THE TELEPHONE ENTERPRISE: THE EVOLUTION OF THE BELL SYSTEM'S HORIZONTAL STRUCTURE, 1876–1909.** Baltimore: Johns Hopkins University Press "AT&T Series in Telephone History," 1985, 210 pp. Explores the technical, legal, operational, and administrative relationships that developed between the central company and its regional affiliates. Based largely on AT&T archival sources. Photo, charts, map, notes, index. For the rest of this series, see 4-106, 7-095, and 7-106.

4-102 Hoddeson, Lillian. "The Emergence of Basic Research in the Bell Telephone System, 1875–1915," *TECHNOLOGY AND CULTURE,* 22:512–544 (July 1981). Primary trends prior to creation of Bell Labs in 1925.

4-103 Page, Arthur W. **THE BELL TELEPHONE SYSTEM.** New York: Harper, 1941, 248 pp. Written in response to the 1939 FCC report (4-108) and Danielian's critique (4-099), this is a broad survey of company management, politics, personnel, activities, and services, with some historical details. There are chapters on research, Bell Labs, and Western Electric. Tables, maps, index.

See Martin, 7-056.

4-104 * Paine, Albert Bigelow. **IN ONE MAN'S LIFE; BEING CHAPTERS FROM THE PERSONAL AND BUSINESS CAREER OF THEODORE N. VAIL.** New York: Harper, 1921 (reissued as THEODORE N. VAIL: A BIOGRAPHY, 1929), 359 pp. The only book-length biography of the longtime head of AT&T (1845–1920) who promoted technical research. Photos, appendix, notes, index.

4-105 Pier, Arthur S. **FORBES: TELEPHONE PIONEER.** New York: Dodd, Mead, 1953, 232 pp. William H. Forbes (1840–1897) served eight years as the first president of AT&T. Photos, index.

See Reich, 4-136.

4-106 ** Smith, George David. **THE ANATOMY OF A BUSINESS STRATEGY: BELL, WESTERN ELECTRIC, AND THE ORIGINS OF THE AMERICAN TELEPHONE INDUSTRY.** Baltimore: Johns Hopkins University Press "AT&T Series in Telephone History," 1985, 237 pp. Study of the AT&T takeover of Western Electric in the 1880s, a manufacturing solution to technical and marketing problems, and the beginning of AT&T's vertical integration. Photos, notes, essay on sources, index. For the rest of this series, see 4-101, 7-095, and 7-106.

4-107 Tosiello, Rosario Joseph. **THE BIRTH AND EARLY YEARS OF THE BELL TELEPHONE SYSTEM, 1876–1880.** New York: Arno Press (reprint of 1971 Boston University dissertation), 1979, 512 pp. Valuable for its details of how the telephone progressed from a solitary invention to the basis for a huge firm. Notes, bibliography (no index).

4-108 ** U.S. Federal Communications Commission. **INVESTIGATION OF THE TELEPHONE INDUSTRY IN THE UNITED STATES.** 78th Cong, 1st Sess., House Document No. 340, 1939 (reprinted by Arno Press "Telecommunications," 1974), 661 pp. The final report of a three-year FCC study of all aspects of AT&T history and operation offers the most detailed study of the company to that time. Focus is more on economics and organization than technology. Tables, charts, index.

Ampex

See Ginsburg, 10-118.

4-109 Rosenbloom, Richard S., and Karen J. Freeze. "Ampex Corp and Video Innovation," in Richard Rosenbloom, ed. **RESEARCH ON TECHNICAL INNOVATION, MANAGEMENT AND POLICY, VOL. 2**. Westport, CT: Greenwood, 1985, pp. 113–186. Describes the company's crucial role in development of video recording tape and equipment and how it lost its lead. Charts, photos, references.

Atwater Kent

4-110 Williams, Ralph O. "Atwater Kent Radio Development," **THE A.W.A. REVIEW**, 1:82–107, (1986); 2:71–93 (1987); 3:6–33 (1988). Careful survey of the company's development and output. Photos.

4-111 _____. "Atwater Kent, Master of Marketing," **THE A.W.A. REVIEW**, 10:7–77 (1996). A review of the firm's full history. Photos.

Bell Telephone Laboratories (of AT&T)
(See also AT&T and Western Electric.)

4-112 *** A HISTORY OF ENGINEERING AND SCIENCE IN THE BELL SYSTEM. Whippany, NJ: Bell Telephone Laboratories, 1975–85, 7 vols. This details the most important work in the premier private research lab, covering the past century—a model of self-published corporate technical history. Each volume focuses on specific area of work with well-researched original papers, heavily illustrated and well indexed. The individual volumes are not numbered, but are listed here in order of publication appearance:

THE EARLY YEARS (1925–1975) edited by M.D. Fagan, 1975, 1,078 pp. Covers all aspects of technical work in AT&T to the formation of Bell Labs in 1926.

NATIONAL SERVICE IN WAR AND PEACE (1925–1975) edited by M.D. Fagan, 1978, 757 pp. Includes detailed chapters on radar, electrical computers for fire control, communication, World War II, air defense, tactical and strategic defense systems, command and control, and operation of the Sandia National Labs.

SWITCHING TECHNOLOGY (1925–1975) edited by G.E. Schindler Jr., 1982, 639 pp. Evolution of electromechanical and manual switching, crossbar switches, preparing for full automation, direct distance dialing, electronic switching, private branch exchanges, etc.

PHYSICAL SCIENCES (1925–1980) edited by S. Millman, 1983, 674 pp. Chapters detail system research in physics (including the transistor, laser, etc.), and materials (semiconductors, fiber optics, etc.).

COMMUNICATION SCIENCES (1925–1980) edited by S. Millman, 1984, 521 pp. Includes chapters on the mathematical foundation of communication, acoustics, picture communication research, vacuum tube electronics, radio systems, fiber optic communications, switching, computer science, digital communications, and behavioral science.

Chapter 4

TRANSMISSION TECHNOLOGY (1925-1975) edited by E.F. O'Neill, 1985, 812 pp. Details such analog technologies as overseas and broadcast radio, coaxial cable, and UHF frequency work, microwave, submarine cable systems, satellites, mobile radio, and the advent of digital systems including the T1 carrier, fiber optic systems, and the technical and business environment for such technological work.

ELECTRONICS TECHNOLOGY (1925-1975) edited by F.M. Smits, 1985, 370 pp. The transistor, integrated chips, electron tubes, magnetic memories, capacitors and resistors, and the like.

4-113 "Bell Labs' First 50 Years: Prelude to Tomorrow," *BELL LABORATORIES RECORD,* 53:l:1–115 (January 1975). Special commemorative issue for the anniversary includes short surveys of major work accomplished over the half century. Photos, diagrams, chronology.

4-114 Not used.

4-115 Bello, Francis. "The World's Greatest Industrial Laboratory," *FORTUNE,* 58:148–157, 208 (November 1958). Pictorial and text survey.

4-116 * Bernstein, Jeremy. **THREE DEGREES ABOVE ZERO: BELL LABS IN THE INFORMATION AGE.** New York: Scribner's, 1984, 241 pp. While informal, this is a useful picture of Bell Labs at the time AT&T was broken up. Photos, bibliography, index.

4-117 Gregor, Arthur. **BELL LABORATORIES: INSIDE THE WORLD'S LARGEST COMMUNICATIONS CENTER.** New York: Scribner's, 1972, 125 pp. Covers creation of the labs, key innovations and then-current operations. Photos, chronology.

See Hanscom, 1-052.

4-118 ** IMPACT: A COMPILATION OF BELL SYSTEM INNOVATIONS IN SCIENCE AND ENGINEERING... Murray Hill, NJ: Bell Telephone Laboratories, 1971, 147 pp.; 1981 [2nd ed.], 132 pp. Subtitle varied between two editions. Sections on basic science and engineering concepts, computer industry, data communication, defense and aeronautics, electronics, entertainment, and telecommunication—each arranged chronologically. Diagrams (second ed. only), index.

4-119 Mabon, Prescott C. **MISSION COMMUNICATIONS: THE STORY OF BELL LABORATORIES.** Murray Hill, NJ: Bell Telephone Laboratories, 1975, 198 pp. Informal corporate history with useful boxed features on key figures over the years. Photos, diagrams, index.

4-120 * Mueser, Roland, ed. **BELL LABORATORIES INNOVATION IN TELECOMMUNICATIONS, 1925-1977.** Murray Hill, NJ: Bell Telephone Laboratories, 1977, 227 pp. Arranged chronologically, this appears in main sections on research and electronics technology, customer systems and services, central office systems, transmission systems, and operation systems. Index.

Institutional and Company Histories

4-121 Noll, A. Michael. "Bell System R&D Activities: The Impact of Divestiture," *TELECOMMUNICATIONS POLICY*, 11:161–178 (June 1987). A one-time member of the Bell Labs staff details the decline in research capability and capacity after the breakup. Tables, notes.

4-122 * Southworth, George C. **FORTY YEARS OF RADIO RESEARCH: A REPORTORIAL ACCOUNT.** New York: Gordon and Breach, 1962, 274 pp. A longtime Bell Labs researcher relates the institution's work in radio transmission systems. Photos, diagrams, index.

Collins Radio Co.

4-123 Braband, Ken C. **THE FIRST 50 YEARS: A HISTORY OF COLLINS RADIO COMPANY AND THE COLLINS DIVISIONS OF ROCKWELL INTERNATIONAL.** Cedar Rapids, IA: Rockwell International, 1983, 219 pp. History of the firm begun in 1933 and taken over by Rockwell in 1973. Photos, product list, bibliography (no index).

Communications Satellite Corporation (COMSAT)
(See also INTELSAT, 4-019 through 4-026.)

4-124 **COMSAT AT 15.** Washington: Comsat, 1978, 56 pp. A useful illustrated chronology covering 1963–78 events.

4-125 **REPORT TO THE PRESIDENT AND THE CONGRESS.** Washington: Comsat, 1964–84, annual. These often provide more detailed technical and service information than Comsat's annual commercial report, at least to about 1980 when they became very brief and perfunctory. Photos, tables, charts.

4-126 Tedeschi, Anthony Michael. **LIVE VIA SATELLITE: THE STORY OF COMSAT AND THE TECHNOLOGY THAT CHANGED WORLD COMMUNICATION.** Washington: Acropolis Books, 1989, 176 pp. Brief institutional history of American member of Intelsat. Photos, index.

DuMont

See DuMont Laboratories, 12-086.

Federal Telephone and Radio Corporation

4-127 Mann, F.J. "Federal Telephone and Radio Corporation: An Historical Review 1909–1946," *ELECTRICAL COMMUNICATION*, 23:376–405 (December 1946). Poulsen arc transmitters, early radio stations, Palo Alto laboratory, company organization, changes, and work during World War II. Photos, notes, bibliography.

General Electric

4-128 Birr, Kendall. **PIONEERING IN INDUSTRIAL RESEARCH: THE STORY OF THE GENERAL ELECTRIC RESEARCH LABORATORY.**

Washington: Public Affairs Press, 1957, 204 pp. See also Hawkins (4-132). Chapter references, note on sources, index.

4-129 Case, Josephine Young, and Everett Needham Case. **OWEN D. YOUNG AND AMERICAN ENTERPRISE: A BIOGRAPHY.** Boston: Godine, 1982, 964 pp. Very detailed biography of the longtime GE leader written by two descendants. See also Tarbell, 4-137.

4-130 * **THE GENERAL ELECTRIC STORY—A PHOTO HISTORY.** Schenectady, NY: GE Elfun Hall of History, 1976–80, 4 vols.; later reissued as single volume. Useful set draws heavily on company archives. Photos, chronologies, selected readings.

> **VOL. 1: THE EDISON ERA, 1876–1892.** (1976), 56 pp.
>
> **VOL. 2: THE STEINMETZ ERA, 1892–1923.** (1977), 74 pp.
>
> **VOL. 3: ON THE SHOULDERS OF GIANTS, 1924–1946.** (1979), 81 pp.
>
> **VOL. 4: PATHWAYS OF PROGRESS, 1947–1978.** (1980), 109 pp.

4-131 ** Hammond, John Winthrop. **MEN AND VOLTS: THE STORY OF GENERAL ELECTRIC.** Philadelphia: Lippincott, 1941, 436 pp. The company story to 1922 with brief epilogue on the next two decades, told in 41 chapters divided into eight major sections. Primarily focused on power and light. Photos, appendices, index. See Miller (4-133) for continuation through World War II.

4-132 Hawkins, Laurence A. **ADVENTURE INTO THE UNKNOWN: THE FIRST FIFTY YEARS OF THE GENERAL ELECTRIC RESEARCH LABORATORY.** New York: William Morrow, 1950, 150 pp. Series of sketches on the character of the laboratory and some of its products with emphasis on personalities. No references or index. See also Birr, 4-128.

4-133 Miller, John Anderson. **MEN AND VOLTS AT WAR: THE STORY OF GENERAL ELECTRIC IN WORLD WAR II.** New York: McGraw-Hill, 1947, 272 pp. Continues Hammond (4-131). Photos, list of war plants and principal war products, index.

4-134 _____. **WORKSHOP OF ENGINEERS: THE STORY OF THE GENERAL ENGINEERING LABORATORY OF THE GENERAL ELECTRIC COMPANY, 1895–1952.** Schenectady, NY: General Electric, 1952, 180 pp. The first 90 pages offer a historical survey while the remainder details then-current activities and developments. Photos, index.

4-135 Nye, David E. **IMAGE WORLDS: CORPORATE IDENTITIES AT GENERAL ELECTRIC, 1890–1930.** Cambridge, MA: MIT Press, 1985, 188 pp. Study of how GE used photography and media to promote itself and its products. Photos, notes, index.

Institutional and Company Histories 77

4-136 ** Reich, Leonard S. THE MAKING OF AMERICAN INDUSTRIAL RESEARCH: SCIENCE AND BUSINESS AND GE AND BELL, 1876–1926. New York: Cambridge University Press, 1985, 309 pp. Comparative study of early industrial research and development at two important electrical firms. Photos, notes, index.

4-137 Tarbell, Ida M. OWEN D. YOUNG: A NEW TYPE OF INDUSTRIAL LEADER. New York: Macmillan, 1932, 353 pp. Biography of the longtime GE leader by a famous investigative journalist. Photos, index. See also Case and Case, 4-129.

General Radio Company

4-138 Sinclair, Donald B. THE GENERAL RADIO COMPANY, 1915–1965. New York: Newcomen Society, 1965, 32 pp. Brief address on the firm's history. Photos.

4-139 Thiessen, Arthur E. A HISTORY OF THE GENERAL RADIO COMPANY. West Concord, MA: General Radio, 1965, 116 pp. General survey. Photos, appendices.

General Telephone and Electronics (GTE)

4-140 McCarthy, Thomas E. THE HISTORY OF GTE: THE EVOLUTION OF ONE OF AMERICA'S GREAT CORPORATIONS. Stamford, CT: GTE, 1990, 224 pp. An in-house history, the only one available of the company which has long operated local telephone services and for a time controlled the Sprint long distance firm. Photos, notes, index.

International Telephone & Telegraph (IT&T)

4-141 "Behn Brothers," and "I.T.&T. Ends a Brilliant Decade," *FORTUNE*, 2:6:35–45, 118–124 (December 1930). Valuable studies of the original company at its height. Photos, maps.

4-142 Deloraine, Maurice. WHEN TELECOM AND ITT WERE YOUNG. New York: Lehigh Books, 1976, 161 pp. English translation of the French work published by Flammarion in 1974, this centers on the activities and observations of the author, a general technical director of IT&T and president of IT&T subsidiaries in France. Chapters discuss company efforts in field telegraphy, early European broadcasting, transatlantic telephone cables, radio research, German developments and telecommunications in World War II. Photos.

4-143 "I.T. &T.'s Nine Lives," *FORTUNE* (September 1945), pp.145–149, 191–207. Review of the firm's many roles and countries of operations—and how it was updating its technology. Photos, map.

4-144 * Kohlhass, H.T. "Milestones of Communication Progress," *ELECTRICAL COMMUNICATION,* 20:143–185 (1942). Illustrated survey of IT&T member company progress in telegraph, telephone, and radio equipment and installations, including early microwave trials. Photos, references (nearly all to previous articles in this IT&T journal).

4-145 Sobel, Robert. **ITT: THE MANAGEMENT OF OPPORTUNITY.** New York: Times Books, 1983, 421 pp. A journalistic history of the firm-become-conglomerate, though focusing more on business and personalities than on technology. Photos, notes, bibliography, index.

Kolster Radio Corporation (formerly Federal-Brandes Inc.)

4-146 **KOLSTER RADIO CORPORATION: THE STRATEGIC EXPLOITATION OF A NEW SCIENCE.** San Francisco: Keib, Keyston & Co., 1928, 48 pp. An informal history of the firm by a stockbroker. Map.

MCI

4-147 ** Cantelon, Philip L. **THE HISTORY OF MCI: THE EARLY YEARS, 1968–1988.** Dallas: Heritage Press, 1993, 719 pp. Privately published narrative history which actually begins in 1963 with the initial formation of the firm and deals in considerable depth with people, legal and policy issues, and technology. Photos, appendices (including a good one on microwave history), index. See also 8-030.

4-148 Kahaner, Larry. **ON THE LINE: THE MEN OF MCI—WHO TOOK ON AT&T, RISKED EVERYTHING, AND WON!** New York: Warner Books, 1986, 344 pp. Largely concerns policy, financial, and legal developments and key company figures, with some discussion of the underlying technology. Photos, index.

4-149 Spurge, Lorraine, Lawrence J. Gitman, and Victor Tabbush. **FAILURE IS NOT AN OPTION—HOW MCI INVENTED COMPETITION IN TELECOMMUNICATIONS.** Encino, CA: Spurge Ink!, 1998, 225 pp. Heavily illustrated popular survey. Photos, graphs, 32-page illustrated time line.

Motorola

4-150 Petrakis, Harry M. **THE FOUNDER'S TOUCH: THE LIFE OF PAUL GALVIN OF MOTOROLA.** New York: McGraw-Hill, 1965, 240 pp. Relates how Galvin Manufacturing, founded in 1928, became Motorola, and provides a history of the first 35 years of the latter. Photos, index.

Philco

4-151 Balderston, William. **"PHILCO": AUTOBIOGRAPHY OF PROGRESS.** New York: Newcomen Society, 1954, 22 pp. Brief survey by the firm's president.

See Ramirez and Prosise, 9-174.

Radio Corporation of America

4-151a* Barnum, Frederick O. **"HIS MASTER'S VOICE" IN AMERICA: NINETY YEARS OF COMMUNICATIONS PIONEERING AND PROGRESS: VICTOR TALKING MACHINE COMPANY, RADIO CORPORATION OF AMERICA, GENERAL ELECTRIC.** Camden, NJ: General Electric, 1991, 390 pp. Lavish corporate history which focuses largely on RCA period of operations in Camden in a series of chronological chapters. Photos (some color), bibliography.

4-152 ** Bilby, Kenneth. **THE GENERAL: DAVID SARNOFF AND THE RISE OF THE COMMUNICATIONS INDUSTRY.** New York: Harper & Row, 1986, 326 pp. Definitive life of the RCA leader and his company by a former public relations official who debunks several long-lived myths. Photos, selected bibliography, index.

See Bitting, 12-078.

4-153 Dreher, Carl. **SARNOFF: AN AMERICAN SUCCESS.** New York: Quadrangle Books, 1977, 282 pp. An informal narrative with a degree of demythologizing. Index.

See Graham, 10-137.

4-154 Lyons, Eugene. **DAVID SARNOFF: A BIOGRAPHY.** New York: Harper & Row, 1966, 372 pp. Written by a first cousin, this must be used with care as it reflects the RCA/Sarnoff public relations machine and many myths—here neither Sarnoff nor RCA can do anything wrong. Photos, index.

4-155 1942–1967: **TWENTY-FIVE YEARS AT RCA LABS.** Princeton, NJ: RCA Laboratories, 1967, 104 pp. Commemorative album. Photos, drawings, charts.

See PIONEERING IN TELEVISION, 12-098.

See Radio Corporation of America, 12-142.

4-156 * RCA—AN HISTORICAL PERSPECTIVE. New York: RCA, 1978 (5th ed.), 49 pp. Useful company promotional piece based on several articles originally appearing in *RCA ENGINEER*, and thus especially strong on technology. Photos.

See "RCA's Television," 12-100.

4-157 Rossi, John P. "An All-American System? Business-Government Relations and the Radio Corporation of America, 1917–1932," *ESSAYS IN ECONOMIC AND BUSINESS HISTORY*, 12:307–318 (1994). Focuses on role of the U.S. Navy and other arms of the federal government.

4-158 * Sarnoff, David. **LOOKING AHEAD: THE PAPERS OF DAVID SARNOFF.** New York: McGraw-Hill, 1968, 313 pp. Includes papers, testimony, reports, and speeches on wireless, radio broadcasting, television, the communications revolution, and other topics. Underwritten by RCA primarily for internal distribution. Index.

4-159 * Schairer, Otto S. **PATENT POLICIES OF RADIO CORPORATION OF AMERICA**. New York: RCA Technical Institutes Press, 1939, 92 pp. The officer in charge of patents details how patents are handled, including licensing and litigation. One of the few such resources available. Photos, patent lists.

4-160 * Sobel, Robert. **RCA**. New York: Stein & Day, 1986, 282 pp. Only book-length history of the company, issued just after it disappeared into GE, this is an informal account with emphasis on people. Tables, bibliography, index.

4-161 25 YEARS OF RADIO PROGRESS WITH RCA. New York: RCA, 1944, 87 pp. Company promotional piece. Photos, color map.

See ULTRAFAX, 13-035.

Raytheon

4-162 Fay, H.J.W. **SUBMARINE SIGNAL LOG**. Portsmouth, RI: Raytheon, Submarine Signal Division, 1963, 37 pp. From the inventions of Reginald Fessenden in 1901 into the 1960s including underwater communication, echo sounding, and submarine detecting. Photos.

4-163 Scott, Otto J. **THE CREATIVE ORDEAL: THE STORY OF RAYTHEON**. New York: Atheneum, 1974, 429 pp. The first 50 years told with considerable feeling for behind-the-scenes developments. Photos, notes, index.

United Fruit Company

4-164 Mason, Roy. "The History and Development of the United Fruit Company's Radio Telegraph System," *RADIO BROADCAST*, 1:377–398 (September 1922). Survey from 1904. Photos.

Varian Associates

4-165 THE KLYSTRON. Palo Alto, CA: Varian Associates, nd, 18 pp. Illustrated booklet on the company's development of the tube and its applications.

4-166 * Varian, Dorothy. **THE INVENTOR AND THE PILOT: RUSSELL AND SIGURD VARIAN**. Palo Alto, CA: Pacific Books, 1983, 314 pp. The combined biography of Russell (1898–1959) and Sigurd (1901–1961) who founded the firm and were cental to development of its key product, the Klystron tube, used in microwaves. Photos, index.

Western Electric
(See also AT&T and Bell Telephone Laboratories.)

4-167 *** Adams, Stephen B, and Orville R. Butler. **MANUFACTURING THE FUTURE: A HISTORY OF WESTERN ELECTRIC**. New York: Cambridge University Press, 1999, 270 pp. Seven chapters trace the full story from before Bell's takeover in 1882 (the company was created in 1869 by Western Union) to AT&T

manufacturing activity after divestiture in 1984 to the spin-off that became Lucent Technologies after 1995. Photos, tables, notes, bibliography, index.

4-168 Gorman, Paul A. **CENTURY ONE...A PROLOGUE.** New York: Newcomen Society, 1969, 24 pp. An address on the company's first century.

4-169 Iardella, Albert B., ed. **WESTERN ELECTRIC AND THE BELL SYSTEM: A SURVEY OF SERVICE.** New York: Western Electric, 1964, 115 pp. A publication designed for new Western Electric managers, includes history and current operations.

4-170 Lovette, Frank H. "Western Electric's First 75 Years: A Chronology," *BELL TELEPHONE MAGAZINE*, 23:271–287 (1944–45). Just that.

4-171 * McKinsey & Co. **A STUDY OF WESTERN ELECTRIC'S PERFORMANCE.** New York: AT&T, 1969, 251 pp. Consultant's report on the manufacturing, maintenance, and other functions of the AT&T subsidiary. Charts.

Western Union

4-172 Not used.

See AMERICAN TELEGRAPHY AFTER 100 YEARS, 6-087

4-173 Brewer, A.R. **WESTERN UNION TELEGRAPH COMPANY: 1851–1901, A RETROSPECT.** New York: J. Kempster for Western Union, 1901, 36 pp. Author was firm's secretary.

4-174 Grant, E.B. **THE WESTERN UNION TELEGRAPH COMPANY: ITS PAST, PRESENT AND FUTURE.** New York: Hotchkiss, Burnham, 1883, 61 pp. A stockbroker's investment analysis, with some discussion of technology.

4-175 NEWCOMEN PAPERS. New York: Newcomen Society in North America. Brief pamphlets based on lectures delivered before meetings of the "American Newcomen" group. Listed chronologically, they include:

> White, R.B. **TELEGRAMS IN 1889—AND SINCE!** 1939, 48 pp. President of the firm reviews its history. Photos.

> Williams, A.N. **"WHAT HATH GOD WROUGHT!" MAY 24, 1844.** (1944, 20 pp.) A later president of the company assesses its historical heritage. Reprinted in AMERICAN TELEGRAPHY, 6-087.

> Marshall, Walter P. **EZRA CORNELL (1807–1874): HIS CONTRIBUTIONS TO WESTERN UNION AND TO CORNELL UNIVERSITY.** (1951, 24 pp.) A brief biography of the pioneer by yet another president of Western Union.

Penrose, Charles. **NEWCOMB CARLTON (1869–1953) OF "WESTERN UNION."** (1956, 36 pp.) A brief biography of the longtime head of the company.

Gallagher, Edward A. **GETTING THE MESSAGE ACROSS: THE STORY OF WESTERN UNION INTERNATIONAL, INC.** (1971, 24 pp.) As told by its president.

4-176 Oslin, G.P. "The Telegraph Industry," in Glover and Cornell, eds. **THE DEVELOPMENT OF AMERICAN INDUSTRIES** (4-060), 1932 [2nd ed.], pp. 687–719. History of telegraph and development of the company and its then-current operations. See also Oslin, 3-048.

4-177 Parsons, Frank. **THE TELEGRAPH MONOPOLY.** Philadelphia: C.F. Taylor "Equity Series," 1890, 239 pp. Reprints a series of muckraking articles attacking most aspects of Western Union operation and arguing for government ownership. Useful for background on technical history. Tables, notes (no index).

4-178 **THE STORY OF WESTERN UNION.** New York: Western Union, 1958, 13 pp. Brief survey including first transcontinental line, systems, services, submarine cables, dates, and some key statistics.

4-179 "Western Union," *FORTUNE*, 10:90–96, 140–150 (November 1935). A careful business-oriented assessment of the company, already playing a secondary role to AT&T's telephone. Photos, map.

Westinghouse Electric Corporation

4-180 Woodbury, David O. **BATTLEFRONTS OF INDUSTRY: WESTINGHOUSE IN WORLD WAR II.** New York: John Wiley, 1948, 342 pp. General account including work with radio and radar. Photos, notes, appendices, index.

Zenith Radio Corporation

See Bryant and Cones, 9-158.

4-181 "Commander McDonald of Zenith," *FORTUNE* (June 1945), pp. 141–143, 209–216. The longtime leader of the company is profiled. Photos.

4-182 ** Cones, Harold N. and John H. Bryant. **ZENITH RADIO: THE EARLY YEARS, 1919–1935.** Atglen, PA: Schiffer, 1997, 222 pp. The first 100 pages offer a history of the firm and its impact on the industry in this period, the next 30 provide color photos of company products, and the remainder focuses on a detailed catalog of Zenith radios. (A continuation volume is in preparation.) Appendices, photos, tables, charts, endnotes.

See Curtis, 3-062.

4-183 **THE ZENITH STORY: A HISTORY FROM 1919.** Chicago: Zenith, 1955, 28 pp. Brief company promotional piece. Photos, charts.

Institutional and Company Histories 83

F. Specific Companies—Foreign

(A few biographies of industry and company leaders are included here rather than in chapter 5 as they deal substantially with business and company affairs.)

Alcatel/Australia

(See also Overseas Telecommunications Commission; Telecom Australia.)

4-184 Murray, James. **CALLING THE WORLD: THE FIRST 100 YEARS OF ALCATEL AUSTRALIA, 1895-1995.** Sydney: Focus Publishing, 1995, 200 pp. An informal history of the carrier and its predecessors.

Bosch Fernseh

4-185 "50 Years of FERNSEH, 1929-1979," *BOSCH TECHNISCHE BERICHTE,* 6:5/6:1-188 (May 25, 1979). Valuable collection of 17 illustrated (English-language) papers, focused particularly on television research. Photos, diagrams, references.

British Broadcasting Corporation (BBC)

4-186 ** **BBC HANDBOOK or YEARBOOK.** London: BBC, 1928-1987 (not published 1953 or 1954), annual. Invaluable for its year-by-year details on the technical basics of radio, and then television service. Title varied. Photos, diagrams, maps, tables, bibliographies.

4-187 * **BBC SOUND BROADCASTING: ITS ENGINEERING DEVELOPMENT.** London: BBC, 1962, 96 pp. Published to mark the 40th anniversary of the BBC. Photos, maps, chronology.

4-188 * **BBC TELEVISION: A BRITISH ENGINEERING ACHIEVEMENT.** London: BBC, 1961 [2nd ed.], 64 pp. Published to mark the 25th anniversary of BBC television. Photos, maps, diagrams, chronology. See also Bishop, 4-191.

4-189 THE BBC TELEVISION CENTRE. London: BBC, 1960, 80 pp. Half devoted to photos (some in color), this is the best description of the original "White City" facilities, including the concept, building, engineering aspects, and use of the studios.

4-190 BBC TELEVISION SERVICE: A TECHNICAL DESCRIPTION. London: BBC, 1951, 32 pp. Useful for its view of early postwar black-and-white 405-line services.

4-191 Bishop, Sir Harold. **TWENTY-FIVE YEARS OF BBC TELEVISION.** London: BBC Engineering Division Monograph No. 39, October 1961, 41 pp. A brief survey of the technical story. Photos, references.

4-192 ** Briggs, Asa. **THE HISTORY OF BROADCASTING IN THE UNITED KINGDOM.** London: Oxford University Press, 1961-1995, 5 vols. Definitive history of the BBC. The first volume, THE BIRTH OF BROADCASTING (1961, 425 pp.), and the second, THE GOLDEN AGE OF WIRELESS

(1965, 688 pp.) include considerable technical material on wireless and radio, while later volumes include somewhat less attention to technical matters. The full set is probably the finest history of any national broadcast system, and takes the story to 1974, covering just over the first half century of the corporation. Photos, diagrams, notes, bibliography, index.

See Burns, 12-113.

4-193 Geddes, Keith. **BROADCASTING IN BRITAIN: 1922–1972—A BRIEF ACCOUNT OF ITS ENGINEERING ASPECTS.** London: HMSO "A Science Museum Booklet," 1972, 63 pp. A summary that focuses primarily on BBC achievements in both radio and television. Photos.

4-194 Higgens, Gavin, ed. **BRITISH BROADCASTING 1922–1982: A SELECTED AND ANNOTATED BIBLIOGRAPHY.** London: BBC, 1982, 279 pp. Some 1,400 items in subject classifications covering radio and television, including technology and history, with good index. Several technical sections. See also Macdonald, 4-195.

See THE LONDON TELEVISION STATION, 12-057.

4-195 Macdonald, Barrie. **BROADCASTING IN THE UNITED KINGDOM: A GUIDE TO INFORMATION SOURCES.** London: Mansell/Cassell, 1988; 1993 [2nd ed.], 316 pp. Invaluable guide to primary and secondary resources on BBC and independent radio and television (including companies, associations, archives) with extensive annotation. Tables, index. See also G. Higgens, 4-194.

4-196 *** Pawley, Edward. **BBC ENGINEERING, 1922–1972.** London: BBC, 1972, 569 pp. Definitive treatment of the first half century of BBC technical development and operations in both radio and television by a senior BBC engineer, covering from the BBC's beginnings to television expansion in the 1970s. A basic source. Photos, tables, maps, references, index.

4-197 * **A TECHNICAL DESCRIPTION OF BROADCASTING HOUSE.** London: BBC, 1932, 105 pp. The full engineering details including acoustic treatment of studios, power supply, and studio and control room equipment as well as architectural features. Useful for its details on then-cutting-edge British technology. Photos, diagrams.

British Telecommunications

4-198 Clark, Jon, et al. **THE PROCESS OF TECHNOLOGICAL CHANGE: NEW TECHNOLOGY AND SOCIAL CHOICE IN THE WORKPLACE.** Cambridge, England: Cambridge University Press "Studies in Management No.11," 1988, 250 pp. Three social scientists and an engineer offer the results of a five-year study of BT's telephone modernization program through three case studies of corporate strategy. Notes, bibliography, index.

See EVENTS IN TELECOMMUNICATIONS HISTORY, 1-051.

See Herbert, 7-134.

See *THE POST OFFICE ELECTRICAL ENGINEERS' JOURNAL*, 7-141.

See Robertson, 7-142.

Cable and Wireless Ltd.

4-199 Baglehole, K.G. **A CENTURY OF SERVICE: A BRIEF HISTORY OF CABLE & WIRELESS LTD., 1868–1968.** Welwyn Garden City and London: Bournehall Press, 1969. A brief company promotional piece. Photos.

4-200 *** Barty-King, Hugh. **GIRDLE ROUND THE EARTH: THE STORY OF CABLE AND WIRELESS AND ITS PREDECESSORS TO MARK THE GROUP'S JUBILEE—1929–1979.** London: Heinemann, 1979, 413 pp. A professional historian divides the history into three 50-year periods: the last half of the 19th century focusing on Far Eastern Telegraph and Latin American cables; the first four decades of the 20th century when wireless joined cables; and the 1939–79 period of the Cable and Wireless conglomerate. Photos, notes, appendices, bibliography, chronology, index.

See THE CABLE AND WIRELESS COMMUNICATIONS OF THE WORLD, 3-084.

4-201 ** **CABLE & WIRELESS: A HISTORY.** London: Cable & Wireless, 1995, 1997 [2nd ed.], CD-ROM. Excellent collection of historical images (including film and audio clips), text and narration allowing study of the company and its technologies from a number of viewpoints. This is one of the first such historical CD-ROMs to be issued by a commercial firm.

L.M. Ericsson

4-202 *** Attman, Artur, et al. **L.M. ERICSSON: 100 YEARS.** Örebro, Sweden: Ljungföretagen, 1977, 3 vols. (translated from the Swedish edition published the year before). A handsome detailed company-sponsored history based substantially on primary documents. Photos, charts (many of these in color), tables, bibliography, index. As follows:

> VOL. 1: **THE PIONEERING YEARS, STRUGGLE FOR CONCESSIONS, CRISIS: 1876–1932,** by Artur Attman, Jan Kuuse, and Ulf Olsson, 369 pp. Divided into three parts: from handicraft to large-scale enterprise (1876–1900); the emerging group (1900–18); and the merger to the Kreuger crash (1918–32).

> VOL. 2: **RESCUE, RECONSTRUCTION, WORLDWIDE ENTERPRISE, 1932–1976,** by Artur Attman, and Ulf Olsson, 389 pp. Includes the period of recovery (1932–40); in the shadow of the war (early 1940s); and the period of expansion (1945–76).

> VOL. 3: **EVOLUTION OF THE TECHNOLOGY, 1876–1976,** by Christian Jacobæus and others, 425 pp. Thirty-one chapters not divided into

parts, and covering lines of development in telephony, cable and line plant, telex systems, broadcasting mobile radio and telephone, railway and other signaling systems, tubes and meters, production techniques, and research and development. Index of persons only.

4-203 Meurling, John, with Richard Jeans. **A SWITCH IN TIME: AN ENGINEER'S TALE**. Chicago: Telephony Publishing, 1985, 181 pp. A key figure in its development relates how Ericsson developed and initially marketed its AXE digital switching system. Tables, diagrams.

Exchange Telegraphy Company

4-204 * Scott, James M. **EXTEL 100: THE CENTENARY HISTORY OF THE EXCHANGE TELEGRAPHY COMPANY**. London: Ernest Benn, 1972, 240 pp. Includes the technology behind the British news agency. Photos, index.

General Electric Company (of Great Britain)

4-205 Clayton, Robert, and Joan Algar. **THE GEC RESEARCH LABORATORIES, 1919–1984**. London: Peter Peregrinus with Science Museum "IEE History of Technology Series 10," 1989, 438 pp. After chapters on institutional growth and direction, specific chapters cover, among other topics, research in valves (tubes), cathode-ray tubes, communications and electronics, and semiconductor materials and devices. Diagrams, photos, chapter references, patent list, statistics, index.

4-206 Whyte, Adam G. **FORTY YEARS OF ELECTRICAL PROGRESS: THE STORY OF THE G.E.C.** London: Ernest Benn, 1930, 166 pp. The rise and progress of the British firm. Photos, index.

Great Northern Telegraph/Great Nordic

4-207 **THE DIARY KEPT BY RASMUS PETERSEN ABOARD THE S.S. GREAT NORTHERN FROM AUGUST 2, 1870 TO JANUARY 4, 1871**. Copenhagen: GN Great Nordic, 1994, 64 pp. Concerns the pioneering undersea cables laid to the Far East. Photos, maps.

4-208 **FROM DOTS AND DASHES TO TELE AND DATACOMMUNICATIONS**. Copenhagen: GN Great Nordic, 1994, 80 pp. Essentially an update of the next entry though primarily given over to historical photographs.

4-209 **THE GREAT NORTHERN COMPANY: AN OUTLINE OF THE COMPANY'S HISTORY 1869–1969**. Copenhagen: Great Northern, 1969, 65 pp. Handsome illustrated album. Photos, maps, drawings.

4-210 Jacobsen, Kurt. "The Great Northern Telegraph Company: A Danish Company in the Service of Globalisation since 1869," in Stein Tønnesson, et al., eds. **BETWEEN NATIONAL HISTORIES AND GLOBAL HISTORY**. Helsingfors, Finland: Hakapaino Oy, 1997, pp. 179–196. A short survey of the company's history and contributions. Notes.

W.T. Henley's Telegraph Works Company

4-211 Anderson, A.F. "William Henley, Pioneer Electrical Instrument Maker and Cable Manufacturer, 1813-1882" *PROCEEDINGS OF THE IEE*, Part A, 132:249-161 (July 1985). Photos, diagrams, maps, bibliography.

4-212 Slater, Ernest. **ONE HUNDRED YEARS: THE STORY OF W.T. HENLEY'S TELEGRAPH WORKS COMPANY.** London: The Company, 1937, 77 pp.

KDD (Kokusai Denshin Denwa, Japan)

4-213 **JAPAN'S PROGRESS IN INTERNATIONAL TELECOMMUNICATIONS: A QUARTER CENTURY OF KDD ACTIVITIES.** Tokyo: KDD, 1980, 121 pp. Focuses on development since World War II. Maps, charts, photos, appendices.

4-214 **TELECOMMUNICATIONS IN JAPAN.** Tokyo: Ministry of Posts and Telecommunications, 1966, 48 pp. About half devoted to domestic (NTT) and broadcasting (NHK) and half to international (KDD) development and activities. Photos, charts, maps.

Marconi Company
(See also under Marconi, Guglielmo, 5-155 through 5-168a.)

4-215 *** Baker, W.J. **A HISTORY OF THE MARCONI COMPANY.** London: Methuen, 1970/New York: St. Martin's Press, 1971, 413 pp. Authoritative account of the origin and progress of the company, but sadly lacking any documentation. Divided into three parts: up to 1918, 1919 to 1939 and World War II into the 1960s. Full details of organization, business matters, people, events, services, products, inventions, engineering, research and development. Many names, dates, and text references. Photos, diagrams, index.

4-216 **CAVALCADE.** London: Marconiphone, n.d., ca 1935. 44 pp. Profusely illustrated history of radio and the company from 1894 to 1935.

4-217 Donaldson, Frances. **THE MARCONI SCANDAL.** New York: Harcourt Brace & World, 1962, 304 pp. Concerns 1912-13 political crisis concerning shareholdings and the building of Britain's Imperial Wireless Chain of stations. Notes, photos, appendices, bibliography, index.

4-218 Godwin, George. **MARCONI 1939-1945: A WAR RECORD.** London: Chatto & Windus, 1946, 130 pp. Broad survey of people and events. Photos.

See Hancock, 3-156.

See Jensen, 9-065.

4-219 **JUBILEE YEAR.** Chelmsford, England: Marconi, 1947, 57 pp. Survey of company development. Photos, list of companies and agents.

See Kreuzer, 5-164.

4-220 Leigh-Bennett, Ernest P. **A CITY OF SOUND**. London: Marconiphone, 1933, 49 pp. Story of the Marconi company and especially its factory in Hayes, Middlesex. Photos.

See Marconi Company, 9-070.

See Marconi International Marine Co., 9-074.

See THE MARCONI WIRELESS TELEGRAPH SYSTEM, 9-015.

NEC (Nippon Electric Co., Japan)

4-221 **SIXTY YEARS OF PROGRESS.** Tokyo: Nippon Electric Co., 1958, 60 pp. Photos, charts. See next entry.

4-222 **THE FIRST 80 YEARS**. Tokyo: NEC Corp., 1984, 120 pp. Two company-produced pieces that briefly survey the past but focus on the present.

NHK (Nippon Hoso Kyokai [Japan Broadcasting Corporation])

4-223 **THE HISTORY OF BROADCASTING IN JAPAN**. Tokyo: History Compilation Room, Radio & TV Culture Research Institute, NHK, 1967, 436 pp. Includes extensive discussion of radio and television engineering developments. Photos, notes, charts, chronology, index.

4-224 **50 YEARS OF JAPANESE BROADCASTING**. Tokyo: History Compilation Room, Radio & TV Culture Research Institute, NHK, 1977, 432 pp. Useful for updating of above. Includes commercial broadcasting organizations as well. Photos, tables, chronology, map, index.

4-225 Not used.

Northern Telecom

4-226 **THE ANATOMY OF A TRANSFORMATION, 1985–1995**. Toronto (?): Northern Telecom, 1995, 33 pp. Through charts and text surveys the change from a traditional manufacturer to a more diversified firm.

4-227 Newman, Peter C. **NORTEL: PAST, PRESENT, FUTURE**. Toronto (?): Northern Telecom, 1995, 120 pp. Six chapter survey, with the first covering to 1956 and the rest dealing with the past four decades. Chronological time line, photos.

Norwegian Telecom (Gyldendal Norsk Forlag or Tele)

4-228 Dahl, Tor Edvin, et al., trans. by Bibbi Lee. **HELLO?! THE HISTORY OF THE TELEPHONE IN NORWAY**. Oslo: Gyldendal Norsk Forlag A/S, 1993, 174 pp. Illustrated album with brief survey text reviews the story. Photos, index.

Institutional and Company Histories 89

Overseas Telecommunications Commission (Australia)
(See also Alcatel Australia, Telecom Australia.)

4-229 ** Harcourt, Edgar. TAMING THE TYRANT: THE FIRST ONE HUNDRED YEARS OF AUSTRALIA'S INTERNATIONAL COMMUNICATION SERVICES. Sydney: Allen & Unwin, 1987, 405 pp. History of the Overseas Telecommunications Commission (OTC) and its predecessors, divided into five parts: beginnings (to 1872); for profit or service (1872–1902); wires and wireless (1902–46); commonwealth partnership (1946–60); and transformation (1960–72). Photos, glossary, charts, sources, index.

Philips

4-230 Bouman, P[ieter] J[an]. ANTON PHILIPS OF EINDHOVEN. London: Weidenfeld & Nicolson, 1958, 255 pp.; revised as GROWTH OF AN ENTERPRISE: THE LIFE OF ANTON PHILIPS. London: Macmillan, 1970, 272 pp. Founder (1874–1951) of the Dutch firm. Revision includes a postscript on the company's history after 1951. Photos, index.

4-231 Heerding, A., trans. by Derek Jordan. THE HISTORY OF N.V. PHILIPS' GLOEILAMPENFABRIEKEN. Cambridge, England: Cambridge University Press, 1982, 2 vols. Primarily devoted to electric light products but sheds some light on the larger company history as well. Photos, references, index.

> VOL. 1: THE ORIGINS OF THE DUTCH INCANDESCENT LAMP INDUSTRY. 341 pp. Takes the story to 1891.

> VOL. 2: A COMPANY OF MANY PARTS. 371 pp. Covers 1891–1922 period.

4-232 Philips, Frederik. 45 YEARS WITH PHILIPS: AN INDUSTRIALIST'S LIFE. Poole, England: Blandford Press, 1978, 280 pp. Biography of firm from 1930 to mid-1970s as seen by a later CEO (1905–). Translation from the Dutch. Photos, map, index.

Siemens

4-233 Bamber, E.F., ed. THE SCIENTIFIC WORKS OF C. WILLIAM SIEMENS. London: John Murray, 1889, 3 vols. The younger brother of Werner von Siemens (see below), and founder of the British branch of the firm.

4-234 FACE TO FACE WITH TECHNOLOGY: ELECTRICAL ENGINEERING, ELECTRONICS, MICROELECTRONICS: THE SIEMENS MUSEUM IN MUNICH. Berlin: Siemens, 1991 [2nd ed.], 147 pp. Concentrates on company products and services. Photos, bibliography.

4-235 Feldenkirchen, Wilfried. WERNER VON SIEMENS: INVENTOR AND INTERNATIONAL ENTREPRENEUR. Columbus: Ohio State University Press "Historical Perspectives on Business Enterprise Series," 1994, 203 pp. Translated from the 1992 German edition. Founder of the company (1816–92). Photos, chronology, appendices, notes, bibliography (no index).

4-236 150 YEARS OF SIEMENS: THE COMPANY FROM 1947 TO 1997. Munich: Wilfried Feldenkirchen for Siemens, 1997, 88 pp. Colorfully illustrated popular survey includes many period illustrations.

4-237 Pole, William. **THE LIFE OF SIR WILLIAM SIEMENS.** London: John Murray, 1888, 412 pp. The founder of the British branch of the company. Portraits, illustrations.

4-238 * Scott, J.D. **SIEMENS BROTHERS 1858–1958: AN ESSAY IN THE HISTORY OF INDUSTRY.** London: Weidenfeld and Nicolson, 1958, 279 pp. The British branch of the firm is detailed, including its work with submarine cables and telephone equipment. Photos, notes, source notes, folding diagram, index.

4-239 * Siemens, Georg, trans. by A.F. Rodger and Lawrence N. Hole. **HISTORY OF THE HOUSE OF SIEMENS.** Freiberg, Germany: Karl Alber, 1957, 2 vols.; 1961 [rev. ed.] (reprinted by Arno Press 1977). Full history of the German firm, translated from the original German edition. Index.

> **VOL. 1: THE ERA OF FREE ENTERPRISE.** 333 pp. The first 19 chapters, covering 1847–1910.

> **VOL. 2: THE ERA OF WORLD WARS.** 326 pp. The final 13 chapters, covering 1910–45.

4-240 Siemens, Werner von. **INVENTOR AND ENTREPRENEUR: RECOLLECTIONS OF WERNER VON SIEMENS.** London: Lund Humphries/Munich: Prestel-Verlag, 1966 [2nd ed.], 314 pp. The author (1816–92), with his brothers, founded the German electronics firm. This book first appeared in English in 1893, translated by W.C. Coupland. This is an English version of the 1956 German second edition which added illustrations, annotations, and an index. Includes discussion of submarine cables and terrestrial telegraph lines. Drawings, photos, index.

4-241 _____. **SCIENTIFIC & TECHNICAL PAPERS OF WERNER VON SIEMENS.** London: John Murray, 1892 and 1895, 2 vols. English translation of the 1889 German work (Berlin: Julius Springer).

4-242 SIR WILLIAM SIEMENS: A MAN OF VISION. London: Siemens PLC, 1993, 80 pp. Founder (1823–83) of the British firm is briefly profiled followed by five additional papers on the development of British electronics industry, the history of Siemens in the U.K., and Siemens today. Illustrations (most colored), references.

4-243 Weiker, Sigfrid von, and Herbert Goetzeler. **THE SIEMENS COMPANY—ITS HISTORICAL ROLE IN THE PROGRESS OF ELECTRICAL ENGINEERING, 1947–1980.** Berlin: Siemens, 1977, 1984 [3rd ed.], 205 pp. Brief historical treatment. Photos, chronology, notes, charts, maps, index.

Sony

4-244 Ibuka, Masaru. "How Sony Developed Electronics for the World Market," *IEEE TRANSACTIONS ON ENGINEERING MANAGEMENT*, EM-22:1 (February 1975). Highlights of the Sony story.

4-245 Lyons, Nick. **THE SONY VISION**. New York: Crown, 1976, 235 pp. An informal company history emphasizing consumer products. Photos, bibliography, index.

4-246 Morita, Akio, et al. **MADE IN JAPAN: AKIO MORITA AND SONY**. New York: Dutton, 1986, 309 pp. Informal history of the firm seen through the life of its founder. Photos, index.

Standard Telephones and Cables

4-247 THE STORY OF S.T.C. 1883–1958. London: Standard Telephones & Cables Ltd., 1958, 108 pp. Issued to commemorate the firm's 75th anniversary. Photos.

4-248 * Young, Peter. **POWER OF SPEECH: A HISTORY OF STANDARD TELEPHONES AND CABLES, 1883–1983**. London: Allen & Unwin, 1983, 221 pp. This is a company-sponsored history—STC began as an arm of Western Electric. Photos, tables, index.

Swedish Telecommunications Administration

4-249 TELE: RETROSPECTIVE SURVEY OF THE LAST TWENTY-FIVE YEARS. Stockholm: Swedish Telecommunications Administration, 1965, 256 pp. Issued in honor of retiring Director General Hokan Sterky, this features 24 technical papers on general development of the administration, telecommunication development, and the development of materials and components. Photos, charts, tables, notes.

Telecom Australia
(See also Alcatel, Overseas Telecommunications Commission.)

4-250 ** Moyal, Ann. **CLEAR ACROSS AUSTRALIA: A HISTORY OF TELECOMMUNICATIONS**. Melbourne: Thomas Nelson, 1984, 437 pp. The first comprehensive history of Telecom Australia and its predecessor organization under the postmaster general. Photos, bibliography, maps, glossary, index.

Telegraph Construction and Maintenance Company

4-251 [Lawford, G.L. and L.R. Nicholson.] **THE TELCON STORY: 1850–1950**. London: Telegraph Construction and Maintenance Company, 1950, 176 pp. Photos, appendix of major cables made by the firm.

Chapter 5

Biography: Inventors, Scientists, and Engineers

Biographical material and sources—reference works, collections, books and articles by or about key individuals—appear in this chapter. The focus is on key inventors and their work, and citations include both primary collections of letters or patents, and secondary biographies or assessments of a given inventor's contributions. (Selected business leader biographies appear with their respective companies in chapter 4.) Arrangement of the major part of this chapter is by specific inventor, and within each such section, alphabetically by author.

After (A) a selective survey of useful general biographical references, the chapter turns to (B) the relatively few sources of collected biographies in this field, and (C) individual inventor entries which include most of the book-length studies devoted to key individuals, as well as the more central periodical articles about them. Most of this is secondary material, but some primary collections of letters or documents are included as well. For related Internet resources, see 15-D.

A. General Biographical Reference

(Often the best sources for specific personal information are the many professional journals noted in chapter 2-D. These contain current notices, memberships, awards, obituaries [some quite extensive] and other personal items. Many serials include brief biographies of authors. Various organizational committee reports, convention records, and news about both organizations and business activities [many described in chapter 4] offer still further choices. Personal information can also be found in society directories, yearbooks, trade magazines, and scholarly journals. More detailed records of the life and work of prominent innovators are collected in the various biographical memoirs and obituaries published annually by professional societies.)

5-001 Abbott, David. **THE BIOGRAPHICAL DICTIONARY OF SCIENTISTS: ENGINEERS AND INVENTORS**. New York: Peter Bedrick, 1985, 188 pp. A column or two on each of several hundred persons, most from Britain or the United States Index.

5-002 * **AMERICAN MEN AND WOMEN OF SCIENCE**. New Providence, NJ: Bowker, 1998 [20th ed.], 8 vols. First issued in 1906, this now includes some 120,000 engineers, scientists, inventors, teachers and others who have made notable contributions to applied science. Each entry includes details of education, vocation, specialty, affiliation, and professional memberships. Last volume is an index by discipline.

5-003 Asimov, Isaac. **BIOGRAPHIC ENCYCLOPEDIA OF SCIENCE AND TECHNOLOGY**. Garden City, NY: Doubleday, 1964, 662 pp.; 1972 [2nd ed.]; 1982 [2nd rev. ed.], 941 pp. Brief biographies of more than 1,500 scientists from the Greek era to modern times. Double-column numbered entries with extensive cross-references provide a virtual roadmap to scientific and technical history. Index.

5-004 Bailey, Martha J. **AMERICAN WOMEN IN SCIENCE: A BIOGRAPHICAL DICTIONARY**. Denver: ABC-Clio, 1994, 463 pp. Includes those who began their careers before 1950. Index.

5-005 ** **BIOGRAPHY INDEX: A QUARTERLY INDEX TO BIOGRAPHICAL MATERIAL IN BOOKS AND MAGAZINES**. New York: H.W. Wilson, 1946–date. Includes annual and three-year cumulations. Available on-line and in CD-ROM format from 1984 to date. Includes material published as books, book chapters, and journal articles.

5-006 Daintith, John, et al., eds. **BIOGRAPHICAL ENCYCLOPEDIA OF SCIENTISTS**. Bristol, England, and Philadelphia: Physics Publishing, 1981, 1983, 1994 [2nd ed.], 2 vols. Some 2,000 entries of about a column each. Chronology, list of institutions, bibliography, and index by name and subject.

5-007 *** **DICTIONARY OF SCIENTIFIC BIOGRAPHY**. New York: Scribner's, 1970–90, 18 vols. Standard reference with signed authoritative essays. Contains short article summaries of the professional careers and scientific accomplishments of figures from all periods, including inventors, engineers and physicists, excluding those still living. Articles of a column or two are signed by contributor and include a brief bibliography of original works and biographical sources. Supplement I and index allowing search by industry appear in vols. 15–16; supplement II in vols. 17–18.

5-008 Elliott, Clark A. **BIOGRAPHICAL INDEX TO AMERICAN SCIENCE: THE SEVENTEENTH CENTURY TO 1920**. Westport, CT: Greenwood, 1979, 360 pp.; 1990 [2nd ed.], 300 pp. Cross references entries for science and technology figures included in over 100 major biographical dictionaries and collections.

5-009 ** Higgins, Thomas J. "A Biographical Bibliography of Electrical Engineers and Electrophysicists," *TECHNOLOGY AND CULTURE*, 2:18–32 (1961); 2:146–165 (1962). First part covers books; second periodicals. In all, some 1,200 items arranged alphabetically by surname. Recollections, surveys, evaluations, anni-

Biography: Inventors, Scientists, and Engineers

versary and obituary notices, reviews, and collections of works by or about individuals who made noteworthy contributions to electrical engineering.

5-010 _____. **BIOGRAPHIES OF ENGINEERS AND SCIENTISTS.** Chicago: Illinois Institute of Technology, 1949, 62 pp. A bibliographic guide.

5-011 Ireland, Norma O. **INDEX TO SCIENTISTS OF THE WORLD FROM ANCIENT TO MODERN TIMES: BIOGRAPHIES AND PORTRAITS.** Boston: F.W. Faxon, 1962, 705 pp. Nearly 7,500 scientists, inventors, and engineers are listed by name with dates, brief identification, and references. Based on 338 collections and ranging from specialized works and multivolume references to monographs, including juvenile and popular books.

5-012 Miller, David, et al. **THE CAMBRIDGE DICTIONARY OF SCIENTISTS.** Cambridge, England: Cambridge University Press, 1996, 387 pp. Some 1,300 brief sketches of people from 38 countries. Useful (and portable!) brief guide. Chronology, list of Nobel winners, index.

5-013 National Academy of Sciences. **BIOGRAPHICAL MEMOIRS.** Various cities and publishers, 1877–date. Contributed articles with portraits and bibliographies, current volumes typically have 12–15 memoirs in about 350 pp. Vol. 40 (1969) includes cumulative index to that point.

5-014 Porter, Roy, ed. **THE BIOGRAPHICAL DICTIONARY OF SCIENTISTS.** New York: Oxford University Press, 1983, vols. 6; 1994 [2nd ed.], 891 pp. Covers thousands of people over time. Appendix on Nobel prize winners, glossary, index.

5-015 Roysdon, Christine, and Linda A. Khatri. **AMERICAN ENGINEERS OF THE NINETEENTH CENTURY: A BIOGRAPHICAL GUIDE.** New York: Garland, 1978, 247 pp. Provides citations for hundreds of figures. Index.

B. Collected Telecommunications Biography

5-016 ** Appleyard, Rollo. **PIONEERS OF ELECTRICAL COMMUNICATIONS.** London: Macmillan, 1930 (reprinted by Books for Libraries, 1968), 347 pp. Reprinted from articles in *ELECTRICAL COMMUNICATION*, Vols. 5–8 (1926–29), this offers chapters on ten key figures: Maxwell (see 5-169), Ampére, Volta, Wheatstone (see 5-079a), Hertz (see 5-138), Oersted, Ohm, Heaviside, Chappe, and Ronalds (see 6-001). No formal references but pertinent names and dates are given in the text. Photos, index.

5-017 * "Audio Pioneers," *AUDIO*, 46:32–60, 72–73 (May 1962). Thirty-five biographical essays with photos.

5-018 ** Bray, John. **THE COMMUNICATIONS MIRACLE: THE TELECOMMUNICATION PIONEERS FROM MORSE TO THE INFORMATION SUPERHIGHWAY.** New York: Plenum Press, 1995, 379 pp. History of telecommunications in form of a useful survey of primarily British and American work, with some emphasis on key individuals. Photos, diagrams, chapter references, index.

5-019 Cortada, James W. **HISTORICAL DICTIONARY OF DATA PROCESSING BIOGRAPHIES.** Westport, CT: Greenwood, 1987, 323 pp. Many of the people briefly profiled played a larger role in telecommunications as well. References, index.

5-020 Dibner, Bern. **TEN FOUNDING FATHERS OF THE ELECTRICAL SCIENCE.** Norwalk, CT: Burndy Library "Publications in the History of Science and Technology, No. 11," 1954, 46 pp. Reprinted from articles in *ELECTRICAL ENGINEERING*, Vols. 73–74 (1954–55), these two-page profiles cover Gilbert, Guericke, Franklin, Volta, Ampére, Ohm, Gauss, Faraday, Henry (see 5-131a), and Maxwell (see 5-169). Engravings.

5-021 ** Dunlap, Orrin E. **RADIO'S 100 MEN OF SCIENCE: BIOGRAPHICAL NARRATIVES OF PATHFINDERS IN RADIO, ELECTRONICS AND TELEVISION.** New York: Harper, 1944 (reprinted by Books for Libraries, 1971), 294 pp. A popular and general treatment with much useful information about the subjects and their contributions, but with very little documentation. The brief personal sketches (one to three pages each) provide numerous quotations and extracts. Arranged by birth year of subject. Photos, notes, index.

5-022 * Garratt, G.R.M. **EARLY HISTORY OF RADIO FROM FARADAY TO MARCONI.** London: IEE "History of Technology Series 20," 1994, 93 pp. Brief biographies of Faraday, Maxwell, Hertz, Lodge, Popov, and Marconi.

5-023 Gray, Prof. Thomas. "The Inventors of the Telegraph and Telephone," **ANNUAL REPORT OF THE BOARD OF REGENTS OF THE SMITHSONIAN INSTITUTION, 1892.** Washington: Government Printing Office, 1893, pp. 639–657. Reprinted in Shiers, 6-120.

5-024 ** Hawks, Ellison. **PIONEERS OF WIRELESS.** London: Methuen, 1927 (reprinted by Arno Press, "Telecommunications," 1974), 304 pp. Based on articles in *WIRELESS WORLD AND RADIO REVIEW*, Vols. 18–19 (1926). A thorough exposition with numerous quotations and extracts, mostly supported by fairly complete references. Ranges from Gilbert up to Alexanderson (see 5-038) and Fessenden (see 5-116). Photos, chronology, notes, index.

5-025 * Hunt, Bruce J. **THE MAXWELLIANS.** Ithaca, NY: Cornell University Press, 1991, 266 pp. Insightful study of G.F. Fitzgerald, Oliver Heaviside, and Oliver J. Lodge (see 5-147) and how by experiments and correspondence, they helped to spread Maxwell's ideas after the latter's early death. Photos, notes, bibliography, index. See also O'Hara and Pritcha, 5-031.

5-026 Jeans, William T. **LIVES OF THE ELECTRICIANS: PROFESSORS TYNDALL, WHEATSTONE AND MORSE.** London: Whittaker, 1887, 327 pp. Though noted as the "first series," only this volume appeared.

5-027 * Lewis, T[homas S.W.]. **EMPIRE OF THE AIR: THE MEN WHO MADE RADIO.** New York: HarperCollins, 1991, 421 pp. A well-written biographical approach to American wireless history, focusing primarily on the progress by and conflicts among Armstrong (5-039), de Forest (5-091), and RCA's Sarnoff (4-152).

Biography: Inventors, Scientists, and Engineers 97

This volume was the basis for a PBS documentary film of the same name. Photos, sources and notes, bibliography, index.

See Lodge, 8-010.

5-028 MacDonald, David K.C. **FARADAY, MAXWELL AND KELVIN**. Garden City, NY: Doubleday, 1964, 143 pp. Brief summary of the three 19th-century giants of electrical research. Photos, notes, index.

5-029 Munro, John. **HEROES OF THE TELEGRAPH**. London: Religious Tract Society, 1891, 288 pp. Popular story with biographies of Wheatstone (see 5-079a), Morse (see 5-187), Thomson, Siemens (see 4-233) and others; includes some references to the telephone. An appendix includes eight more shorter biographies. Full text is available on the web as Project Gutenberg Release No. 979.

5-030 Nebeker, Frederik, ed. **SPARKS OF GENIUS: PORTAITS OF ELECTRICAL ENGINEERING EXCELLENCE**. New York: IEEE, 1994, 268 pp. The eight chapters, each devoted to a single figure, include "Amos Joel and the Advent of Electronic Telephone Switching," Charles Townes as an electrical engineer, Harold Alden Wheller and applied electronics. Based on the IEEE oral history program. Photos, diagrams, references.

5-031 O'Hara, J[ames] G., and W[illibald] Pritcha. **HERTZ AND THE MAXWELLIANS: A STUDY AND DOCUMENTATION OF THE DISCOVERY OF ELECTROMAGNETIC WAVE RADIATION, 1873-1894**. London: Peter Peregrinus and the Science Museum "IEE History of Technology Series 8," 1987, 154 pp. Focuses on documenting the inter-relationships between Hertz and a number of British scientific figures, and the repetition of his pioneering experiments in Britain and Ireland. Photos, references, sources, index. See also Hunt, 5-025.

5-032 "Personnel of the Telegraph," in Twenty-Fifth Anniversary Number of *TELEGRAPH AGE* (January 1, 1908), pp. 15–97. Useful resource for biographies of managers and other key figures of all operating telegraph companies in the U.S. and Canada at the time. Photos.

5-033 Perucca, Eligio, and Vittorio Gori. "Pioneers in Electrical Communications," *JOURNAL OF THE FRANKLIN INSTITUTE*, 261:61–79 (January 1956). A brief survey with highlights of progress to 1956 on the electric telegraph, telegraph lines, telephone, wireless and radio communication. Numerous names and dates in the text. Notes.

5-034 Potamian, Brother [Michael F. O'Reilly], and James J. Walsh. **MAKERS OF ELECTRICITY**. New York: Fordham University Press, 1909, 408 pp. Chapters devoted to Peregrinus and Columbus, Norman and Gilbert, Franklin and some contemporaries, Galvani, Volta, Coulomb, Oersted, Ampére, Ohm, Faraday, Maxwell (see 5-169), and Lord Kelvin. Engravings, notes, index.

5-035 * Reid, James D. **THE TELEGRAPH IN AMERICA: ITS FOUNDERS, PROMOTERS AND NOTED MEN**. New York: Derby Bros, 1879 (reprinted by Arno Press "Telecommunications," 1974), 846 pp.; 1886 [2nd ed.], 894 pp. First 100 pages are a memorial to Samuel F.B. Morse which, curiously,

98   Chapter 5

continues for the last 150 pages (and includes a memorial to William Orton, longtime head of Western Union, who died the year before). The majority of the volume details the history of specific firms and their key figures. Engravings.

5-036 Taltavall, John B. **TELEGRAPHERS OF TO-DAY: DESCRIPTIVE, HISTORICAL, BIOGRAPHICAL.** New York: Telegraph Age, 1894, 354 pp. A large album survey of late-19th-century figures in the industry—operators, company officers, and telegraph "personalities"—with their service histories and often a photo.

5-037 Towers, Walter K. **MASTERS OF SPACE: MORSE AND THE TELEGRAPH, THOMPSON AND THE CABLE, BELL AND THE TELEPHONE, MARCONI AND THE WIRELESS TELEGRAPHY, CARTY AND THE WIRELESS TELEPHONE.** New York: Harper, 1917, 300 pp.; revised as FROM BEACON FIRE TO RADIO (MASTERS OF SPACE): THE STORY OF LONG-DISTANCE COMMUNICATION, 1924, 303 pp. General survey relating telecommunications development through key inventors and innovators and their work. Photos, appendices, index.

See Van Deusen, 3-022.

C. Individuals

(Includes primarily publications about key inventors, plus a few primary papers or collections of papers by inventors (more appear in the subject chapters). See Shiers, 1-017, for further minor entries which are largely superceded by items noted below.)

ALEXANDERSON, Ernst Fredrik Werner (1878–1975)

See Alexanderson, 9-045.

5-038 *** Brittain, James E. **ALEXANDERSON: PIONEER IN AMERICAN ELECTRICAL ENGINEERING.** Baltimore: Johns Hopkins University Press "Studies in the History of Technology," 1992, 381 pp. Definitive biography of the General Electric engineer who originated the high-frequency alternator, many radio innovations, experimented with mechanical television systems, and took part in many other innovations—and took out some 340 patents. Photos, appendix, bibliographic note, notes, index.

ARMSTRONG, Edwin Howard (1890–1954)

See Armstrong, 9-130 through 9-132, and 9-140 through 9-143.

5-039 Erickson, D.H.V. **ARMSTRONG'S FIGHT FOR FM BROADCASTING: ONE MAN vs BIG BUSINESS AND BUREAUCRACY.** University: University of Alabama Press, 1973, 226 pp. Though based on the author's dissertation (Illinois, 1969) and drawing heavily on Lessing (5-041), this is more of a polemic than a scholarly analysis of the man and his work. Must be used carefully. Notes, bibliography, index.

Biography: Inventors, Scientists, and Engineers

5-040 * Lessing, Lawrence. "Armstrong of Radio," *FORTUNE*, 37: 88–91, 198–210 (February 1948). Survey of Armstrong's life and contributions to radio engineering, with special reference to FM. Photos.

5-041 ** _____. MAN OF HIGH FIDELITY: EDWIN HOWARD ARMSTRONG. Philadelphia: Lippincott, 1956, 320 pp. Reprinted by Bantam Books, 1969 with added chapter, 272 pp. A solid and colorful story, popular and nontechnical with little documentation. Diagrams, photos, bibliography of Armstrong's papers, index.

5-041a ** Morrisey, John W., ed. "The Legacies of Edwin Howard Armstrong," *PROCEEDINGS OF THE RADIO CLUB OF AMERICA*, 63:3: 1–321 (November 1990). Extensive anthology of news reports, technical articles by and about Armstrong, and other material arranged in sections on the regenerative, superheterdyne, superregenerative, and FM circuits. Photos, diagrams, references, appendices.

5-042 Ragazzini, John R. "Creativity in Radio: Contributions of Major Edwin H. Armstrong," *JOURNAL OF ENGINEERING EDUCATION*, 45:112–119 (October 1954). A personal study set against the historical background by a close colleague.

5-043 "Revolution in Radio," *FORTUNE*, 20:86–88, 116–121 (October 1939). One of the first in-depth discussions for a general audience of Armstrong's FM innovation. Photos.

BAIRD, John Logie (1888–1946)

5-044 ** Baird, John Logie. SERMONS, SOAP AND TELEVISION: AUTOBIOGRAPHICAL NOTES. London: Royal Television Society, 1988, 147 pp. Useful for its first-person account and illustrations of the struggle for mechanical television success. Photos, diagrams.

5-045 * Baird, Margaret. TELEVISION BAIRD. Capetown: HAUM, 1973, 160 pp. A fond memoir by the inventor's wife (she joined the firm as secretary in 1928; they married in 1931). Photos.

5-046 * Bridgewater, T.H. "Baird and Television," *JOURNAL OF THE BRITISH KINEMATOGRAPHY, SOUND AND TELEVISION SOCIETY*, 49:60–68 (March 1967). Then chief engineer of BBC Television reviews Baird's life and accomplishments.

5-047 Burns, R.W. "The First Demonstration of Television," *ELECTRONICS & POWER*, 26:953–956 (October 9, 1975). The public demonstrations of late 1925 and early 1926 are reviewed. Photos, references. See related letters in following issues: 21:1193 (December 11, 1975); 22:40 (January 1976); 22:113, 114 (February 1976); 22:246, 247 (April 1976); and 22:310 (May 1976). See also Burns, 12-113.

5-048 _____. "J.L. Baird: Success and Failure," *PROCEEDINGS OF THE IEE*, 126:921–928 (September 1979). A survey of his accomplishments compared with those of Marconi. Photos, references.

5-049 * Exwood, Maurice. **JOHN LOGIE BAIRD: 50 YEARS OF TELEVISION**. London: Institution of Electronic and Radio Engineers, 1976, 31 pp. One of the better surveys from an engineering point of view. Photos, bibliography.

5-050 Hallett, Michael. **JOHN LOGIE BAIRD AND TELEVISION**. Hove, England: Priory Press "Pioneers of Science and Discovery," 1978, 95 pp. Intended for younger readers, this is useful largely for its illustrations. Photos, chronology, glossary, bibliography, index.

5-051 Herbert, Ray. "J.L. Baird's Colour Television 1937–1946," *TELEVISION*, 27:1:23–29 (January–February 1990). Color photos of the inventor and his late-life work.

5-052 _____. **SEEING BY WIRELESS: THE STORY OF BAIRD TELEVISION**. Sanderstead, England: Author, 1996, 27 pp. Valuable photographic record assembled from the collections of 34 former Baird company employees, of whom the author was one. Includes both prewar and wartime efforts. Photos.

5-053 ** McArthur, Tom, and Peter Waddell. **THE SECRET LIFE OF JOHN LOGIE BAIRD**. London: Century Hutchinson, 1986 (reprinted as VISION WARRIOR: THE HIDDEN ACHIEVEMENT OF JOHN LOGIE BAIRD, by Scottish Falcon Books, 1990, 315 pp.). A sometimes breathless account of the inventor's life, this focuses especially on his wartime secret work for the British government on radar and related technologies. Photos, notes, index.

See McLean, 10-021 through 10-123.

5-054 * Moseley, Sydney. **JOHN BAIRD: THE ROMANCE AND TRAGEDY OF THE PIONEER OF TELEVISION**. London: Odhams, 1952, 256 pp. Nontechnical account giving the "inside story" by Baird's staunchest supporter (Moseley was a business writer and television publicist). Photos, index.

See Moseley and Chapple, 12-012.

5-055 * Percy, J.D. **JOHN L. BAIRD: THE FOUNDER OF BRITISH TELEVISION**. London: Television Society, 1952, 16 pp. Chronological review of developments from 1926 to 1936. Photos, diagram. Reprinted in Shiers, 12-128.

5-056 Rowland, John. **THE TELEVISION MAN: THE STORY OF JOHN L. BAIRD**. New York: Roy, 1966, 144 pp. Written for younger readers, this is a narrative of Baird's life and work. Bibliographic note.

5-057 * Shiers, George. "Television 50 Years Ago," *JOURNAL OF BROADCASTING*, 19:387–400 (Fall 1975). Account of Baird's activities from 1923 to the end of 1926 with mention of Jenkins work at the same time. References. Reprinted in *TELEVISION*, 16:1: 6–9 (January–February 1976), adding photos.

5-058 * Tiltman, Ronald F. **BAIRD OF TELEVISION: THE LIFE STORY OF JOHN LOGIE BAIRD**. London: Seeley, Service, 1933 (reprinted by Arno Press, "Telecommunications," 1974), 220 pp. A popular account by a reporter and friend of the inventor. Noncritical and vague except for a few dates in the text. Photos, index.

Biography: Inventors, Scientists, and Engineers

See "The World's Earliest Television Recordings," 15-062.

BELL, Alexander Graham (1847–1922)

See "Alexander Graham Bell Family Papers," 15-034.

See "Bell's Path to the Telephone," 15-035.

See Bell, 7-047.

5-059 *** Bruce, Robert V. BELL: ALEXANDER GRAHAM BELL AND THE CONQUEST OF SOLITUDE. Boston: Little, Brown, 1973 (reprinted by Cornell University Press, 1990), 564 pp. Magisterial and definitive biography. Roughly 200 pages focus on Bell's work with the telephone; the remainder deals with his life before and after that work, including aeronautical research. Photos, bibliography, references, index.

5-060 Gorman, Michael E. "Alexander Graham Bell, Elisha Gray and the Speaking Telegraph: A Cognitive Comparison," *HISTORY OF TECHNOLOGY*, 15:1–56 (1993). Modern review of the classic face-off, comparing the conflicting claims.

5-061 ** Grosvenor, Edwin S., and Morgan Wesson. ALEXANDER GRAHAM BELL: THE LIFE AND TIMES OF THE MAN WHO INVENTED THE TELEPHONE. New York: Harry Abrams, 1997, 304 pp. Handsome photographic album dealing about equally with the man and his primary invention. Includes many examples of Bell's research notebook pages and early apparatus. Photos, bibliography, index.

See Hounshell, 5-129 and 5-130.

5-062 Mackay, James. SOUNDS OUT OF SILENCE: A LIFE OF ALEXANDER GRAHAM BELL. Edinburgh, Scotland: Mainstream Publishing / New York: Wiley, 1997, 320 pp. Using both primary documents and secondary resources (especially Bruce, 5-059, who forced the American edition's withdrawal under threat of legal action for plagiarism), the author provides a brief modern assessment of the inventor's life. Photos, notes, bibliography, index.

5-063 Mackenzie, Catherine D. ALEXANDER GRAHAM BELL, THE MAN WHO CONTRACTED SPACE. Boston: Houghton Mifflin, 1928 (reprinted by Grosset & Dunlap, and again by Books for Libraries, 1971), 382 pp. His former secretary (1914–22) relates the inventor's life with numerous extracts, but no documentation. Until replaced by Bruce (5-059), this was the standard biography. Photos, index.

5-064 Stevenson, O.J. THE TALKING WIRE: THE STORY OF ALEXANDER GRAHAM BELL. London: Bodley Head, 1954, 158 pp. A brief survey biography.

BERLINER, Emile (1851-1929)

5-065 * Wile, Frederick William. **EMILE BERLINER, MAKER OF THE MICROPHONE.** Indianapolis: Bobbs-Merrill, 1926 (reprinted by Arno Press "Telecommunications," 1974) 353 pp. Still the only book-length study of a key developer of the microphone, phonograph, and telephone and radio apparatus. Photos, appendices, index.

BOSE, Jagadish Chandra (1858-1937)

5-066 Bhattacharyya, P., and M. Engineer, eds. **ACHARYA J.C. BOSE—A SCIENTIST AND A DREAMER.** Calcutta: Bose Institute, 1996, 2 vols. The major work on his life, including the wireless research to about 1900.

5-067 Bondynpadhay, Probir K. "Under the Glare of a Thousand Suns—The Pioneering Work of Sir J.C. Bose," *PROCEEDINGS OF THE IEEE*, 86:218-224 (January 1998). Useful brief summary of a nearly forgotten Indian experimenter and inventor.

5-068 ** _____. "Sir J.C. Bose's Diode Detector Received Marconi's First Transatlantic Wireless Signal of December 1901 (The 'Italian Navy Coherer' Scandal Revisited," *PROCEEDINGS OF THE IEEE*, 86:259-285 (January 1998). Historic detective work showing Bose created the device, not J.A. Fleming or Marconi himself as often claimed. Photos, references. See Phillips, 5-071.

5-069 Dasgupta, Subrata. "Forgotten History: Sir Jagadish Bose and the Origin of Radio," *TRANSACTIONS OF THE NEWCOMEN SOCIETY*, 67:207-219 (1995/96).

5-070 Geddes, Sir Patrick. **THE LIFE AND WORK OF SIR JAGADIS C. BOSE, AN INDIAN PIONEER OF SCIENCE.** New York: Longmans, Green, 1920 (reprinted by Benjamin Blom, 1971), 259 pp. Long the only available biography. Photos, diagrams.

5-071 * Phillips, Vivian J. "The 'Italian Navy Coherer' Affair: A Turn-of-the-Century Scandal," *PROCEEDINGS OF THE IEE*, Series A, Vol. 140:175-185 (May 1993); reprinted *PROCEEDINGS OF THE IEEE*, 86:248-258 (January 1998). Concerns primacy of the coherer invention.

5-072 Sengupta, Dipak L., et al. "Centennial of the Semiconductor Diode Detector," *PROCEEDINGS OF THE IEEE*, 86:235-247 (January 1998). Bose's invention was announced in 1899.

BRAUN, Karl Ferdinand (1850-1918)

5-073 ** Kurylo, Friedrich, and Charles Susskind. **FERDINAND BRAUN: A LIFE OF THE NOBEL PRIZEWINNER AND INVENTOR OF THE CATHODE-RAY OSCILLOSCOPE.** Cambridge, MA: MIT Press, 1981, 289 pp. Only book in English on the inventor, this is a revision of a German work first pub-

Biography: Inventors, Scientists, and Engineers 103

lished in 1965. Diagrams, photos, appendices (including Braun's Nobel speech of 1909), chronology, bibliography of works by and about Braun, index.

5-074 Lewis, G., and F.J. Mann. "Ferdinand Braun—Inventor of the Cathode-Ray Tube," *ELECTRICAL COMMUNICATION,* 25:319–327 (December 1948). Long the only account in English. Notes.

5-075 * Shiers, George. "Ferdinand Braun and the Cathode Ray Tube," *SCIENTIFIC AMERICAN,* 230:3:92–101 (March 1974). Brief discussion of how the tube was developed and how it fit into what was then known—and what has happened since. Photos, charts.

BRIGHT, Charles Tilston (1832–1888)

5-076 ** Bright, Edward B., and Charles Bright. **THE LIFE STORY OF THE LATE SIR CHARLES TILSTON BRIGHT, CIVIL ENGINEER, WITH WHICH IS INCORPORATED THE STORY OF THE ATLANTIC CABLE, AND THE FIRST TELEGRAPH TO INDIA AND THE COLONIES.** London: Constable, 1899, 2 vols, 506 and 692 pp.; 1908 [Rev. and abridged], 478 pp. The subject was chief engineer for the pioneering cable expeditions. This was written by his brother and son (the latter an authority on the subject generally—see 6-218 and 6-219). The second edition of one volume eliminates much detail on the telegraphs themselves. Photos, maps, notes, appendices, index.

CARTY, John Joseph (1861–1932)

5-077 Jewett, Frank B. "John J. Carty, Telephone Engineer," *BELL SYSTEM QUARTERLY,* 16:160–177 (April 1937).

5-078 _____. "John Joseph Carty, 1861–1932," *NATIONAL ACADEMY OF SCIENCES, BIOGRAPHICAL MEMOIRS.* 18:69–91 (1938).

5-079 ** Rhodes, Frederick Leland. **JOHN J. CARTY—AN APPRECIATION.** New York: privately printed, 1932, 280 pp. The subject rose to become chief engineer of AT&T. Photos, notes, appendices, index.

COOKE, William Fothergill (1806–1879) and WHEATSTONE, Charles (1802–1875)

5-079a ** Adams, W.G., et al., eds. **THE SCIENTIFIC PAPERS OF SIR CHARLES WHEATSTONE.** London: Physical Society of London, Taylor & Francis, 1879, 380 pp. Includes papers on sound, electricity and telegraphy, optics, and his patents. 21 plates.

5-080 *** Bowers, Brian. **SIR CHARLES WHEATSTONE, FRS, 1802–1875.** London: HMSO, for the Science Museum, 1975, 239 pp. The only full biography, this was based on the author's Ph.D. thesis. Photos, diagrams, genealogical table, bibliographical references, index.

Chapter 5

5-081 Clarke, L. "Sir William Fothergill Cooke," *SOCIETY OF TELEGRAPH ENGINEERS JOURNAL*, 8:361-397 (1879). A survey at the time of his death.

5-082 Cooke, Rev. Thomas Fothergill. **AUTHORSHIP OF THE PRACTICAL ELECTRIC TELEGRAPH OF GREAT BRITAIN, OR THE BRUNEL AWARD VINDICATED...** Bath, England: Simpkin, Marshall, 1868, 131 pp. Includes letters and arbitration evidence from the 1841 case, supporting the author's brother.

5-083 * Cooke, William Fothergill. **TELEGRAPHIC RAILWAYS: OR, THE SINGLE WAY RECOMMENDED BY SAFETY, ECONOMY, AND EFFICIENCY, UNDER THE SAFEGUARD AND CONTROL OF THE ELECTRIC TELEGRAPH; WITH PARTICULAR REFERENCE TO RAILWAY COMMUNICATION WITH SCOTLAND AND TO IRISH RAILWAYS.** London: Simpkin, Marshall, 1842, 39 pp.

5-084 *_____. **THE ELECTRIC TELEGRAPH: WAS IT INVENTED BY PROFESSOR WHEATSTONE?** London: printed for the author by W.H. Smith, 1854, 48 pp.; 1856 [2nd ed.], 152 pp.; 1856-57 [3rd ed.], 2 vols.:

> **PART 1: MR. COOKE'S FIRST PAMPHLET, MR WHEATSTONE'S ANSWER, MR COOKE'S REPLY.** 282 pp. The publications of 1854-56. Includes 5-085, 5-089.

> **PART 2: ARBITRATION PAPERS AND DRAWINGS.** 100 pp. 1866 [4th ed.], 268 pp. Offers many documents in support of his argument backing his own primacy in the invention.

5-085 _____. **THE ELECTRIC TELEGRAPH: WAS IT INVENTED BY PROFESSOR WHEATSTONE? A REPLY TO MR. WHEATSTONE'S ANSWER.** London: W.H. Smith, 1856, 152 pp.

5-086 _____. **INVENTION OF THE ELECTRIC TELEGRAPH: THE CHARGE AGAINST SIR CHARLES WHEATSTONE OF TAMPERING WITH THE PRESS.** London and Bath: Simpkin, Marshall, 1869.

5-087 ** Hubbard, Geoffrey. **COOKE AND WHEATSTONE AND THE INVENTION OF THE ELECTRIC TELEGRAPH.** London: Routledge & Kegan Paul, 1965, 158 pp. Places the work of the two telegraph English inventors in context. Perhaps the best modern survey of what they did and the controversies they created. Photos, notes, index.

5-088 Webb, F.H., ed. **EXTRACTS FROM THE PRIVATE LETTERS OF SIR WILLIAM FOTHERGILL COOKE, 1836-1839, RELATING TO THE INVENTION AND DEVELOPMENT OF THE ELECTRIC TELEGRAPH.** London and New York: Spon, 1895, 95 pp. Includes a memoir by Latimer Clark.

5-089 Wheatstone, Charles. **A REPLY TO MR. COOKE'S PAMPHLET, "THE ELECTRIC TELEGRAPH; WAS IT INVENTED BY PROFESSOR WHEATSTONE?"** London: Taylor & Co., 1855, 74 pp.

Biography: Inventors, Scientists, and Engineers 105

5-090 "Wheatstone and Cooke's Electric Telegraph," *MECHANIC'S MAGAZINE* (August 1, 1840), pp. 161–170.

DE FOREST, Lee (1873–1961)

5-091 * Arvin, W.B. "The Life and Work of Lee de Forest," *RADIO NEWS*, Vols. 6–7 (October 1924 to November 1925). Fourteen-part biography with considerable technical information, this is interesting for its contemporary viewpoint. Photos, diagrams.

5-092 Carneal, Georgette. **A CONQUEROR OF SPACE: AN AUTHORIZED BIOGRAPHY OF THE LIFE AND WORK OF LEE DE FOREST.** New York: Horace Liveright, 1930, 296 pp. An authorized life, popular and nontechnical. Photos.

5-093 ** de Forest, Lee. **FATHER OF RADIO.** Chicago: Wilcox and Follett, 1950, 502 pp. An autobiographical record of 50 active years compiled from memory and notebooks—a breezy, poetic and often flamboyant (and ego-filled) revelation of the man—his personal life, inner sentiments, philosophy, triumphs, and disasters. Many extracts from newspapers and correspondence. Photos, notes (though little documentation) appendix on audion's development, a patent list, some of his poems and an index.

5-094 Hijiya, James A. **LEE de FOREST AND THE FATHERHOOD OF RADIO.** Cranbury, NJ: Lehigh University Press, 1993, 182 pp. Focuses primarily on the inventor's personal life and beliefs rather than technology. Useful balance to the inventor's autobiography. Photos, notes, bibliography, index.

5-095 * Lubell, Samuel. "Magnificent Failure," *SATURDAY EVENING POST*, 17 Jan: pp. 9–11, 75, 76, 78, 80; 24 Jan: pp. 20–21, 35, 36, 38, 43; 31 Jan: pp. 27, 38, 40–42, 46, 48, 49. Highly readable, colorful, and revealing story of de Forest's tumultuous life and activities in business and electrical communication inventing. Photos.

DRAWBAUGH, Daniel (1827–1911)

5-096 * Harder, Warren J. **DANIEL DRAWBAUGH: THE EDISON OF THE CUMBERLAND VALLEY.** Philadelphia: University of Pennsylvania Press, 1960, 228 pp. Only biography of one of the numerous claimed inventors of the telephone. Photos, diagrams, patent list, index.

EDISON, Thomas Alva (1847–1931)

(Given his huge output and central role, there are probably more publications on Edison than on any other single figure in electronic history. These are among the best and most useful.)

5-097 * Baldwin, Neil. **EDISON: INVENTING THE CENTURY.** New York: Hyperion, 1995, 531 pp. Excellent survey of the inventor's life and inventions, focusing on the personal side including his relationships with others and his driving competitiveness. Photos, notes, index.

5-098 Clark, Ronald W. **EDISON: THE MAN WHO MADE THE FUTURE.** New York: Putnam, 1977, 256 pp. Largely based on secondary sources. Photos, references, index.

5-099 * Conot, Robert. **A STREAK OF LUCK: THE LIFE AND LEGEND OF THOMAS ALVA EDISON.** New York: Seaview Books, 1979, 565 pp. This was acclaimed on publication as offering a balanced view of the man and his work. Photos, chart of patents granted, chronology chart, reference guide, source notes, index.

5-100 Davidson, George E. **BEEHIVES OF INVENTION: EDISON AND HIS LABORATORIES.** Washington: Government Printing Office "National Park Service History Series," 1973, 70 pp. Brief guide for the Edison National Historic Site in Orange, NJ. Photos, bibliography.

5-101 Dyer, Frank L., and Thomas C. Martin. **EDISON: HIS LIFE AND INVENTIONS.** New York: Harper, 1910, 2 vols. 989 pp.; 1929 [2nd ed.]. This authorized biography lacks references except for items noted in the text. Appendices detail and list major patents in U.S. and overseas; photos, index.

5-102 ** Israel, Paul. **EDISON: A LIFE OF INVENTION.** New York: John Wiley, 1998, 560 pp. The managing editor of the project publishing the inventor's papers (see 5-106) draws on them for the best recent "life." Notes, photos, index.

5-103 *** Jehl, Francis. **MENLO PARK REMINISCENCES.** Dearborn, MI: Edison Institute, 1937–41, 3 vols. (Vol. I [only] reprinted by Dover, 1990), 1,156 pp. A thorough record of events at Edison's Menlo Park, NJ laboratory up to the 1880s, written by a close associate. Photos, references, notes, index (to each volume).

5-104 * Josephson, Matthew. **EDISON.** New York: McGraw-Hill, 1959, 511 pp. A thorough, serious, and valuable study—long the standard modern biography, now supplanted by Israel (5-102). Photos, notes, references, index.

5-105 * Millard, Andre. **EDISON AND THE BUSINESS OF INNOVATION.** Baltimore: Johns Hopkins University Press "Studies in the History of Technology," 1990, 387 pp. Focuses primarily on the second half of the inventor's life and the rise, operation, and function of his companies. Photos, tables, figures, notes, index.

5-106 *** **PAPERS OF THOMAS A. EDISON.** Baltimore: Johns Hopkins University Press, 1989–date, continuing. Massive multivolume project with many editors which, when completed sometime in the 21st century, will make available most of the inventor's papers. See also on-line information: *http://edison.rutgers.edu/*

See "Thomas Edison Papers," 15-036.

5-107 Wachhorst, Wyn. **THOMAS ALVA EDISON: AN AMERICAN MYTH.** Cambridge, MA: MIT Press, 1981, 328 pp. Not a biography per se, but a study of the public image of Edison as seen through media coverage in 62 books, 21 pamphlets, 326 chapters and excerpts, 936 periodical articles, 3,200 newspaper items, 148 book reviews, four plays, five films, and four TV documentaries. Photos, tables (of Edison content in various media) references, bibliography, index.

ERLANG, Agner Krarup (1878–1929)

5-108 Brokmeyer, E., et al. **THE LIFE AND WORKS OF A.K. ERLANG.** Copenhagen: Copenhagen Telephone Company, 1948, 278 pp. A short biography of the mathematical theorist and head of the telephone laboratory leader is followed by further information on his statistical, mathematical, and electrotechnology works and a collection of his papers.

FARNSWORTH, Philo Taylor (1906–1971)

5-109 *** Abramson, Albert. "Pioneers of Television—Philo Taylor Farnsworth," *SMPTE JOURNAL,* 101:770–784 (November 1992). The best concise survey of exactly what the American innovator developed and when, with photos, diagrams, and documentation.

5-110 **DEDICATION OF THE STATUE OF PHILO T. FARNSWORTH.** 101st Congress, 2nd Sess., House Document 101-188. Washington: Government Printing Office, 1991, 72 pp. Includes proceedings of May 2, 1990 as well as paper on Farnsworth's role and a list of his patents. Photos.

5-111 * Everson, George. **THE STORY OF TELEVISION: THE LIFE OF PHILO T. FARNSWORTH.** New York: Norton, 1949 (reprinted by Arno Press "History of Broadcasting," 1971), 266 pp. A nontechnical story ("dramatic and absorbing") designed for the layman by a former backer and strong supporter of the inventor. Disappointingly vague on technical facts, dates, and events. Photos, index.

5-112 * Farnsworth, Elma G. **DISTANT VISION: ROMANCE AND DISCOVERY ON AN INVISIBLE FRONTIER.** Salt Lake City: Pemberly Kent Publishing, 1990, 375 pp. The inventor's wife tells of their life but book suffers from a lack of technical detail and basic editing. Valuable largely for its first-person account. Photos, list of patents, index.

5-113 Farnsworth, Philo T., and Harry R. Lubcke. "The Transmission of Television Images," *CALIFORNIA ENGINEER,* 8:5:12–33 (February 1930). The early work with image dissectors.

See Farnsworth, 12-088.

5-114 Godfrey, Donald G., and Alf Pratte. "Elma Pem Gardner Farnsworth: The Pioneering of Television," *JOURNALISM HISTORY,* 20:2:74–79 (Summer 1994). The inventor's wife and partner in their early work.

5-115 * Hofer, Stephen F. "Philo Farnsworth: Television's Pioneer," *JOURNAL OF BROADCASTING,* 23:153–165 (Spring 1979). First scholarly survey of his contributions. Notes.

FESSENDEN, Reginald Aubrey (1866–1932)

5-116 *** Fessenden, Helen. **FESSENDEN—BUILDER OF TOMORROWS.** New York: Coward-McCann, 1940 (reprinted by Arno Press "History of Broadcasting," 1971), 362 pp. A personal record of the inventor (best known for his high-frequency alternator, heterodyne theory, and early broadcasting) by the inventor's wife that reveals domestic and business affairs. Many extracts from correspondence and diary entries. Bibliography, no index (except in Arno reprint).

See Fessenden, 9-060.

5-117 Krauter, David W. "A New Bibliography of Reginald A. Fessenden," THE A.W.A. REVIEW, 8:55–66 (1993). Includes two parts—a reprint of the bibliography from Fessenden (5-116), and new materials not included there. Not annotated.

5-118 * Raby, Ormond. **RADIO'S FIRST VOICE: THE STORY OF REGINALD FESSENDEN**. Toronto: Macmillan of Canada, nd, but ca 1970, 161 pp. Written for younger readers, this volume claimed that Fessenden, a Canadian, was the true inventor of radio. Photos.

5-119 Webb, Michael. **REGINALD FESSENDEN: RADIO'S FORGOTTEN VOICE**. Toronto: Copp Clark, Pitman, 1991. For younger readers.

FIELD, Cyrus West (1819–1892)

5-120 *** Carter, Samuel III. **CYRUS FIELD: MAN OF TWO WORLDS**. New York: Putnam, 1968, 380 pp. The best modern biography of the key figure behind the first Atlantic telegraph cables. Photos, bibliography, index.

5-121 * Judson, Isabella Field. **CYRUS W. FIELD, HIS LIFE AND WORK [1819–1892]**. New York: Harper, 1896 (reprinted by Garrett Press, 1969), 332 pp. By his daughter and drawing heavily on (and reproducing) many of his papers. Illustrations, map.

5-122 McDonald, Philip B. **A SAGA OF THE SEAS: THE STORY OF CYRUS W. FIELD AND THE LAYING OF THE FIRST ATLANTIC CABLE**. New York: Wilson-Erickson, 1937, 288 pp. The man and his key work. Photos, bibliography, index.

5-123 **THE STORY OF CYRUS FIELD: THE PROJECTOR OF THE ATLANTIC TELEGRAPH**. London: Thomas Nelson, 1878, 120 pp. Useful for contemporary viewpoint.

FLEMING, John Ambrose (1849–1945)

5-124 ** Fleming, John A[mbrose]. **FIFTY YEARS OF ELECTRICITY: THE MEMORIES OF AN ELECTRICAL ENGINEER**. London: Wireless Press, 1921, 371 pp. Provides an overview from 1870 to 1920 including chapters on telegraphs and telephones, and wireless. A valuable account by a foremost engineer and educator (and author). Some names, dates, and references in the text but otherwise no documentation. Diagrams, photos, appendix of Fleming papers and publications, index.

5-125 **_____. **MEMORIES OF A SCIENTIFIC LIFE**. London: Marshall, Morgan & Scott, 1934, 244 pp. On the early days of the telephone and electric lighting, the author's work as a professor and consultant, his scientific researches, lectures, and books along with personal events. Some names and dates appear in the text. Photos, no references or index.

5-126 *** MacGregor-Morris, John T. **THE INVENTOR OF THE VALVE: A BIOGRAPHY OF SIR AMBROSE FLEMING**. London: Television Society, 1954, 141 pp. A concise biography by an early associate with some discussion of technical issues. Photos, appendices, bibliography, index.

See Shiers, 11-022.

GOLDMARK, Peter C. (1906–1977)

5-127 Goldmark, Peter C. **MAVERICK INVENTOR: MY TURBULENT YEARS AT CBS**. New York: Saturday Review Press, 1973, 278 pp. Longtime head of CBS Laboratories tells his story and that of the LP record, color television, and electronic video recording. Index.

GRAY, Elisha (1835–1901)

See Gorman, 5-060.

5-128 * Hounshell, David A. "Elisha Gray and the Telephone: On the Disadvantages of Being an Expert," *TECHNOLOGY AND CULTURE,* 16:133–161 (April 1975). Argues that Gray, as an engineer, relied too heavily on business associates, and did not see the telephone as a useful proposition. References.

5-129 _____. "Bell and Gray: Contrasts in Style, Politics, and Etiquette," *PROCEEDINGS OF THE IEEE,* 64:1305–1314 (September 1976). Points out the many ways that Bell was a more accomplished—and thus successful—innovator of his invention. References.

5-130 _____. "Two Paths to the Telephone," *SCIENTIFIC AMERICAN,* 244:156–163 (January 1981). Underlines the same argument as items immediately above.

5-131 Not used.

HENRY, Joseph (1797–1878)

5-131a * Coulson, Thomas. **JOSEPH HENRY: HIS LIFE AND WORK**. Princeton, NJ: Princeton University Press, 1949, 280 pp. Standard biography. Photos, bibliography, notes, index.

5-132 Dickerson, Edward N. **JOSEPH HENRY AND THE MAGNETIC TELEGRAPH: AN ADDRESS...** New York: Charles Scribner's Sons, 1885, 65 pp. Lecture reviewing the Smithsonian director's role.

5-133 * Moyer, Albert E. **JOSEPH HENRY: THE RISE OF AN AMERICAN SCIENTIST.** Washington: Smithsonian Institution Press, 1997, 348 pp. Written largely from primary documents, this covers the first 50 years of Henry's life—to the beginning of his three decades as secretary of the Smithsonian. This early era contained most of his own scientific work, including that on the telegraph. Notes, index.

5-134 Pope, F.L. **LIFE AND WORK OF JOSEPH HENRY.** New York: American Electrical Society, 1879.

5-135 ** Reingold, Nathan, and Marc Rothenberg, et al., eds. **THE PAPERS OF JOSEPH HENRY.** Washington: Smithsonian Institution Press, 1972–date. At the time of writing, seven volumes took the story to 1850, with several more to come. Photos, notes, index.

5-136 * Taylor, William B. **AN HISTORICAL SKETCH OF HENRY'S CONTRIBUTION TO THE ELECTRO-MAGNETIC TELEGRAPH WITH AN ACCOUNT OF THE ORIGIN AND DEVELOPMENT OF PROF. MORSE'S INVENTION.** Washington: Government Printing Office, 1879, 103 pp. Taken from the SMITHSONIAN REPORT FOR 1878 (pp. 262–360), this surveys the key work by many innovators in the late 18th and early 19th centuries. Notes, tables. (Reprinted in Shiers, 6-120.)

5-137 U.S. Congress. **A MEMORIAL OF JOSEPH HENRY.** Washington: Government Printing Office, 1880, 528 pp. Largely devoted to proceedings observing his death around the country. Includes William B. Taylor's "The Scientific Work of Joseph Henry," (pp. 205–364), notes, list of his papers.

HERTZ, Heinrich Rudolph (1857–1894)

5-138 * Baird, Davis, R.I.G. Hughes, and Alfred Nordmann, eds. **HEINRICH HERTZ: CLASSICAL PHYSICIST, MODERN PHILOSOPHER.** Boston: Kluwer Academic "Boston Studies in the Philosophy of Science, Vol. 198," 1998, 318 pp. Collection of essays by scientists and historians of science on the role and impact of Hertz, especially his 1894 PRINCIPLES OF MECHANICS. Notes, bibliography, index.

5-139 Blanchard, Julian. "Hertz, the Discoverer of Electric Waves," *PROCEEDINGS OF THE IRE,* 26:505–515 (May 1938). Also appears in *BELL SYSTEM TECHNICAL JOURNAL,* 17:327–337 (July 1938). No references.

5-140 *** Buchwald, Jed Z. **THE CREATION OF SCIENTIFIC EFFECTS: HEINRICH HERTZ AND ELECTRIC WAVES.** Chicago: University of Chicago Press, 1994, 482 pp. First of two volumes (the second not yet published), this definitive life, based on many primary documents, is focused on "Hertz himself—his thoughts, his practice, and his career before he achieved fame." Photos, tables, diagrams, appendices, notes, bibliography, index.

5-141 Hertz, Heinrich, trans. by Daniel E. Jones. **ELECTRIC WAVES: BEING RESEARCHES ON PROPAGATION OF ELECTRIC ACTION WITH FINITE VELOCITY THROUGH SPACE.** London: Macmillan, 1893, (reprinted by Dover, 1962), 278 pp.; 1900 [2nd ed.], 278 pp. With a preface by Lord Kelvin. Includes the key papers.

5-141a* Hertz, Johanna (arranger), updated and translated by Lisa Brinner, Mathilde Hertz, and Charles Susskind. **MEMOIRS, LETTERS, DIARIES: HEINRICH HERTZ/ERINNERUNGEN, BRIEFE, TAGEBUCHER: HEINRICH HERTZ.** San Francisco: San Francisco Press/Weinheim, West Germany: Physik, 1977 [2nd ed.], 361pp. Updates, expands, and translates original German publication (Leipzig, 1927). Bilingual German/English text on alternate pages collects his personal papers. Includes a biographical introduction by Max von Laue. Appendix, bibliography of Hertz' work, photos, name index.

See Lodge, 8-010.

5-142 * Morrison, Philip, and Emily Morrison. "Heinrich Hertz," *SCIENTIFIC AMERICAN*, 197: 98–106 (December 1957). A concise review of his accomplishments. Photo, diagrams.

5-143 Mulligan, Joseph F., ed. **HEINRICH RUDOLPH HERTZ (1857–1894): A COLLECTION OF ARTICLES AND ADDRESSES.** New York: Garland "Reference Library of the Humanities, No. 1697," 1994, 442 pp. After a brief biography, this includes 18 papers: 11 of the inventor's most important papers, and seven contemporary accounts of his life. Bibliography of his scientific papers, a general bibliography of books and articles, index.

IVES, Herbert Eugene (1882–1953)

5-144 Buckley, Oliver E., and Karl K. Darrow, "Herbert Eugene Ives," *NATIONAL ACADEMY OF SCIENCE BIOGRAPHICAL MEMOIRS*, 29:145–189 (1956). Bell Labs television pioneer's life is outlined. Includes bibliography and list of patents.

JENKINS, Charles Francis (1867–1934)

See Jenkins, 12-007, 13-024.

5-145 * Jenkins, Charles Francis. **THE BOYHOOD OF AN INVENTOR.** Washington: Author, 1931, 273 pp. Revealing record by a man who regarded himself as an inventor above all else. He was a founder of the Society of Motion Picture Engineers and a pioneer in phototelegraphy and radio pictures. Appendix of eight broadcast talks by Jenkins, hundreds of photos (but no index).

KORN, Arthur (1870–1945)

5-146 Korn, Terry, and Elizabeth P. Korn. **TRAILBLAZER TO TELEVISION.** New York: Scribner's, 1950, 144 pp. While written for younger readers, this is the only biography of a key developer of facsimile by his wife and daughter-in-law, unfortunately lacking references or index.

LODGE, Oliver Joseph (1851–1940)

5-147 ** Jolly, W.P. **SIR OLIVER LODGE: PHYSICAL RESEARCHER AND SCIENTIST.** Cranbury, NJ: Associated University Presses, 1975, 256 pp. Only book-length biography of the British physicist and wireless pioneer, helping to place him in context with his (now) better-known contemporaries. Photos, sources and references, index.

5-148 Lodge, Oliver J. **TALKS ABOUT RADIO, WITH SOME PIONEERING HISTORY AND SOME HINTS AND CALCULATIONS FOR RADIO AMATEURS.** London: Cassell/New York: George H. Doran, 1925, 267 pp. Part one (to page 120) is largely historical. A few references in text. Index.

5-149 *_____. **PAST YEARS, AN AUTOBIOGRAPHY.** London: Hodder & Stoughton, 1931/ New York: Scribner's, 1932, 364 pp. Photos, references, index. Well over half of this informal memoir concerns his scientific work in the late 19th century, including pioneering wireless experimentation. Photos, index.

5-150 _____. **MY PHILOSOPHY, REPRESENTING MY VIEWS ON THE MANY FUNCTIONS OF THE ETHER OF SPACE.** London: Ernest Benn, 1933, 318 pp. Combines his views on science and what some would call pseudoscience.

See Lodge, 8-010, 8-020, and 8-021.

5-151 Rowlands, Peter, and J. Patrick Wilson, eds. **OLIVER LODGE AND THE INVENTION OF RADIO.** Liverpool, England: PD Publications, 1994, 241 pp. Photos, index.

LOOMIS, Mahlon (1826–1886)

5-152 ** Appeby, Thomas. **MAHLON LOOMIS, INVENTOR OF RADIO.** Washington: Loomis Publications, 1967, 160 pp. The standard treatment of the Washington, D.C., dentist and wireless telegraphy experimenter. Copious extracts from contemporary sources with some references. Photos, diagrams, index.

5-153 Winters, S.R. "The Story of Mahlon Loomis, Pioneer of Radio," *RADIO NEWS*, 4:836–837, 966–981 (November 1922). Includes text of original 1872 patent and copies of inventor's drawings plus extracts from articles and records, with text references from 1868. Photos.

5-154 * Young, Otis B. "Mahlon Loomis, the Discover and Inventor of Radio: A Report by the Chairman of the Radio Discovery Committee of the Illinois State Academy of Science," *TRANSACTIONS, ILLINOIS STATE ACADEMY OF SCIENCE*, 60:3–8 (1967). Very useful for its chronology and list of references and past attempts to gain recognition for Loomis. (The same author wrote numerous other papers in the mid-1960s largely repeating the same information found here.)

Biography: Inventors, Scientists, and Engineers 113

MARCONI, Guglielmo (1874–1937)

5-155 Armstrong, Edwin H. "The Sprit of Discovery: An Appreciation of the World of Marconi," *ELECTRICAL ENGINEERING,* 72:670–676 (August 1953). A study of Marconi's three great discoveries in the field of radio transmission and reception (discovery of the grounded wave, the "daylight" wave, and microwaves) by another electronics inventor. Photos, references.

5-156 Bucciante, Giuseppe, ed. **MARCONI**. Rome: Dedado Editore, 1963, 431 pp. Combines selected reprints of Marconi's lectures and articles with original essays on his life and work, the development of radiotelegraphy, wireless at sea, wireless at the Pole, and the man and scientist. Photos, diagrams, maps.

5-157 Coe, Douglas [pseud. for Samuel Epstein]. **MARCONI, PIONEER OF RADIO**. New York: Julian Messner, 1943, 272 pp. Written for younger readers, this includes a list of Marconi's awards and brief bibliography, diagrams, index.

5-158 * Dunlap, Orrin E. **MARCONI, THE MAN AND HIS WIRELESS**. New York: Macmillan, 1937 (reprinted by Arno Press "History of Broadcasting," 1971), 360 pp.; 1938 [rev. ed.], 362 pp. A popular and somewhat romantic survey with which the inventor cooperated. Based on much original material, but lacks specific references. Photos, notes, list of honors, degrees and awards, index.

5-159 Geddes, Keith. **GUGLIELMO MARCONI: 1874–1937**. London: Her Majesty's Stationery Office, "Science Museum Booklet," 1974, 40 pp. A brief telling of the story, issued on the centenary of the inventor's birth, focusing almost entirely on wireless. Photos.

5-160 **GUGLIELMO MARCONI**. Chelmsford, England: The Marconi Company, 1984, 24 pp. Company brochure on the 100th anniversary of the inventor's birth. Photos.

5-161 Gunston, David. **GUGLIELMO MARCONI**. Geneva, Switzerland, Heron Books "The Great Nobel Prizes," 1970, 350 pp. Divided into three parts: a brief biography, the inventor's major wireless discoveries, and a more general history of radio. Historical appendix with brief chronology, the 1909 Nobel address, brief biographies of ten other figures, glossary, photos, index.

5-162 Jacot, Bernard L., and D.M.B. Collier. **MARCONI—MASTER OF SPACE, AN AUTHORIZED BIOGRAPHY OF THE MARCHESE MARCONI**. London: Hutchinson, 1935, 287 pp. Popular and nontechnical survey of the inventor's life. Photos, list of honors, index.

5-163 ** Jolly, W.P. **MARCONI**. New York: Stein & Day, 1972, 292 pp. The standard modern biography covering both his personal life and his work with wireless. Photos, references, index.

5-164 * Kreuzer, James H., and Felicia A. Kreuzer "Marconi—The Man and His Apparatus," **THE A.W.A. REVIEW**, 9:7–96 (1995). Wonderful collection of photos with a connected narrative that surveys the inventor and his companies and many of their products—a fine supplement to the biographies noted above. Photos, chart, references.

5-165 MacLeod, Mary K. **WHISPER IN THE AIR: MARCONI—THE CANADA YEARS, 1902–1946.** Hantsport, Nova Scotia: Lancelot Press, 1992, 235 pp. Focus is on work in the Maritime provinces early in the century.

5-166 Marconi, Degna. **MY FATHER MARCONI.** London: Muller; New York: McGraw-Hill, 1962, 321 pp.; Ottawa: Balmuir Book Publishing, 1982 [2nd ed.], 258 pp. A family record based on personal notes, journals, and scientific records by the inventor's eldest daughter. Interesting and useful for background material. Includes a variety of extracts but has few references. Revision includes some corrections and more material on the Canadian experience. Photos, index [1st ed. only].

5-167 * Masini, Giancarlo. **MARCONI.** New York: Marsilio Publishers, 1995 (translated from 1975 Italian edition), 370 pp. Insightful biography by an Italian science writer using archival and interview sources—but lacking any documentation. Photos, index.

5-168 Read, Leslie. **MARCONI AND THE DISCOVERY OF WIRELESS.** London: Faber "Men and Events," 1963, 166 pp. For younger readers. Photos, map, bibliography, index.

5-168a * Simons, R.W. "Guglielmo Marconi and Early Systems of Wireless Communication," *GEC REVIEW*, 11:1 (1996), 19 pp. Useful survey, based largely on original documentation, of just what Marconi accomplished and the context of the 1895–1905 period. Photos, diagrams, references. Also available on-line: see 15-029.

MAXWELL, James Clerk (1831–1879)

5-169 * Campbell, Lewis, and William Garnett. **THE LIFE OF JAMES CLERK MAXWELL WITH A SELECTION FROM HIS CORRESPONDENCE AND OCCASIONAL WRITINGS AND A SKETCH OF HIS CONTRIBUTIONS TO SCIENCE.** London: Macmillan, 1882 (reprinted by Johnson Reprint "The Sources of Science No. 85," 1969), 662 pp.; 1884 [rev. and abridged ed.], 421 pp. The official biography, by a lifelong friend. Portraits, engravings, diagrams.

5-170 Dibner, Bern. "James Clerk Maxwell," *IEEE SPECTRUM*, 1:50–56 (December 1964). A tribute on the centennial of Maxwell's historic paper, "A Dynamical Theory of the Electromagnetic Field." Photos, references.

5-171 Domb, C[yril], ed. **CLERK MAXWELL AND MODERN SCIENCE: SIX COMMEMORATIVE LECTURES.** London: Athlone Press, 1963, 188 pp. Talks by Sir Edward Appleton and others highlighting major contributions by Clerk Maxwell.

5-172 Everitt, C.W.F. **JAMES CLERK MAXWELL: PHYSICIST AND NATURAL PHILOSOPHER.** New York: Charles Scribner's "DSB Editions," 1974, 205 pp. Expanded from author's entry in the *Dictionary of Scientific Biography*, this covers Maxwell's brief life and major scientific work. Photos, notes, bibliography, index.

Biography: Inventors, Scientists, and Engineers

5-173 Glazebrook, Sir Richard Tetley. **JAMES CLERK MAXWELL AND MODERN PHYSICS**. London: Cassell "Century Science Series," 1896, 224 pp. Useful late-19th-century assessment.

5-174 Goldman, Martin. **THE DEMON IN THE AETHER: THE STORY OF JAMES CLERK MAXWELL**. Edinburgh: Paul Harris, 1983, 224 pp. A brief summary of his life and work. Diagrams, photos, references, index.

5-175 ** Harmon, P[eter] M., ed. **THE SCIENTIFIC LETTERS AND PAPERS OF JAMES CLERK MAXWELL**. Cambridge, England: Cambridge University Press, 1990–95, 2 vols. Standard collection, the first volume covering 1846–62; the second 1862–73. The selections, few of which have been fully published before, are annotated. Notes, index. See Niven (5-180) for an earlier collection.

5-176 Hendry, John. **JAMES CLERK MAXWELL AND THE THEORY OF THE ELECTROMAGNETIC FIELD**. Bristol, England: Adam Hilger, 1986, 305 pp. Photos, bibliography, index. A detailed modern survey of his chief work.

5-177 JAMES CLERK MAXWELL: A COMMEMORATION VOLUME: 1831–1931. Cambridge, England: Cambridge University Press/New York: Macmillan, 1931, 146 pp. Issued on the centenary of his birth, this includes essays by Sir Joseph Thomson and a series of tributes by 20th century figures.

5-178 Larmor, Sir James, ed. **ORIGINS OF CLERK MAXWELL'S ELECTRICAL IDEAS AS DESCRIBED IN FAMILIAR LETTERS TO WILLIAM THOMSON**. Cambridge, England: Cambridge University Press, 1937, 56 pp. Useful collection of primary documentation. Photos, diagrams. Reprints 8-005.

5-179 Maxwell, James Clerk. **AN ELEMENTARY TREATISE ON ELECTRICITY**. Oxford: Clarendon Press, 1881, 208 pp., edited by W. Garnett; 1888 [2nd ed.], 108 pp.; and as A TREATISE ON ELECTRICITY AND MAGNETISM. 1892 [3rd ed.], 2 vols. Revised and annotated by Joseph John Thomson. Reprinted by Oxford University Press "Oxford Classic Texts in the Physical Sciences No. 1," 1998, 2 vols. The major published work.

5-180 Niven, William D., ed. **THE SCIENTIFIC PAPERS OF JAMES CLERK MAXWELL**. Cambridge, England: Cambridge University Press, 1890, 2 vols. (reprinted by Dover in single vol., 1965), 607 and 806 pp. Plates, diagrams, tables, index (in both vols.). See Harmon (5-175) for a newer collection.

5-181 ** Simpson, Thomas K. **MAXWELL ON THE ELECTROMAGNETIC FIELD: A GUIDED STUDY**. New Brunswick, NJ: Rutgers University Press "Masterworks of Discovery: Guided Studies of Great Texts in Science," 1997, 440 pp. Analysis for the general reader of what Maxwell said in his three classic 1855–64 papers with an introductory chapter on Maxwell and Faraday and the state of electrical understanding in the 19th century. Diagrams, references, selected readings (an annotated guide, by chapter), notes, index.

5-182 Tolstoy, Ivan. **JAMES CLERK MAXWELL: A BIOGRAPHY**. Edinburgh: Canongate / Chicago: University of Chicago Press, 1981, 184 pp. Modern survey based largely on secondary sources. Photo, bibliography, index.

5-183 Tricker, R. A. R. **THE CONTRIBUTIONS OF FARADAY AND MAXWELL TO ELECTRICAL SCIENCE.** Oxford, England: Pergamon "The Commonwealth and International Library: Selected Readings in Physics," 1966, 289 pp. Mixture of biographical material and scientific assessment. Photos, bibliography, index.

MEUCCI, Antonio (1808–1889)

5-184 Bargellini, Pier L. "An Engineer's Review of Antonio Meucci's Pioneer Work in the Invention of the Telephone," *TECHNOLOGY IN SOCIETY*, 15:409–421 (1993). Concludes Meucci was clearly one of several telephone pioneers.

5-185 * Nese, Marco, and Francesco Nicotra. **ANTONIO MEUCCI, 1808–1889.** Rome: Italy Italy Magazine, 1989, 173 pp. Lavish bilingual pictorial on the inventor's life and claims to have invented the telephone before Bell. Photos, bibliography.

5-186 ** Schiavo, Giovanni E. **ANTONIO MEUCCI: INVENTOR OF THE TELEPHONE.** New York: The Vigo Press, 1958, 288 pp. First biography in English of the Italian-born claimant to the telephone invention. Photos, notes, bibliography (of primary and secondary material), appendix of documents including list of his inventions.

MORSE, Samuel Finley Breese (1791–1872)

5-187 Kendall, Amos. **MORSE'S PATENT: FULL EXPOSURE OF DR. CHAS. T. JACKSON'S PRETENSIONS TO THE INVENTION OF THE AMERICAN ELECTRO-MAGNETIC TELEGRAPH.** Washington: J.T. Towers, 1852, 64 pp. Jackson (1805–1880) was one of many claimants to the telegraph's invention.

5-188 Larkin, Oliver W. **SAMUEL F.B. MORSE AND AMERICAN DEMOCRATIC ART.** Boston: Little, Brown "Library of American Biography," 1954, 215 pp. While emphasizing the artistic output of Morse as a painter, this includes about 80 pages on the telegraph portion of his life. Source notes, index.

5-189 *** Mabee, Carleton. **THE AMERICAN LEONARDO: THE LIFE OF SAMUEL F.B. MORSE.** New York: Knopf, 1943 (reprinted by Octagon Books, 1969 and again by Easton Press, 1990), 435 pp. This Pulitzer Prize-winner (in 1944, for biography) remains the definitive treatment of the painter and inventor's life and covers both his artistic and scientific careers. Photos, references, index.

5-190 MEMORIAL OF SAMUEL FINLEY BREESE MORSE INCLUDING APPROPRIATE CEREMONIES OF RESPECT AT THE NATIONAL CAPITOL, AND ELSEWHERE. Washington: Government Printing Office, 1875, 359 pp. A 16-page biographical sketch is followed by proceedings of the many memorial services on the inventor's death. Useful for contemporary views about Morse and the telegraph.

5-191 ** Morse, Edward Lind, ed. **SAMUEL F.B. MORSE, HIS LETTERS AND JOURNALS.** Boston: Houghton Mifflin, 1914, 2 vols., (reprinted in one volume by Da Capo "Library of American Art," 1973), 440 and 548 pp. The inventor's son assembled these two volumes, the first focused on art and the second dealing with the telegraph. The editor notes that many letters included in Prime (5-194) are not duplicated here. Photos, index.

5-192 Morse, S.F.B. **THE LEADING TELEGRAPH PATENTS, INCLUDING ORIGINAL AND REISSUED PATENTS, OF S.F.B. MORSE.** New York: Polhemus, 1876, 157 pp. Useful compilation, growing out of the Paris exposition.

5-193 Nicolay, Helen. **WIZARD OF THE WIRES: A BOY'S LIFE OF SAMUEL F.B. MORSE.** New York: Appleton Century, 1938, 326 pp. As title suggests, for younger people. Index.

5-194 ** Prime, Samuel Irenaeus. **THE LIFE OF SAMUEL F.B. MORSE, LL.D., INVENTOR OF THE ELECTRO-MAGNETIC RECORDING TELEGRAPH.** New York: Appleton, 1875 (reprinted by Arno Press "Telecommunications," 1974), 776 pp. Based on available papers and extensive interviews, the author was asked by Morse's family to undertake this biography. Engravings, diagrams, document excerpts (no index).

POPOV, Alexander Stepanovich (1859–1905)

5-195 Howe, G.W.O. "Alexander S. Popov," *WIRELESS ENGINEER*, 25:1–5 (January 1948). Review of Popov's life and work with observations on the invention of radio communication and the rival claims of Marconi. See subsequent letters: 25: 135–137 (May 1948); 26:141–142, 249–150 (April, August 1949).

5-196 * Radovsky, M., trans. by G. Yankovsky. **ALEXANDER POPOV, INVENTOR OF RADIO.** Moscow: Foreign Languages Publishing House "Men of Russian Science," 1957, 130 pp. This biography of Russia's pioneer provides useful and interesting material not found elsewhere, but must be used carefully. The third of four chapters focuses on radio. Photos, notes (no index).

5-197 ** Süsskind, Charles. "Popov and the Beginnings of Radiotelegraphy," *PROCEEDINGS OF THE IRE*, 50:2036–2047 (October 1962). Critical review of the Marconi–Popov controversy. Thoroughly documented. Photos, references. See subsequent letters 51:473–474 (March 1963); 53:162–164 (February 1965). See also the author's book of same title, San Francisco Press, 1962, 30 pp.; 1970 [2nd ed.]. Photos, references.

PREECE, William Henry (1834–1913)

5-198 ** Baker, E.C. **SIR WILLIAM PREECE F.R.S.: VICTORIAN ENGINEER EXTRAORDINARY.** London: Hutchinson, 1976, 377 pp. Well-written biography of a central figure in British telegraphy, telephony, and early wireless development. Photos, appendix of lectures and patents, index.

5-199 Tucker, D. Gordon. "Sir William Preece (1834–1913)," *NEWCOMEN SOCIETY TRANSACTIONS*, 53:119–138 (1981–82). Useful brief account. References.

PUPIN, Michael Idvorsky (1858–1935)

5-200 David, R. "Michael Idvorsky Pupin, 1858–1935," *NATIONAL ACADEMY OF SCIENCE BIOGRAPHICAL MEMOIRS*, 19:307–323 (1938). Includes bibliography, portrait.

5-201 Pupin, Michael I. **FROM IMMIGRANT TO INVENTOR.** New York: Scribner's, 1922, 396 pp. A popularly-written (and commercially successful) autobiography lacking references and nontechnical in tone. Photos, index.

REIS, Johann Philipp (1834–1874)

5-202 Thompson, Silvanus P. **PHILIPP REIS: INVENTOR OF THE TELEPHONE. A BIOGRAPHICAL SKETCH, WITH DOCUMENTARY TESTIMONY, TRANSLATIONS OF ORIGINAL PAPERS OF THE INVENTOR, AND CONTEMPORARY PUBLICATIONS.** London: E. & F.N. Spon, 1883 (reprinted by Arno Press "Telecommunications," 1974), 184 pp. Still the only complete life, this argues Reiss' German work in the 1860s preceded Bell and he should be so recognized. Engravings, notes, appendices, references.

SCHOENBERG, Isaac (1880–1963)

5-203 McGee, James D. "The Life and Work of Sir Isaac Shoenberg, 1880–1963," *THE ROYAL TELEVISION SOCIETY JOURNAL* 13: 207–216 (May-June 1971). Schoenberg headed the team that developed the British 405-line television system. Photos.

5-204 Tucker, John D. "Schoenberg," *INTERNATIONAL BROADCAST ENGINEERING SUPPLEMENT*, (December 1971), pp. 5–9. Brief survey of the prewar work.

SIEMENS, Werner von (1816–1892)

See Feldenkirchen, 4-235.

See Siemens, 4-240 and 4-241.

STONE, John Stone (1869–1943)

5-205 Clark, George H. **THE LIFE OF JOHN STONE STONE: MATHEMATICIAN, PHYSICIST, ELECTRICAL ENGINEER AND GREAT INVENTOR.** San Diego: Frye & Smith, 1946, 200 pp. The only published life. Photos, appendix on Stone patents.

STUBBLEFIELD, Nathan (1859–1928)

5-206 Fawcett, Waldon. "The Latest Advance in Wireless Telephony," *SCIENTIFIC AMERICAN*, 86:21 (May 24, 1902). Contemporary report of Stubblefield demonstration—the first in a scientific magazine.

5-207 * Hoffer, Thomas W. "Nathan B. Stubblefield and his Wireless Telephone," *JOURNAL OF BROADCASTING*, 15:317-329 (Summer 1971). The Kentucky inventor, what he may have invented, and when. Notes.

5-208 Horton, L.J. "Did He Invent Radio?" *BROADCASTING* (March 19, 1951), pp. 90-93. Brief journalistic biography. Photos.

SWINTON, Alan Archibald Campbell (1863-1930)

5-209 ** Bridgewater, T.H. **A.A. CAMPBELL SWINTON**. London: Royal Television Society "Monograph 1," 1982, 32 pp. The man who foresaw electronic television is detailed by a one-time BBC television chief engineer. Photos, references.

5-210 McGee, James D. "Campbell Swinton and Television," *NATURE*, 138:674-676 (October 17, 1936). A tribute to Swinton's 1908 proposal for electronic television issued just as the BBC began regular television transmissions. Extracts, references.

5-211 _____. "The Contribution of A.A. Campbell Swinton, F.R.S. to Television," *NOTES AND RESEARCHES OF THE ROYAL SOCIETY*, 32:91-105 (July 1977).

5-212 Swinton, A.A. Campbell. **AUTOBIOGRAPHICAL AND OTHER WRITINGS**. London: Longmans, Green, 1930, 181 pp. About 50 pages offers the inventor's view of his own life, though with little technical insight. A modest story of a man who accomplished far more than the text indicates. Photos.

TESLA, Nikola (1857-1943)

5-213 Anderson, Leland I., ed. **NIKOLA TESLA: ON HIS WORK WITH ALTERNATING CURRENTS AND THEIR APPLICATION TO WIRELESS TELEGRAPHY, TELEPHONE, AND TRANSMISSION OF POWER**. Denver: Sun Publishing, 1992, 237 pp. Largely reprints Tesla's own patents and related material. Illustrated, bibliography, index.

5-214 * Cheney, Margaret. **TESLA: MAN OUT OF TIME**. Englewood Cliffs, NJ: Prentice-Hall, 1981, 320 pp. The first attempt at a balanced and somewhat documented biography. Photos, bibliographic essay, notes, index.

5-215 Hunt, Inez, and Wanetta W. Draper. **LIGHTNING IN HIS HAND: THE LIFE STORY OF NIKOLA TESLA**. Denver: Sage Publications, 1964 (reprinted by Omni Publications, 1981), 269 pp. Appears to concentrate primarily on the inventor's personality and how that affected his inventions and work with others. Focuses on his brief stay in Colorado Springs. Photos, notes, bibliography, index.

5-216 *** Martin, Thomas Commerford. **THE INVENTIONS, RESEARCHES AND WRITINGS OF NIKOLA TESLA, WITH SPECIAL REFERENCE TO HIS WORK IN POLYPHASE CURRENTS AND HIGH POTENTIAL LIGHTING**. New York: The Electrical Engineer, 1894 (reprinted by Angriff Press, 1981; reprinted again by Barnes & Noble, 1995), 495 pp. After a brief biography, four main sections include polyphase currents (24 chapters), the Tesla effects with high-frequency and high-potential currents (including reprints of three Tesla lectures), miscellaneous inventions and writings (nine chapters); and an appendix on early phase motors and the Tesla oscillators. Descriptive rather than critical analysis. Photos, diagrams, index.

See "Nikola Tesla," 15-040, 15-041.

5 217 O'Neill, John J. **PRODIGAL GENIUS: THE LIFE OF NIKOLA TESLA.** New York: Ives, Washburn, 1944 (reprinted by David McKay, 1972 and again by Angriff Press, 1981), 326 pp. The first American biography, published just months after the inventor's death, by a journalist who knew him, it suffers from lack of references and generality, so must be used with care. Partial list of patents, index.

5-218 Ratzlaff, John T., comp. **DR. NIKOLA TESLA: COMPLETE PATENTS.** Millbrae, CA: Tesla Book Co., 1979, 2 vols., 500 pp. Reprints original patent specifications.

5-219 _____. **DR. NIKOLA TESLA: SELECTED PATENT WRAPPERS FROM THE NATIONAL ARCHIVES.** Millbrae, CA: Tesla Book Co., 1980, 4 vols., 922 pp. Facsimile reprints.

5-220 **Ratzlaff, John T., and Leland I. Anderson, comps. **DR. NIKOLA TESLA BIBLIOGRAPHY, 1884–1878.** Palo Alto, CA: Ragusan Press, 1979, 237 pp. Includes some 3,000 citations by and about the inventor, most with brief annotations. Includes list of Tesla's U.S. patents.

5-221 Raucher, Elisabeth, and Toby Grotz, eds. **TESLA 1984: PROCEEDINGS OF THE TESLA CENTENNIAL.** Colorado Springs, CO: Tesla Society, 1984.

5-222 ** Seifer, Marc J. **WIZARD: THE LIFE AND TIMES OF NIKOLA TESLA.** Secaucus, NJ: Birch Lane, 1996, 542 pp. Perhaps the definitive (though too-positive) biography to date, this is also the most detailed and best documented, much of it based on primary materials never used before. Photos, bibliography, notes, index.

5-223 * Tesla, Nikola, edited by Ben Johnston. **MY INVENTIONS: THE AUTOBIOGRAPHY OF NIKOLA TESLA.** Williston, VT.: Hart Bros., 1982 (reprinted by Barnes & Noble, 1995), 111 pp. A book made up of six articles written by Tesla and published in *Electrical Experimenter* in 1919. Appendix on Tesla's two-phase induction motor. Photos, map, diagrams.

5-224 _____, edited by Alexander Marincic. **COLORADO SPRINGS NOTES AND COMMENTARY, 1899–1900.** Belgrade, Yugoslavia: Tesla Museum, 1979, 440 pp. Reprints of primary material.

VARIAN, Russell (1898–1959)

See Varian, 4-166, 11-064, and 11-065.

WHEATSTONE, Charles, see under Cooke, William.

ZWORYKIN, Vladimir Kosma (1889–1982)

5-225 *** Abramson, Albert. **ZWORYKIN: PIONEER OF TELEVISION.** Urbana: University of Illinois Press, 1995, 317 pp. Definitive biography of the Westinghouse and then RCA television researcher with considerable material on subsequent patent battles with Farnsworth and others. Photos, notes (90 pages), index.

5-226 *_____. "Pioneers of Television—Vladimir Kosma Zworykin," *JOURNAL OF THE SMPTE,* 90:579–590 (July 1981). Largely subsumed in book above, this includes useful diagrams, a list of patents, and a bibliography of Zworykin's published writings. Photos, references.

5-227 Zworykin, Vladimir K. "The Early Days: Some Recollections," *TELEVISION QUARTERLY,* 1:69–73 (November 1962). On his association with Boris Rosing, early work on camera tubes at Westinghouse, and later events up to 1931.

See Zworykin, 11-054, 12-108, and 12-109.

Chapter 6

Telegraphy

Electrical engineering began with the advent of the wire telegraph, first in Britain and then the U.S., around 1840. The laying of the first Atlantic cables (1857–1866) was a high point of engineering achievement in the mid-19th century. Stories of the telegraph and submarine telegraph abound throughout both technical and popular literature—indeed the literature on the first means of telecommunications seems to considerably outpace that available on the telephone.

This chapter begins with (A) a chronological listing of pioneering contemporary books covering the first half century of the telegraph's implementation. This is followed by sections with (B) a selection of later telegraph works (1896–1950); (C) modern histories of U.S. land telegraphy; (D) foreign telegraph systems, subdivided into General and Europe, and Britain and the Commonwealth; (E) the first halfcentury of books on submarine telegraphy; (F) modern histories of telegraph (and telephone) submarine cables; and (G) guides to collecting old telegraph equipment and ephemera. For related Internet resources, see 15-E.

A. Contemporary Books on Telegraphy (1823–1895)

(Arranged chronologically [and alphabetically by author or title within a given year], the first 70 years of contemporary telegraph books provides a valuable historical record—as well as period engravings of apparatus and installations. Excluded from this list are commercial telegraph codebooks and works devoted to legal matters (including a substantial number of works on potential government ownership of the telegraph issued in and after the 1870s). Biographies, including material on Cooke and Wheatstone, and Field appear in chapter 5-C. This listing draws heavily from Weaver [1-023]—quotations appearing below are from entries therein—Higgins [1-008], and Shiers [1-017]. Unfortunately, a handful of citations are incomplete or annotations are lacking due to the scarcity of many of these works.)

Chapter 6

6-001 * Ronalds, Francis. **DESCRIPTIONS OF AN ELECTRIC TELEGRAPH, AND OF SOME OTHER ELECTRICAL APPARATUS.** London, Hunter, 1823, 83 pp; 1871 [2nd ed.]. Appears to be first monograph in English on the telegraph. The author's work was "the prototype of electric dial-telegraphs" (Weaver, 1-023, I:803). See also Ronalds, 1-015, for the catalogue of the author's own collection of books and papers on electricity.

See Cooke, 5-083.

6-002 Finlaison, John **AN ACCOUNT OF THE REMARKABLE APPLICATIONS OF THE USEFUL ARTS BY MR. ALEXANDER BAIN WITH A VINDICATION OF HIS CLAIM TO BE THE FIRST INVENTOR OF THE ELECTRO-MAGNETO** *PRINTING* **TELEGRAPH AND ALSO OF THE ELECTROMAGNETIC CLOCK.** London: Chapman & Hall, 1843, 127 pp. Four foldout illustrations.

6-003 *** Vail, Alfred, and J. Cummings Vail. **EYEWITNESS TO EARLY AMERICAN TELEGRAPHY.** New York: Arno Press "Telecommunications," 1974, 260 pp. Facsimile reprints of three important, early publications by or about Alfred Vail (1807–1859), assistant superintendent of the first telegraph service in the country. Engravings, diagrams. See also Vail 6-009.

THE AMERICAN ELECTRO MAGNETIC TELEGRAPH: WITH THE REPORTS OF CONGRESS, AND A DESCRIPTION OF ALL TELEGRAPHS KNOWN, EMPLOYING ELECTRICITY OR GALVINISM by Alfred Vail. Philadelphia: Lea & Blanchard, 1845, 1847, 208 pp. Appears to be the first American book devoted to the electric telegraph. 81 wood engravings.

DESCRIPTION OF THE AMERICAN ELECTRO MAGNETIC TELEGRAPH: NOW IN OPERATION BETWEEN THE CITIES OF WASHINGTON AND BALTIMORE by Alfred Vail. Washington: J. and G. S. Gideon, 1845, 1847, 24 pp. Fourteen wood engravings.

EARLY HISTORY OF THE ELECTRO-MAGNETIC TELEGRAPH FROM LETTERS AND JOURNALS OF ALFRED VAIL, ARRANGED BY HIS SON, J. CUMMINGS VAIL. New York: Hine Brothers, 1914, 36 pp.

6-004 Hatcher, W[illiam]. H. **AN ACCOUNT OF THE ELECTRIC TELEGRAPH NOW IN USE FOR RAILWAYS AND OTHER PURPOSES: EXTRACTED FROM THE ENGINEER'S AND CONTRACTOR'S POCKETBOOK FOR 1847 AND 1848.** London: n.p. 1847, 20 pp. Also appeared in another edition without subtitle.

6-005 Phipson, T[homas] L. **MANUAL OF THE VARIOUS ELECTROMAGNETIC TELEGRAPHS AT PRESENT IN USE.** London (?): n.p., 1847, 28 pp. Includes Brett's printing telegraph.

6-006 Progress, Peter (pseud.). **ELECTRIC TELEGRAPH: COMPRISING A BRIEF HISTORY OF FORMER MODES OF TELEGRAPHIC COMMUNI-**

CATION; AN ACCOUNT OF THE ELECTRIC CLOCK, BRETT AND LITTLE'S ELECTRIC-TELEGRAPHIC CONVERSER; BAIN'S PRINTING TELEGRAPH, ETC. WITH ILLUSTRATIVE ANECDOTES. London: R. Yorke Clarke, 1847, 84 pp. Short descriptions of Ronald's telegraph, the printing telegraph, Bain's chemical telegraph, and the electric clock.

6-007 HANDBOOK OF THE ELECTRIC TELEGRAPH: BEING A TREATISE ON THE CONSTRUCTION, NATURE AND POWERS, OF THIS INSTRUMENT, WITH A FULL ACCOUNT OF ITS ORIGIN AND PROGRESS. London: n.p., 1848 [3rd ed.], 30 pp. Popular account of the needle telegraph.

6-007a ANECDOTES OF THE TELEGRAPH. London: David Bogue, 1849.

See Head, 6-158.

6-008 Park, Andrew H. A POPULAR EXPLANATION OF THE ELECTRIC TELEGRAPH AND OF SOME OTHER WONDERFUL APPLICATIONS OF ELECTRICITY. Boston: n.p., 1849, 34 pp.

6-009 Vail, Alfred, comp. THE TELEGRAPH REGISTER OF THE ELECTRO-MAGNETIC TELEGRAPH COMPANIES IN THE UNITED STATES AND THE CANADAS, USING PROF. MORSE'S PATENT, CONTAINING THE RATES OF CHARGES FOR TRANSMISSION OF MESSAGES. Washington: J.T. Towers, 1849, 61pp. One of the earliest published telegraph tariff schedules—useful in showing early telegraph network routes. See also Vail, 6-003.

6-010 Walker, Charles V. ELECTRIC TELEGRAPH MANIPULATION: BEING THE THEORY AND PLAIN INSTRUCTIONS IN THE ART OF TRANSMITTING SIGNALS TO DISTANT PLACES, AS PRACTICED IN ENGLAND... London: George Knight, 1850. Author was superintendent of electric telegraphs for the South Eastern Railway. (Not in Weaver.)

6-011 Alexander, William. PLAN AND DESCRIPTION OF THE ORIGINAL ELECTRO-MAGNETIC TELEGRAPH. London: Longman, Brown, Green & Longmans, 1851, 30 pp. "Letters written in 1837 by Lord John Russell relating to the author's proposal to establish telegraphic communication between Edinburgh and London by underground conductors; description of apparatus" (Weaver, 1-023, I:1194).

6-012 Davis, Daniel Jr. BOOK OF THE TELEGRAPH. Boston: D. Davis, 1851, 44 pp. (Not in Weaver.)

6-013 Archer, Charles Maybury. GUIDE TO THE ELECTRIC TELEGRAPH SHOWING THE EVERY-DAY PRACTICAL UTILITIES, WITH A SCALE OF CHARGES FOR MESSAGES, LIST OF COMMUNICATION STATIONS, ETC., WITH ILLUSTRATIVE MAP OF THE ENTIRE NETWORK OF TELEGRAPHIC COMMUNICATION IN GREAT BRITAIN. London: W.H. Smith, 1852, 62 pp. Collection of anecdotes relating to the early days of the electric telegraph. Map refers to facilities of the Electric Telegraph Co.

6-014 Channing, William F. THE MUNICIPAL ELECTRIC TELEGRAPH: ESPECIALLY IN ITS APPLICATION TO FIRE ALARMS. New Haven: B.L.

Hamlen, 1852, 28 pp. Focuses on this alarm system as developed by Moses Farmer for the city of Boston. See also the author's later lecture reprinted in SMITHSONIAN INSTITUTION [THIRTEENTH] ANNUAL REPORT FOR 1854 (Washington: Government Printing Office, 1855), pp. 147–155.

6-015 * Highton, Edward. THE ELECTRIC TELEGRAPH: ITS HISTORY AND PROGRESS. London: John Weale, 1852, 179 pp. Author was a telegraph engineer with the London and North Western Railway. Includes electric clocks and regulators of time.

6-016 ** Jones, Alexander. HISTORICAL SKETCH OF THE ELECTRIC TELEGRAPH; INCLUDING ITS RISE AND PROGRESS IN THE UNITED STATES. New York: Putnam, 1852 (reprinted by Echo Press, 1998), 194 pp. "Chronology of electric telegraphy and brief account of methods and apparatus used; determination of longitude, communication of time" (Weaver, 1-023, I:1232).

See Kendall, 5-187.

6-017 Turnbull, Laurence, M.D. LECTURES ON THE ELECTRO-MAGNETIC TELEGRAPHY, WITH AN HISTORICAL ACCOUNT OF ITS RISE AND PROGRESS, CONTAINING A LIST OF THE NUMBER OF TELEGRAPHIC LINES OF THE WORLD, WITH AN APPENDIX CONTAINING THE DECISIONS OF JUDGES WOODBURY AND KANE IN THE CELEBRATED TELEGRAPHIC TRIALS. Philadelphia: Franklin Institute, 1852, 186 pp.

6-018 Wilson, Dr. George. ELECTRICITY AND THE ELECTRIC TELEGRAPH, TOGETHER WITH THE CHEMISTRY OF THE STARS. London: Longman "Traveller's Library No. 26," 1852. Includes 77 pp. specifically on the telegraph.

6-019 Russell, Robert. W. HISTORY OF THE INVENTION OF THE ELECTRIC TELEGRAPH; ABRIDGED FROM THE WORKS OF LAURENCE TURNBULL M.D., AND EDWARD HIGHTON C.E., WITH REMARKS ON ROYAL E. HOUSE'S AMERICAN PRINTING TELEGRAPH AND THE CLAIMS OF S.F.B. MORSE, AS AN INVENTOR. New York: W.C. Bryant, 1853, 130 pp. Abridgement of 6-020.

6-020 ** Turnbull, Laurence. THE ELECTRO-MAGNETIC TELEGRAPH WITH AN HISTORICAL ACCOUNT OF ITS RISE, PROGRESS AND PRESENT CONDITION, ALSO PRACTICAL SUGGESTIONS IN REGARD TO INSULATION, AND PROTECTION FROM THE EFFECTS OF LIGHTENING, WITH AN APPENDIX CONTAINING SEVERAL IMPORTANT TELEGRAPHIC DECISIONS AND LAWS. Philadelphia: A. Hart, 1853, 264 pp.; [2nd ed. "revised and improved], 294 pp. "Detailed history of the development of the electric telegraph with numerous illustrations. The appendix contains a brief account of important telegraph decisions" (Weaver, 1-023, I:1271).

6-021 Not used.

6-022 Adley, Charles C. THE ELECTRIC TELEGRAPH: ITS HISTORY, THEORY, AND PRACTICAL APPLICATIONS," Reprinted from *PROCEEDINGS OF THE INSTITUTE OF CIVIL ENGINEERS* (Vol. 11, 1854), 33 pp.

See Cooke, 5-084.

6-023 ** Shaffner, Tal[iaferro] P., ed. *SHAFFNER'S TELEGRAPH COMPANION: DEVOTED TO THE SCIENCE AND ART OF THE MORSE TELEGRAPH*. New York: Pudney & Russell, 1854–1855, 2 vols. in one, 422 pp. The author's early and short-lived quarterly journal (10 issues: January–June 1854, monthly; not published July–December 1854; 1855, quarterly) on people, events, and technical developments. "The first number of Vol. 2 consists of Morse's defense against charges of Prof. Henry, with index" (Weaver, 1-023, II:403). Much of what is here, including valuable patent news and comment, is subsumed in the later integrated treatment from the same author, 6-029.

See Wheatstone, 5-089.

6-024 Window, Frederick Richard. **ON THE ELECTRIC TELEGRAPH AND THE PRINCIPAL IMPROVEMENTS IN ITS CONSTRUCTION**. Reprinted from *PROCEEDINGS OF THE INSTITUTE OF CIVIL ENGINEERS* (Vol. 2, 1854), 62 pp. "General account of electric telegraphy: the double needle instrument, Brett's printing, Blakewell's copying and Siemen's printing telegraphs" (Weaver, 1-023, I:1307)

6-025 Channing, William F. **THE AMERICAN FIRE-ALARM TELEGRAPH: A LECTURE DELIVERED BEFORE THE SMITHSONIAN INSTITUTION, MARCH 1855**. Reprinted from THE [THIRTEENTH] ANNUAL REPORT OF THE SMITHSONIAN INSTITUTION. Boston: Redding Co., 1855, 19 pp.

6-026 Lardner, Dr. Dionysus. **THE ELECTRIC TELEGRAPH POPULARIZED**. London: Walton & Maberly, 1855, 250 pp. Revised and rewritten by Edward B. Bright; London: J. Walton "Lardner's Museum of Science, No. 31 and No. 41," 1867 ["A new edition"], 272 pp. Includes systems developed by Morse, Bain, Steinheil, Wheatstone, Brett, House, Froment, Siemens, and Henley. Many engravings.

See Cooke, 5-085.

See Andrew, 6-144.

6-027 * Henry, Joseph. **THE ELECTRO-MAGNETIC TELEGRAPH**. Reprinted from PROCEEDINGS OF THE BOARD OF REGENTS, SMITHSONIAN INSTITUTION. Washington: Government Printing Office, 1857, 39 pp. "Morse's charges against the author, with appendix on the history of the electromagnetic telegraph" (Weaver, 1-023, I:1392).

6-028 Richards, William C. **ELECTRON, OR THE PRANKS OF THE MODERN PUCK: A TELEGRAPHIC EPIC FOR THE TIMES**. New York: D. Appleton, 1858, 84 pp. Written by a Baptist minister, an epic poem celebrating progress through the 1858 trans-Atlantic telegraph. (Not in Weaver.)

See Hamel, 6-157.

6-029 *** Shaffner, Taliaferro P. **THE TELEGRAPH MANUAL: A COMPLETE HISTORY AND DESCRIPTION OF THE SEMAPHORIC, ELECTRIC AND MAGNETIC TELEGRAPHS OF EUROPE, ASIA, AFRICA, AND AMERICA, ANCIENT AND MODERN**. New York: Pudney and Russell, 1859 (reprinted by van

Nostrand, 1867), 850 pp. The basic source for early telegraph history. This first comprehensive survey includes semaphore telegraphs, principles of the prominent systems in each major country, submarine telegraphs, telegraph construction and maintenance, organization and administration of national systems, instruments, apparatus. Engravings, index. A considerable part of material is drawn from the author's earlier telegraph periodical, 6-023. Chapters 9, 12, 13, and 27 reprinted in Shiers, 6-120.

6-030 Wilson, George. **THE PROGRESS OF THE TELEGRAPH**. Cambridge, England: Macmillan, 1859, 60 pp. "Lecture of literary merit on the electric telegraph" (Weaver, 1-023, I:1473).

6-031 ** Prescott, George B. **HISTORY, THEORY, AND PRACTICE OF THE ELECTRIC TELEGRAPH**. Boston: Ticknor and Fields, 1860, 468 pp.; 1863 [2nd ed.], 468 pp; 1866 [stated as 4th ed. but really 3rd; reprinted by Frank Jones, 1972], 508 pp. Later "editions" appear unchanged. Offers much historical background on the invention and progress of the technology including submarine telegraphy, instruments and apparatus, line construction, and practical working. Various systems are described in detail: needle, Morse, Bain, Hughes, dial, printing, and others. Engravings, index. See also Prescott, 6-059, which is something of a continuation.

6-032 Brown, Charles. **THE TELEGRAPHIC TRAIL OF THE TRANSCONTINENTAL TELEGRAPH**. n.p. 1861. Concerns construction of the first transcontinental telegraph completed in the year of publication.

6-033 Burnett, William Hickling. **THE ELECTRIC TELEGRAPH; AND THE PATENTED IMPROVEMENT THEREON**. London: n.p. 1861, 27 pp. Account of the author's own work.

6-034 Dodwell, Robert. **ILLUSTRATED HANDBOOK TO THE ELECTRIC TELEGRAPH**. London: Lemare, 1861, 80 pp.; 1862 [2nd ed.]; 80 pp. Popular treatment. Plates.

6-035 Bond, R. **HANDBOOK OF THE TELEGRAPH: BEING A MANUAL OF TELEGRAPHY, TELEGRAPH CLERK'S REMEBRANCER, AND GUIDE TO CANDIDATES FOR EMPLOYMENT IN THE TELEGRAPH SERVICE**. London: J. Weale, 1862, 68 pp.; and 1873 [2nd ed. revised and enlarged as "Weale's Rudimentary Series No. 138,"], 178 pp. has appended "Questions on magnetism, electricity, and practical telegraphy by W. McGregor."

6-036 * Culley, R[ichard] S[pellman]. **A HANDBOOK OF PRACTICAL TELEGRAPHY**. London: Longmans, Green, 1863, 191 pp.; 1867 [2nd ed.]; 1868 [3rd ed.]; 1870 [4th ed.], 302 pp.; 1871 [5th ed.], 408 pp.; 1874 [6th ed.], 443 pp.; 1878 [7th ed.], 468 pp.; 1885 [8th ed.], 442 pp. A standard and detailed treatment of electrical and magnetic theory, telegraph apparatus, line construction, submarine cables, apparatus and methods of testing, and fault-finding. The 7th ed. expanded to include the telephone and quadruplex telegraph systems. Engravings, diagrams (several foldout), tables, index.

See Preece, 6-166.

See Collins, 6-128.

Telegraphy

6-037 WOOD'S PLAN OF TELEGRAPHIC INSTRUCTION. Syracuse, NY: Morses's Telegraphic, 1864, 116 pp.

6-038 Smith, J[ohn] S. **AN OUTLINE OF THEORETICAL TELEGRAPHY.** Oberlin, OH: Calkins, Griffin, 1866, 66 pp.

6-038a Pond, Chester. **MANUAL OF TELEGRAPHY, DESIGNED FOR BEGINNERS.** Poughkeepsie, NY: I. Platt, 1865; New York: L.G. Tillotson and Co., 1868 [2nd ed.], 48 pp.; 1874 [17th ed.]. Largely concerned "reading by sound" (Weaver, 1-023, I:1733).

6-039 TELEGRAPH SECRETS: BY A STATIONMASTER. London: C.H. Clarke, 1866; 1868 [4th ed.], 122 pp. "Series of curious incidents" (Weaver, 1-023, I:1737).

See Adley, 6-142.

6-040 * Dodd, George. **RAILWAYS, STEAMERS, AND TELEGRAPHS: A GLANCE AT THEIR RECENT PROGRESS AND PRESENT STATE.** Edinburgh: W. and R. Chambers, 1867, 326 pp. Early work on the unity between the new modes of land and sea transport and the electric telegraph. Third chapter of 100 pages traces the telegraph, primarily the Atlantic cable. Engravings, chronology, index.

See Johnston, 6-161.

6-041 Morse, S.F.B. **MODERN TELEGRAPHY: SOME ERRORS OF DATES OF EVENTS AND OF STATEMENTS IN THE HISTORY OF TELEGRAPHY EXPOSED AND RECTIFIED.** Paris, 1867, 50 pp.; 1868 [2nd ed.], 38 pp. "Reply to personal criticisms" (Weaver, 1-023, I:1687-88). See also Morse, 6-044.

6-042 ** Sabine, Robert. **THE ELECTRIC TELEGRAPH.** London and New York: Virtue, 1867, 428 pp.; Van Nostrand, 1869 [2nd ed. adding subtitle: **HISTORY AND PROGRESS OF THE ELECTRIC TELEGRAPH, WITH DESCRIPTIONS OF SOME OF THE APPARATUS**], 280 pp.; London: Lockwood, 1872 [3rd ed. "with additions"], 280 pp. A standard technical history by Sir Charles Wheatstone's son-in-law. Engravings.

6-043 Varley, Cromwell F. **REPORT ON THE CONDITION OF THE LINES OF THE WESTERN UNION TELEGRAPH CO.** New York: Western Union, 1867, 129 pp.

6-044 Morse, S.F.B. **EXAMINATION OF THE TELEGRAPHIC APPARATUS AND THE PROCESSES IN TELEGRAPHY.** Washington: Government Printing Office "Reports of the U.S. Commissioners, Universal Exposition, Paris, 1867," 1869, 166 pp. "Comprehensive description of the telegraphic apparatus exhibited in Paris" (Weaver, 1-023, I:1749). See also Morse, 6-041.

6-045 Clark, [Josiah] Latimer. **ELEMENTARY TREATISE ON ELECTRICAL MEASUREMENT FOR THE USE OF TELEGRAPHIC INSPECTORS AND OPERATORS.** London, 1868, 175 pp. A "handbook valuable both to student and practical telegraphicist" (Weaver, 1-023, I:1715).

6-046 POLES, WIRES AND CABLES; OR, ELECTRIC TELEGRAPHS: THEIR PAST AND THEIR FUTURE, THEIR COMMERCIAL ADVANTAGES AND CHRISTIAN ASPECTS. London: E.J. Francis, 1869, 60 pp.; 1870 [2nd ed.], 60 pp. "Early history of the electric telegraph; ocean telegraphy; government acquisition of telegraphs" (Weaver, 1-023, I:1762).

6-047 Pope, Franklin L. **MODERN PRACTICE OF THE ELECTRIC TELEGRAPH: A HANDBOOK FOR ELECTRICIANS AND OPERATORS.** New York: Russell Bros., 1869, 128 pp.; Van Nostrand, 1870; 1871 [3rd ed.] through 1881 [11th ed.], 160 pp.; 1891 [14th ed., "rewritten and enlarged"] though 1902 [17th ed.], 234 pp. Most "editions" appear to be reprints, with pagination unchanged from earlier versions. Engravings showing instruments, apparatus, and circuit diagrams. Batteries, circuits, the Morse system, insulators, testing, construction, hints to learners, recent improvements. The full text of the 11th edition of 1881 is available free on-line—see 15-045.

See Sauer, 6-138.

See Kennan, 6-136.

6-048 Anderson, Sir James. **STATISTICS OF TELEGRAPHY.** London: Waterlow & Sons, 1872, 121 pp. "Discussion of the principles which affect the price of messages; also type of cable best suited for submarine telegraphy" (Weaver, 1-023, I:1816). Map.

6-049 Hoskiaer, Capt. [Otto] V. **GUIDE FOR THE ELECTRIC TESTING OF TELEGRAPH CABLES.** London: E. and F.N. Spon, 1873, 54 pp.; 1879 [2nd ed.], 72 pp.; 1889 [3rd ed.], 75 pp. "The usual tests for conductivity, charge, insulation and faults with useful formulae" (Weaver, 1-023, I:1869). Includes 10 plates. See also Hoskiaer, 6-061.

6-050 Mattison, I.W. **A HANDBOOK OF THE ELECTRIC TELEGRAPH: DESIGNED FOR THE USE OF STUDENTS AND OPERATORS.** Oberlin, OH: Union Telegraph College, 1873, 150 pp.

See Goldsmid, 6-153.

6-051 Schwendler, [Carl] Louis. **ON THE THEORY OF DUPLEX TELEGRAPHY—CONFERENCE PROCEEDING.** London: n.p., 1874, 48 pp.

6-052 Churchill, Charles H. **THEORY AND PRACTICE OF THE ELECTRIC TELEGRAPH: A MANUAL FOR OPERATORS AND STUDENTS.** Oberlin, OH: Shearman & Bros., 1875, 144 pp.

6-053 Douglas, John. C. **MANUAL OF TELEGRAPH CONSTRUCTION; THE MECHANICAL ELEMENTS OF ELECTRIC TELEGRAPH ENGINEERING.** London and Glasgow: Charles Griffin, 1875, 421 pp.; 1877 [2nd ed.], 468 pp. Author was with East India Government Telegraph Dept. "Handbook of the principles and practice of civil engineering necessary for the telegraph engineer" (Weaver, 1-023, I:1938). Second edition adds appendices and an index.

6-054 Huntington, F.M. **THE TELEGRAPHER'S SOUVENIR: A WORK COMPRISING COMPILATIONS AND ORIGINAL ARTICLES...**

INTENDED TO BE INSTRUCTIVE, INTERESTING AND AMUSING, NOT ONLY TO THE FRATERNITY, BUT TO STRANGERS AS WELL. Paterson, NJ: Lyon & Halsted, 1875, 49 pp.

See Prime, 5-194.

See Morse, 5-192.

See MEMORIAL OF S.F.B. MORSE, 5-190.

6-055 Beechey, Frederick S. **ELECTRO-TELEGRAPHY.** London: E. and F.N. Spon, 1876, 126 pp.; 1881 [2nd ed.], 126 pp. Basic primer.

6-056 Davis, Charles H. and Frank B. Rae. **HANDBOOK OF ELECTRICAL DIAGRAMS AND CONNECTIONS.** New York: Graphic Co, 1876, 46 pp., 29 plates, folding charts, map. See also Terry and Finn, 6-074.

6-057 * Preece, W[illiam] H., and Sir J[ames] Sivewright. **TELEGRAPHY.** London: Longmans, Green "Text Books of Science," 1876, 293 pp; 1884 [3rd ed.] through 1888 [7th ed.], 328 pp.; 1892 [9th ed.], 396 pp.; 1895 [11th ed.], 417 pp.; 1899 [15th ed.]; 1905 [18th impression, new ed.], 504 pp. Revised and partly rewritten by W. Llewellyn Preece, 1914, 422 pp. A standard text.

6-058 * Johnston, William J. **LIGHTNING FLASHES AND ELECTRIC DASHES: A VOLUME OF CHOICE TELEGRAPHIC LITERATURE, HUMOR, FUN, WIT AND WISDOM, CONTRIBUTED TO BY ALL THE PRINCIPAL WRITERS IN THE RANKS OF TELEGRAPHIC LITERATURE, AS WELL AS SEVERAL WELL-KNOWN OUTSIDERS.** New York: W.J. Johnston, 1877, 141 pp.; 1882 [2nd ed.], 160 pp. Preface: "This is...published with a view of giving telegraphy a literature of its own." Collection of stories, cartoons, and poetry reflecting in some way on telegraph operations and impact. On p. 97 of the first edition is a poem concerning the telephone—less than a year after Bell's patent. Woodcuts. See also the author's 6-068.

6-059 ** Prescott, George B. **ELECTRICITY AND THE ELECTRIC TELEGRAPH.** New York: D. Appleton, 1877, 978 pp.; 1881[4th ed.], 980 pp.; 1885 [6th ed.], 2 vols., 1,121 pp. The first 280 pages or so are on electricity in general while the rest focuses on the telegraph. Engravings, appendix, glossary, tables, index. Continuation of the author's 6-031.

6-060 Bolton, F., and J[ames] Sivewright. **TELEGRAPH POCKET BOOK, DIARY AND TELEGRAPH CODE FOR THE YEAR 1878. CONTAINING TABLES AND FORMULAE FOR USE IN THE CONSTRUCTION AND MAINTENANCE OF TELEGRAPHS, TOGETHER WITH REGULATIONS, TARIFFS AND NUMEROUS OTHER PARTICULARS OF THE BRITISH AND FOREIGN TELEGRAPH SYSTEMS, AND STATISTICS OF THE PRINCIPAL TELEGRAPH COMPANIES OF THE WORLD.** London: Letts, 1878.

6-061 Hoskiaer, Capt. [Otto] V. **LAYING AND REPAIRING OF ELECTRIC TELEGRAPH CABLES.** London: E. & F.N. Spon, 1878, 71 pp.; 1879 [2nd ed.],

72 pp.; 1889 [3rd ed.], 75 pp. Includes both land and sea cables. Diagrams. See also the author's 6-049.

6-062 Loring, A.E. **A HAND-BOOK OF THE ELECTRO-MAGNETIC TELEGRAPH.** New York: D. Van Nostrand "Science Series No. 39," 1878, 98 pp.; 1883 [2nd ed.], 98 pp.; 1900 [4th ed.], 116 pp. A standard text.

6-063 Samson [pseud. for J.A Clippinger]. **SAM JOHNSON; THE EXPERIENCE AND OBSERVATIONS OF A RAILROAD TELEGRAPH OPERATOR.** New York, 1878, 176 pp. "Work of wit and humor" (Weaver, 1-023, I: 2101).

6-064 Schwendler, [Carl] Louis. **INSTRUCTIONS FOR TESTING TELEGRAPH LINES AND THE TECHNICAL ARRANGEMENT OF OFFICES.** London and Edinburgh: Trubner, 1878–80 [2nd ed.], 2 vols. "Details of the practical methods for testing in general use; their mathematical theory" (Weaver, 1-023, I:2104). Written on behalf of the director-general of telegraphs of India.

6-065 * Tegg, William. **POSTS AND TELEGRAPHS, PAST AND PRESENT; WITH AN ACCOUNT OF THE TELEPHONE AND PHONOGRAPH.** London: T. Tegg, 1878, 318 pp. Includes both land and cable telegraphy, with a chapter each on the telephone and phonograph.

6-066 Craig, [Daniel H.]. **CRAIG'S MANUAL OF THE TELEGRAPH, ILLUSTRATING THE ELECTRO-MECHANICAL SYSTEM OF THE AMERICAN "RAPID" TELEGRAPH COMPANY OF NEW YORK.** New York: John Polhemus, 1879.

See Reid, 5-035.

6-067 Sumner, Charles A. **A PLAIN TALK ABOUT THE POSTAL TELEGRAPH.** San Francisco, Bacon & Co, 1879, 23 pp. A lecture by the famous congressman. (For a later version, see the same title as published by Government Printing Office; 48th Cong. "Speeches, Vol. 5, No. 53," 1884, 35 pp.)

See Taylor, 5-136.

6-068 Johnston, William J. **TELEGRAPHIC TALES AND TELEGRAPHIC HISTORY: A POPULAR ACCOUNT OF THE ELECTRIC TELEGRAPH, ITS USES, EXTENT AND OUTGROWTHS.** New York: Author, 1880, 254 pp.; 1882 [2nd ed.], 286 pp. See also the author's 6-058.

6-069 Abernathy, J[ohn] P[atterson]. **THE MODERN SERVICE OF COMMERCIAL AND RAILWAY TELEGRAPHY, IN THEORY AND PRACTICE, INCLUDING THE RAILWAY STATION AND EXPRESS SERVICE, ARRANGED IN QUESTIONS AND ANSWERS.** Cleveland (later St. Louis): Author, 1882 (with title preceded by AN OUTLINE OF COMMERCIAL...); 1883 [2nd ed.], 318 pp.; 1884 [3rd ed.], 333 pp.; 1886 [4th ed.], 355 pp.; 1887 [6th ed.], 423 pp.; 1891 [7th ed.], 424 pp.; and 1894 [8th ed.], 424 pp. A standard text.

See Plum, 3-140.

Telegraphy

6-070 Lockwood, Thomas D. ELECTRICITY, MAGNETISM, AND ELECTRIC TELEGRAPHY: A PRACTICAL GUIDE AND HAND-BOOK OF GENERAL INFORMATION FOR ELECTRICAL STUDENTS, OPERATORS, AND INSPECTORS. New York: D. Van Nostrand, 1883, 1888 [2nd ed.], 377 pp.

6-071 Swift, Lt. James A. THE PRACTICAL TELEGRAPHER: A MANUAL OF PRACTICAL TELEGRAPHY AND TELEGRAPHIC CONSTRUCTION. New York: W.J. Johnston, 1883, 189 pp.

6-072 *** Fahie, J[ohn] J. A HISTORY OF ELECTRIC TELEGRAPHY TO THE YEAR 1837, CHIEFLY COMPILED FROM ORIGINAL SOURCES, AND HITHERTO UNPUBLISHED DOCUMENTS. London: E. and F.N. Spon, 1884 (reprinted by Arno Press, "Telecommunications," 1974), 542 pp. Invaluable and largely definitive tracing of the beginnings of telegraphy, arranged chronologically and including developments in many countries. Engravings, diagrams, notes, bibliography of sympathetic telegraphy, index.

6-073 Lynd, William. PRACTICAL TELEGRAPHIST AND GUIDE TO THE TELEGRAPH SERVICE. London: Wyman, 1884, 227 pp. Includes submarine telegraphy. "Useful compilation of telegraph matter" (Weaver, 1-023, I:2367).

6-074 Terry, Astley C., and William Finn. ILLUSTRATIONS AND DESCRIPTION OF TELEGRAPHIC APPARATUS. Buffalo, NY: A.C. Astley, 1884, 94 pp.; 1889 [2nd ed.], 100 pp. See also Davis and Rae, 6-056

See Dickerson, 5-132.

6-075 Clark, L[atimer], and Robert Sabine. ELECTRIC TABLES AND FORMULAE FOR THE USE OF TELEGRAPH INSPECTORS AND OPERATORS. London: E. & F.N. Spon, 1885.

6-076 Maver, William Jr., and Minor M. Davis. THE QUADRUPLEX: WITH CHAPTERS ON THE DYNAMO-ELECTRIC MACHINE IN RELATION TO THE QUADRUPLEX, THE PRACTICAL WORKING OF THE QUADRUPLEX, TELEGRAPH REPEATERS, AND THE WHEATSTONE AUTOMATIC TELEGRAPH. New York: W.T. Johnston, 1885, 128 pp. "Description of the Edison quadruplex-system of telegraphy, written for operators" (Weaver, 1-023, I:2388).

6-077 Pangborn, Joseph G. ELECTRIC B & O. New York: American Bank Note Co., 1885. A history of the Baltimore and Ohio Telegraph Co. Illustrated.

6-078 Williams, W. MANUAL OF TELEGRAPHY. London: Longmans, Green, 1885, 327 pp. "Book of reference for telegraph matters" (Weaver, 1-023, I:2399).

6-079 Nicol, Donald. THE TELEGRAPH AND TELEPHONE CONSIDERED IN RELATION TO ECONOMY AND EFFICIENCY. London, 1887, 32 pp. "The paper advocates the use of bitumen for insulation purposes" (Weaver, 1-023, I:2424).

6-080 Maver, William Jr. **PRACTICAL SYSTEMS OF ELECTRICAL TELEGRAPHY.** Newport, RI: Torpedo Station Print, 1888, 46 pp. A lecture before the class of 1888.

See Munro, 5-029.

6-081 ** Maver, William Jr. **AMERICAN TELEGRAPHY: SYSTEMS, APPARATUS, AND OPERATION.** New York: J.H. Bunnell, 1892, 563 pp.; Maver, 1897 [2nd ed.], 563 pp.; continued as **AMERICAN TELEGRAPHY AND ENCYCLOPEDIA OF THE TELEGRAPH: SYSTEMS, APPARATUS, OPERATION.** 1903 [5th ed.], 656 pp.; 1909 [7th ed.], 668 pp.; 1912 (reprinted by Lindsay Publications, 1997), 695 pp. A full treatment of all kinds of systems and apparatus, including writing telegraphs, time services, ticker systems, burglar and fire alarms, policy and railway signaling. Descriptions of individual instruments are given in detail with illustrations of mechanical parts, construction, assembly, operation, and circuits. Diagrams, index.

6-082 Thom, Charles, and Willis H. Jones. **TELEGRAPHIC CONNECTIONS: EMBRACING RECENT METHODS IN QUADRUPLEX TELEGRAPHY.** New York: Van Nostrand, 1892, 59 pp.

6-083 Not used.

6-084 Cornell, Alonzo B. **HISTORY OF THE ELECTRO-MAGNETIC TELEGRAPH.** Schnectady, NY: Union College "Butterfield Lectures," 1894, 24 pp.

6-085 Flegel, George C. **THE TELEGRAPHERS' HANDBOOK.** Vinton, IA: The Telegraphers' Publishing Co., 1894, 112 pp. Diagrams, tables.

See Taltavall, 5-036.

6-086 Routledge, Robert (trans. from French of A.L. Ternant). **THE TELEGRAPH.** London: G. Routledge, 1895, 289 pp. First 77 pages treat optical, acoustic, and pneumatic systems while the remainder focuses on electric telegraph: history, land lines, submarine cables, apparatus, and applications. Engravings.

B. Later Telegraphy Works (1896–1950)

(This is a very selective listing of works offering substantial historical intent or value. Arrangement is alphabetical by author.)

6-087 ** **AMERICAN TELEGRAPHY AFTER 100 YEARS.** New York: Western Union, 1944, ca 65 pp. Compilation of six articles issued to celebrate the centennial, all previously published in technical journals and most listed elsewhere here, assembled by a company committee on technical publication. Original to this publication is I.S. Coggelshall's "A Bibliography of the Telegraph" with 414 un-annotated items arranged by subject with emphasis on technical citations from engineering periodicals from 1920 to 1944. Photos, diagrams, maps, references.

6-088 Bell, John H. "Printing Telegraph Systems," *TRANSACTIONS OF THE AIEE* 39:167–230 (1920). A full survey. Diagrams, references.

See Bright, 6-148.

6-089 * D'Humy, Ferdinand E., and P.J. Howe. "American Telegraphy after 100 Years," *TRANSACTIONS OF THE AIEE* 63:1014–1032 (December 1944). Good survey of then-current operating conditions and practices. Photos, diagrams, maps, references. Reprinted in AMERICAN TELEGRAPHY, 6-087; and Shiers, 6-120.

6-090 * Duncan, J.A., et al. "Telegraphy in the Bell System," *TRANSACTIONS OF THE AIEE* 63:1032–1044 (December 1944). Useful survey of then-current operations. Photos, diagrams, maps, tables, charts, references. Reprinted in AMERICAN TELEGRAPHY, 6-087.

6-091 Fowle, Frank. "Telegraph Transmission," *TRANSACTIONS OF THE AIEE* 19:1683–1741 (1911).

See Harlow, 3-038.

6-092 Harrison, H[arry] H. **PRINTING TELEGRAPH SYSTEMS AND MECHANISMS.** London: Longmans, Green "Manuals of Telegraph and Telephone Engineering," 1923, 435 pp. Photos, diagrams, index.

See Herbert, 6-159.

6-093 Hillis, George C. "Telegraphy—Pony Express to Beam Radio," **SMITHSONIAN INSTITUTION ANNUAL REPORT FOR 1947**. Washington: Government Printing Office, 1948, pp. 191–205. Includes discussion of microwave (see also 8-D for more). Photos, references. Reprinted in Shiers, 6-120.

6-094 Houston, Edwin J., and Arthur E. Kennelly. **ELECTRIC TELEGRAPHY.** New York: W.J. Johnston, 1897, 488 pp.; McGraw, 1906 [2nd ed.], 480 pp. Diagrams, folding map.

6-095 * Maver, William, and Donald McNicol. "American Telegraph Engineering—Notes on History and Practice," *TRANSACTIONS OF THE AIEE* 20:1303–1339; discussion 1339–1356 (1910). History from 1840, sources of emf, printers, repeaters, superimposed systems, inductive disturbances, relays, engineering details, the telephone in testing, and line construction.

6-096 * McNicol, Donald. **AMERICAN TELEGRAPH PRACTICE: A COMPLETE TECHNICAL COURSE IN MODERN TELEGRAPHY INCLUDING SIMULTANEOUS TELEGRAPHY AND TELEPHONY.** New York: McGraw-Hill, 1913, 507 pp. A standard text. Diagrams, photos, appendices, tables, index.

See Meyer, 6-164.

6-097 Murray, John. **A STORY OF THE TELEGRAPH.** Montreal: John Lovell & Son, 1905, 269 pp. Includes more coverage of British and Canadian developments than most American works.

6-098 Perkins, W.T. **MODERN TELEGRAPH SYSTEMS AND EQUIPMENT.** London: George Newnes, 1946, 215 pp. Useful example of a survey (there were several) of postwar equipment and services, concentrating on developments after 1920. Includes details of American, British, and German systems. Photos, diagrams, tables, index.

6-099 Rhoads, Stanley. "Some Phases of Railroad Telegraph and Telephone Engineering," *TRANSACTIONS OF THE AIEE* 40:301–380 (1921).

6-100 Roberts, A.H. "The Romance and History of the Electric Telegraph," *POST OFFICE ELECTRICAL ENGINEERS JOURNAL* 16:121–130, 207–224, 301–321; 17:1–19, 91–96, 171–120 (1923–25).

6-101 *** Thompson, Robert Luther. **WIRING A CONTINENT: THE HISTORY OF THE TELEGRAPH INDUSTRY IN THE UNITED STATES, 1832–1866.** Princeton, NJ: Princeton University Press, 1947 (reprinted by Arno Press "Technology and Society," 1972), 544 pp. Intended as the first of a definitive two-volume history, only this volume appeared. It details the technical and corporate story up to the formation of Western Union's near-monopoly after the Civil War. Notes, tables, sources, index.

C. Modern Histories of U.S. Land Telegraphy

6-102 Ault, Phil. **WIRES WEST: THE STORY OF THE TALKING WIRES.** New York: Dodd, Mead, 1974, 176 pp. Written for younger readers, this traces the expansion of the telegraph into the Mississippi and into the American West. Diagrams, photos, bibliography, index.

6-103 Blondheim, Menaham. **NEWS OVER THE WIRES: THE TELEGRAPH AND THE FLOW OF PUBLIC INFORMATION IN AMERICA, 1844–1897.** Cambridge, MA: Harvard University Press "Studies in Business History 42," 1994, 305 pp. History of the news wire services (mainly the Associated Press of New York and its relationship with Western Union) focusing on the effect of the new technology. Tables, selected primary sources, notes, index.

6-104 * Coe, Lewis. **THE TELEGRAPH: A HISTORY OF MORSE'S INVENTION AND ITS PREDECESSORS IN THE UNITED STATES.** Jefferson, NC: McFarland, 1993, 184 pp. Useful survey history of the pre-telephone era including its many applications. Parallels author's later studies of the telephone (7-073) and wireless (9-096). Photos, diagrams, appendix of biographical sketches, bibliography, index.

6-105 * Dawson, Keith. "Electromagnetic Telegraphy: Early Ideas, Proposals, and Apparatus," in A. Rupert Hall and Norman Smith, eds. **HISTORY OF TECHNOLOGY: FIRST ANNUAL VOLUME, 1976.** London: Mansell Information Publishing, 1976, pp. 113–141. The work done prior to Wheatstone and Morse, based on the author's dissertation at University of London (1973).

6-106 DuBoff, Richard B. "Business Demand the Development of the Telegraph in the United States, 1844–1860," *BUSINESS HISTORY REVIEW* 54: 459–479 (Winter 1980). Pre–Civil War impact of the telegraph on business operations.

6-107 _____. "The Telegraph and the Structure of Markets in the United States, 1845–1890," *RESEARCH IN ECONOMIC HISTORY* 8:253–277 (1982). Interesting analysis of the many business impacts of telegraphy, some positive and others not so. References.

6-108 _____. "The Telegraph in Nineteenth-Century America: Technology and Monopoly," *COMPARATIVE STUDIES IN SOCIETY AND HISTORY* 26: 571–586 (October 1984).

6-109 Field, Alexander James. "The Magnetic Telegraph: Price and Quantity Data, and the New Management of Control," *THE JOURNAL OF ECONOMIC HISTORY* 52:401–413 (June 1992). Impact of the innovation on business management.

6-110 Gabler, Edwin. **THE AMERICAN TELEGRAPHER: A SOCIAL HISTORY, 1860–1900**. New Brunswick, NJ: Rutgers University Press, 1988, 264 pp. Focuses on those who worked the keys and other parts of the business. Notes, index. See also Ulriksson, 6-123.

6-111 * Harris, Robert Dalton, and Diane DeBlois. **U.S. POSTAL HISTORY DOCUMENTS No. 2: CONGRESS & THE TELEGRAPH**. Wynantskill, NY: A Gatherin', no date, 34 pp. Annotated listing of hearings and related documents from 1837 to 1902 concerning land and sea telegraphy. Annotations are extensive and provide full particulars of telegraph events discussed. See also Brightbill, 1-002.

6-112 *** Israel, Paul. **FROM MACHINE SHOP TO INDUSTRIAL LABORATORY: TELEGRAPHY AND THE CHANGING CONTEXT OF AMERICAN INVENTION, 1830–1920**. Baltimore: Johns Hopkins University Press "Studies in the History of Technology No.14," 1992, 255 pp. Unique study exploring the relationship of research and the telegraph—and how the former became more corporate over time. Photos, diagrams, notes, bibliography, index.

6-113 Israel, Paul, and Keith A. Nier. "The Transfer of Telegraph Technologies in the Nineteenth Century," in David J. Jeremy, ed. **INTERNATIONAL TECHNOLOGY TRANSFER: EUROPE, JAPAN AND THE USA, 1700–1914**. Aldershot, England: Edward Elgar Publishing, 1991, pp. 95–121.

6-114 Klein, Maury. "What Hath God Wrought?" *AMERICAN HERITAGE OF INVENTION AND TECHNOLOGY* 8:34–43 (Spring 1993). Popular treatment of early telegraph.

6-115 Lindley, Lester G. **THE CONSTITUTION FACES TECHNOLOGY: THE RELATIONSHIP OF THE NATIONAL GOVERNMENT TO THE TELEGRAPH, 1866–1884**. New York: Arno Press "Dissertations in American Economic History," 1975, 280 pp. Based largely on Western Union documents, this study (reprinting a 1971 Rice University dissertation) focuses on the early years of the company's monopoly operation. Footnotes, bibliography (no index).

6-116 Lipartito, Kenneth. "A Comparative Analysis of the Early History of Southern and Northern Telegraph Systems," *BUSINESS AND ECONOMIC HISTORY* 14:156–176 (1985).

6-117 * Lubrano, Annteresa. **THE TELEGRAPH: HOW TECHNOLOGICAL INNOVATION CAUSED SOCIAL CHANGE.** New York: Garland "Garland Studies on Industrial Productivity 49," 1997, 182 pp. Based on a dissertation, this reviews telegraphy's impact on formal organization, politics, culture, social stratification, law, and nationalism. Diagrams, notes, bibliography, index.

See Marland, 3-043.

6-118 Morus, Iwan R. "Telegraphy and the Technology of Display: The Electricians and Samuel Morse," *HISTORY OF TECHNOLOGY* 13:20–40 (1991).

6-119 Nakagawa, Yasuo. "The Development of Early Practical Electromagnetic Telegraphs and the Mechanization of Skilled Operation," *HISTORIA SCIENTIARUM* 27: 77–89 (1984). Details both English and American instrument developments—and how they affected system operation. Photos, charts, references.

See Oslin, 3-048.

6-120 *** Shiers, George, ed. **THE ELECTRIC TELEGRAPH: AN HISTORICAL ANTHOLOGY.** New York: Arno Press "Historical Studies in Telecommunications," 1977, ca 500 pp. Valuable collection includes 18 articles originally published from 1859 to 1948 (many listed separately here), plus introductory overview essay and bibliography. Photos, diagrams, notes, references.

6-121 * Standage, Tom. **THE VICTORIAN INTERNET: THE REMARKABLE STORY OF THE TELEGRAPH AND THE NINETEENTH CENTURY'S ON-LINE PIONEERS.** New York: Walker, 1998, 227 pp. A popular and readable history of the telegraph in Britain and the United States described as an early version of the Internet. Photos, sources, index.

6-122 Tarr, Joel A., with Thomas Finholt and David Goodman. "The City and the Telegraph: Urban Communications in the Pre-Telephone Era," *JOURNAL OF URBAN HISTORY* 14:38–80 (November 1987).

6-123 * Ulriksson, Vidkunn. **THE TELEGRAPHERS, THEIR CRAFT AND THEIR UNIONS.** Washington: Public Affairs Press, 1953, 218 pp. The people behind the keys. See also Gabler, 6-110.

6-124 Yates, Joanne. "The Telegraph's Effects on Nineteenth-Century Markets and Firms," *BUSINESS AND ECONOMIC HISTORY,* 15:149–193 (1986). Useful study of early impact on business organization, markets, and trade.

D. Foreign Telegraphy Systems

(See also chapter 3-E for general foreign telecommunications; 4-F for foreign carriers and administrations; 6-A for many early British works; 7-E for works on foreign telephone systems, and 9-D for works on radio.)

General and Europe

6-125 *** Ahvenainen, Jorma. THE FAR EASTERN TELEGRAPHS: THE HISTORY OF TELEGRAPHIC COMMUNICATIONS BETWEEN THE FAR EAST, EUROPE, AND AMERICA BEFORE THE FIRST WORLD WAR. Helskinki: Suomalainen Tiedeakatemia, Series B, No. 216, 1981, 226 pp. An important study, this focuses on the Danish Great Northern and British Eastern companies, mainly on business and politics. References, map.

6-126 ** Baark, Erik. LIGHTNING WIRES: THE TELEGRAPH AND CHINA'S TECHNOLOGICAL MODERNIZATION, 1860–1890. Westport, CT: Greenwood "Contributions in Asian Studies No. 6," 1997, 216 pp. The varied impacts of the telegraph in late Imperial China—and the lessons drawn from this case study. A pioneering study. Notes, bibliography, index.

6-127 Butricica, Andrew J. "Telegraphy and the Genesis of Electrical Engineering Institutions in France, 1845–1895," *HISTORY AND TECHNOLOGY* 3:365–380 (1987).

6-128 Collins, Perry M. OVERLAND EXPLORATIONS IN SIBERIA, NORTHERN ASIA, AND THE GREAT AMOOR RIVER COUNTRY: INCIDENTAL NOTICES OF MANCHOORIA, MONGOLIA, KASCHATKA, AND JAPAN WITH MAP AND PLAN OF AN OVERLAND TELEGRAPH AROUND THE WORLD, VIA BEHRING'S STRAIT AND ASIATIC RUSSIA TO EUROPE. New York: Appleton, 1864. Title largely sums up this telegraph-related travelogue.

6-128a de Cogan, Donard. "International Cable Telegraph and the Austro-Hungarian Monarchy," *BLATTER FUR TECHNIKGESCHICHTE*, 50:7–44 (1988–90). In matters of international communications, Austro-Hungary had its own cable network in the Adriatic which was largely laid by German cable ships. Notes.

6-129 Field, Alexander J. "French Optical Telegraphy, 1793–1855: Hardware, Software, Administration," *TECHNOLOGY AND CULTURE*, 35:315–347 (April 1994).

6-130 Flichy, Patrice. "The Birth of Long Distance Communication: Semaphore Telegraphs in Europe (1790–1840)," *RÉSEAUX: THE FRENCH JOURNAL OF COMMUNICATION*, 1:81–101 (1993). Really "pre" telecommunications.

6-131 Hodge, John E. "The Role of the Telegraph in the Consolidation and Expansion of the Argentine Republic," *AMERICAS*, 41:59–80 (July 1984). For a parallel work on the telephone in the same country, see 7-119.

6-132 * Holzmann, Gerard J., and Björn Pehrson. THE EARLY HISTORY OF DATA NETWORKS. Los Alamitos, CA: IEEE Computer Society Press, 1995, 290 pp. An interesting history of optical telegraphs, primarily in Europe, including an English translation of a 1796 "Treatise on Telegraphs." Photos, diagrams, appendices, notes, bibliography, index. See also Wilson, 6-139.

6-133 James, John Alton. THE FIRST SCIENTIFIC EXPLORATION OF RUSSIAN AMERICA AND THE PURCHASE OF ALASKA. Evanston: Northwestern University Press "Studies in the Social Sciences No. 4," 1942, 444 pp.

The mid-1860s expedition to develop a land telegraph connection to Europe. Photos, maps. See also Neering, 6-137.

6-134 Japanese Ministry of Communications. **OUTLINE OF THE HISTORY OF TELEGRAPHS IN JAPAN.** Tokyo: Japanese Ministry of Communications, 1892, 56 pp. Official story. Tables. (A revised edition adding coverage of postal services appeared in 1902.)

6-135 Johnson, John J. **PIONEER TELEGRAPHY IN CHILE, 1852–1876.** New York: AMS Press "Stanford University Publications, University Series: History, Economics and Political Science, 6:1," 1968, 159 pp.

6-136 Kennan, George. **TENT LIFE IN SIBERIA, AND OTHER ADVENTURES AMONG THE KORAKS AND OTHER TRIBES IN KAMCHATKA AND NORTHERN ASIA.** New York: G.P. Putnam, 1870, 425 pp.; 1910 [2nd ed.]; many later reprints. One of the better-known first-person accounts of the ill-fated Siberian telegraph, but heavier on travel and sights than on the telegraph. Folding map. See also Neering, 6-137.

6-137 ** Neering, Rosemary. **CONTINENTAL DASH: THE RUSSIAN-AMERICAN TELEGRAPH.** Ganges, British Columbia: Horsdal & Schubart Publishers, 1989, 230 pp. Modern history of the early 1860s attempt to develop a land telegraph connection to Europe via Alaska and Siberia. Photos, notes, maps, bibliography, index. See also Adams, 6-141; James, 6-133; Kennan, 6-136; Lawrence, 6-163.

6-138 * Sauer, George. **THE TELEGRAPH IN EUROPE: A COMPLETE STATEMENT OF THE RISE AND PROGRESS OF TELEGRAPHY IN EUROPE, SHOWING THE COST OF CONSTRUCTION AND WORKING EXPENSES OF TELEGRAPHIC COMMUNICATIONS IN THE PRINCIPAL COUNTRIES, FROM OFFICIAL RETURNS.** Paris, printed for private circulation, 1869, 177 pp. "The introduction gives a sketch of the rise and progress of telegraphy; the rest of the work treats of general telegraphic statistics in the various countries of Europe" (Weaver, 1-023, I:1754). Important given its early date.

6-139 ** Wilson, Geoffrey. **THE OLD TELEGRAPHS.** London: Phillimore / Totowa, NJ: Rowman and Littlefield, 1976, 252 pp. Well-researched history of the various mechanical systems of visual communication developed before the electric telegraph. Arranged geographically, it begins with Britain and then covers Europe and other major countries. Photos, diagrams, maps, bibliography, index. See also Holzmann and Pehrson, 6-132.

6-140 Yarotsky, A.V. "150th Anniversary of the Electromagnetic Telegraph," *TELECOMMUNICATION JOURNAL,* 49:709–715 (October 1982). Claims that Paval Schilling demonstrated a working system for the czar in October 1832. Photos, diagrams, bibliography.

British Commonwealth

6-141 Adams, George R. **LIFE ON THE YUKON, 1865–1867.** Kingston, ON: Limestone Press "Alaska History No. 22," 1982, 219 pp. Endpaper maps, list of related works. See also Neering, 6-137.

Telegraphy

6-142 Adley, Charles C. **THE STORY OF THE TELEGRAPH IN INDIA.** London: E. and F.N. Spon, 1866, 86 pp. "Criticism of the instruments and methods used on the Indian telegraph lines; also plea for the emancipation of telegraphs from Government control" (Weaver, 1-023, I:1635).

6-143 ALICE SPRINGS TELEGRAPH STATION NATIONAL PARK. Alice Springs, NT: Northern Territory Reserves Board, 1969, 24 pp. The history, development and significance of this "outback" site. Photos. See also Blackwell and Lockwood, 6-147.

6-144 Andrew, Sir W[illiam] P. **MEMOIR ON THE EUPHRATES VALLEY ROUTE TO INDIA; WITH OFFICIAL CORRESPONDENCE AND MAPS.** London: publisher not given, 1857, 267 pp. "This work was written to show the importance of establishing telegraphic communication between England and India via the Euphrates Valley" (Weaver, 1-023, I:1378).

6-145 Atkinson, Hugh. **THE LONGEST WIRE.** Sydney: Angus & Robertson, 1982, 232 pp. Building the telegraph in Australia is the setting for this novel. See also Clune, 6-149.

6-146 Bennett, R. **REMINISCENCES OF THE CAPE GOVERNMENT TELEGRAPHS.** Cape Town: S.A. Newspaper Col, no date, but ca 1908, 100 pp. Memoirs of 40 years of development of the electric telegraph service in South Africa. Includes chapters on its uses in the Kimberly gold fields and during the Boer War. Photos.

6-147 Blackwell, Doris, and Douglas Lockwood. **ALICE ON THE LINE.** Adelaide: Rigby, 1965, 204 pp. The first author's father was in charge of the Alice Springs telegraph station from 1899 to 1908. See also ALICE SPRINGS TELEGRAPH STATION, 6-143.

6-148 Bright, Charles. **IMPERIAL TELEGRAPHIC COMMUNICATION.** London: P.S. King, 1911, 212 pp. As with many books of this period, "communication" includes chapters on various modes of transport.

6-149 * Clune, Frank. **OVERLAND TELEGRAPH: THE STORY OF A GREAT AUSTRALIAN ACHIEVEMENT AT THE LINK BETWEEN ADELAIDE AND PORT DARWIN.** Sydney: Angus and Robertson, 1955, 238 pp. Concerns the 1870–72 construction of a line connecting the south coast with Darwin. Reprinted in 1984 with different subtitle: "An Epic Feat of Endurance and Courage." Photos, maps, references. See also Atkinson, 6-145, and Giles, 6-152..

6-150 Durham, John C.E. **TELEGRAPHS IN VICTORIAN LONDON.** Cambridge, England: Golden Head Press, 1959, 30 pp. Traces developments of two competing firms in 1859–60 as both strung wires overhead.

6-151 Ghose, Saroj. "Commercial Needs and Military Necessities: The Telegraph in India," in Roy Macleod and Deepak Kumar, eds. **TECHNOLOGY AND THE RAJ: WESTERN TECHNOLOGY AND TECHNICAL TRANSFERS TO INDIA, 1700–1947.** New Delhi: Sage, 1995, pp. 153–176. British colonial need for and uses of the telegraph.

6-152 Giles, Alfred. **EXPLORING IN THE SEVENTIES AND THE CONSTRUCTION OF THE OVERLAND TELEGRAPH LINE**. Adelaide: Friends of the State Library of South Australia, 1995, 172 pp. A facsimile reprint of a contemporary book (date not given). Illustrations, folding map. See also Clune, 6-149.

6-153 * Goldsmid, Sir Frederick John. **TELEGRAPH AND TRAVEL: A NARRATIVE OF THE FORMATION AND DEVELOPMENT OF TELEGRAPHIC COMMUNICATION BETWEEN ENGLAND AND INDIA, UNDER THE ORDERS OF HER MAJESTY'S GOVERNMENT, WITH INCIDENTAL NOTICES OF THE COUNTRIES TRAVERSED BY THE LINES**. London: Macmillan, 1874, 673 pp. "The Persian Gulf cable also the Russo-Persian lines. The volume contains a map showing the telegraph lines to India, 1874" (Weaver, 1-023, I:1811). Engravings, maps, diagrams.

6-154 Gorman, Me. "Sir William O'Schaughnessy, Lord Dalhousie, and the Establishment of the Telegraph System in India," *TECHNOLOGY AND CULTURE*, 12:581-601 (October 1971). Establishment of the colonial service in British India.

6-155 Graves, E. "A Decade in the History of the English Telegraph," *SOCIETY OF TELEGRAPH ENGINEERS JOURNAL*, 9:249–277 (1880).

6-156 * Gribble, P.J. **WHAT HATH GOD WROUGHT: THE STORY OF THE QUEENSLAND TELEGRAPH SERVICE FROM 1861**. Brisbane: Telecom, 1981, 711 pp. Photos, appendices, index. Detailed study of the telegraph's development in western Australia.

6-157 Hamel, Dr. J[oseph]. **HISTORICAL ACCOUNT OF THE INTRODUCTION OF THE GALVANIC AND ELECTRO-MAGNETIC TELEGRAPH INTO ENGLAND WITH COMMENTS THEREON BY W. F. COOKE**. London: n.p., 1859, 79 pp. "Chronological data about the electrical work of Soemmering, Schilling, Zamboni and Romagnosi" short account of the introduction of the electric telegraph into England by Cooke and Wheatstone" (Weaver, 1-023, I:1457).

6-158 [Head, Sir Francis Bond]. **STOKERS AND POKERS: OR, THE LONDON AND NORTH-WESTERN RAILWAY, THE ELECTRIC TELEGRAPH, AND THE RAILWAY CLEARING-HOUSE**. London: John Murray, 1849 (reprinted by Frank Cass, 1968), 208 pp. Based on articles from the *QUARTERLY REVIEW*, chapter 13 offers material on the telegraph. Lengthy appendix concerns rules of the railway.

6-159 * Herbert, Thomas E. **TELEGRAPHY: A DETAILED EXPOSITION OF THE TELEGRAPH SYSTEM OF THE BRITISH POST OFFICE**. London: Whittaker, 1901; 1907 [2nd ed.], 509 pp; 1916 [3rd ed.], 916 pp.; London: Pitman "The Specialists' Series," 1920 [4th ed.], 1,019 pp.; 1930 [5th ed.], 1,199 pp. Standard descriptive information. Photos, diagrams. For the author's parallel volume on British telephony, see 7-134.

6-160 Holmes, T.W. **THE SEMAPHORE: THE STORY OF THE ADMIRALTY-TO-PORTSMOUTH SHUTTER TELEGRAPH AND SEMAPHORE LINES, 1796 TO 1847**. Ilfracombe, England: Stockwell, 1983, 223 pp.

Pre-electromagnetic telegraphy, important as many of the same routes were converted to the electrical means as it became available.

6-161 Johnston, Robert W. **THE TELEGRAPH AND ITS PROPOSED ACQUISITION BY THE GOVERNMENT.** Edinburgh: W.P. Nimmo, 1867, 43 pp. Author was with the Chief Metropolitan Post Office and assesses the advantages of government ownership.

6-162 *** Kieve, Jeffrey. **THE ELECTRIC TELEGRAPH IN THE U.K.: A SOCIAL AND ECONOMIC HISTORY.** Newton Abbot, England / New York: Barnes & Noble, 1973, 310 pp. The best modern history which, despite the subtitle, includes technical development of the system as well. Photos, maps, tables, notes and references, bibliography, index.

6-163 Lawrence, Guy. **40 YEARS OF THE YUKON TELEGRAPH.** Vancouver: Mitchell Press, 1965. The line that ran 1,900 miles from Vancouver up to Atlin, Whitehorse, and Dawson City. See also Neering, 6-137.

See Livingston, 3-079.

6-164 Meyer, Richard Hugo. **THE BRITISH STATE TELEGRAPHS: A STUDY OF THE PROBLEM OF A LARGE BODY OF CIVIL SERVANTS IN A DEMOCRACY.** London: Macmillan, 1907, 408 pp. Largely administrative policy in focus.

6-165 Morus, Iwan Rhys. "The Electric Ariel: Telegraphy and Commercial Culture in Early Victorian England," *VICTORIAN STUDIES,* 39:339–378 (Spring 1996). Impacts of the technology.

6-166 Preece, William H. **RAILWAY TELEGRAPHS AND THE APPLICATION OF ELECTRICITY TO THE SIGNALING AND WORKING OF TRAINS.** London, 1863, 86 pp. "Details of the system of signalling used on the London and South-Western Railway" (Weaver, 1-023, I:1580).

6-167 Shridharani, Krishnalal G. **STORY OF THE INDIAN TELEGRAPHS: A CENTURY OF PROGRESS.** New Delhi: Posts and Telegraphs Department, 1953, 172 pp. Photos. From the colonial telegraph systems through operations in the early years of independence.

6-168 Taylor, Peter. **AND END TO SILENCE: THE BUILDING OF THE OVERLAND TELEGRAPH LINE FROM ADELAIDE TO DARWIN.** Sydney: Methuen, 1980, 192 pp. Photos, endpaper maps. See also Clune, 6-149.

E. Contemporary Books on Submarine Telegraphy (1855–1905)

(Arranged chronologically, this first half century of contemporary submarine telegraphy books and pamphlets includes material now of considerable historical value. Parts of this listing originated with Higgins, 1-008. See also biographies of Cyrus Field, 5-120.)

Chapter 6

6-169 * Mullaly, John. **A TRIP TO NEWFOUNDLAND: ITS SCENERY AND FISHERIES: WITH AN ACCOUNT OF THE LAYING OF THE SUBMARINE TELEGRAPH CABLE.** New York: T.W. Strong, 1855, 108 pp. Appears to be first book-length treatment on the subject in English. "Narrative of the laying of the cable between Port au Basque and North Sydney, 1855; the writer was a member of the expedition. He was also on board the U.S. steam frigate *Niagara* as secretary to Prof. Morse and afterwards to Cyrus W. Field, while the cables of 1857 and 1858 were being laid" (Weaver, 1-023, I:1329). Engravings.

6-170 Allen, Thomas. **OCEAN TELEGRAPHY.** London: n.p., 1857, 26 pp. "The author's system of light cables for submarine telegraphy; telegraphic communication with America" (Weaver, 1-023, I:1377).

6-171 * [Mann, Robert J.] **THE ATLANTIC TELEGRAPH: A HISTORY OF PRELIMINARY EXPERIMENTAL PROCEEDINGS AND A DESCRIPTIVE ACCOUNT OF THE PRESENT STATE AND PROSPECTS OF THE UNDERTAKING.** London: Jarrold & Sons, 1857, 69 pp. First booklet on the Atlantic project. "Mechanical and electrical difficulties of making, laying and working a cable under the Atlantic; remarks on Maury's telegraphic plateau; induction coil used in transmitting signals through cables" (Weaver, 1-023, I:1396).

6-172 Window, Frederick Richard. **ON SUBMARINE ELECTRIC TELEGRAPHS.** London: Institute of Civil Engineers, 1857, 40 pp. "Review and discussion of results, principally from an engineering point of view" (Weaver, 1-023, I:1402).

6-173 Allan, Thomas. **ALLAN'S SYSTEMS OF INLAND AND SUBMARINE TELEGRAPHY.** London: n.p., 1858, 79 pp. "Short communications to various papers on telegraphic matters from 1853–1858" (Weaver, 1-023, I:1402bis).

6-174 * Brett, John W. **ON THE ORIGIN AND PROGRESS OF THE OCEAN ELECTRIC TELEGRAPH: WITH A FEW BRIEF FACTS AND THE OPINION OF THE PRESS.** London: W.S. Johnson, 1858, 104 pp.; later edition in same year, with revised title, 175 pp. "History of the author's pioneer efforts in submarine telegraphy with original documents" (Weaver, 1-023, I:1411,1412).

6-175 THE CRUISE OF THE "AGAMEMNON" EXTRACTED FROM THE "TIMES" NEWSPAPER. Plymouth: printed for private circulation, July 1858, 43 pp. Stories concerning the 1857–58 expeditions.

6-176 ** Briggs, Charles F., and Augustus Maverick. **THE STORY OF THE TELEGRAPH, AND A HISTORY OF THE GREAT ATLANTIC CABLE; A COMPLETE RECORD OF THE INCEPTION, PROGRESS, AND FINAL SUCCESS OF THAT UNDERTAKING. A GENERAL HISTORY OF LAND AND OCEANIC TELEGRAPHS. DESCRIPTIONS OF TELEGRAPHIC APPARATUS AND BIOGRAPHICAL SKETCHES OF THE PRINCIPAL PERSONS CONNECTED WITH THE GREAT WORK.** New York: Rudd and Carleton, 1858, 255 pp. Origins, finance, science, and history of the Atlantic telegraph, cable construction and experiments, expeditions of 1857 and 1858. Engravings, folding map, no index.

Telegraphy

6-177 Longbridge, J[ames] A., and C.H. Brooks. **ON THE SUBMERGING OF TELEGRAPHIC CABLES**. London: Institute of Civil Engineers, 1858, 43 pp. "Mathematical discussion" (Weaver, 1-023, I:1421).

6-178 * Mullaly, John. **THE LAYING OF THE CABLE, OR THE OCEAN TELEGRAPH: BEING A COMPLETE AND AUTHENTIC NARRATIVE OF THE ATTEMPT TO LAY THE CABLE ACROSS THE ENTRANCE TO THE GULF OF ST. LAWRENCE IN 1855 AND OF THE THREE ATLANTIC TELEGRAPH EXPEDITIONS OF 1857 AND 1858**. New York: D. Appleton, 1858, 329 pp. Just that.

6-179 Rowett, William **THE NEW SUBMARINE TELEGRAPH CABLE; THE REGULATION OF ITS SPECIFIC GRAVITY AND ITS TRUE CONSTRUCTION AND SUBMERSION EXPLAINED, SHOWING ITS EASY ADAPTATION TO DEEP SEA AS WELL AS TO SHALLOW WATERS, AND AT A GREAT DIMINUTION OF EXPENSE**. London, 1858, 43 pp. "Ocean currents and density of sea-water; difficulties of laying cables, specific gravity of cable" (Weaver, 1-023, I:1428).

6-180 "Sunnyside" (pseud.). **THE TRUE ALLIANCE; OR, THE HISTORY OF THE TRANSATLANTIC CABLE UNITING BRITAIN WITH AMERICA**. London, 1858, 48 pp. 1 map.

6-181 Van Rensselaer, Cortlandt. **SIGNALS FROM THE ATLANTIC CABLE**. Philadelphia: J.M. Wilson "Technological Pamphlets, Vol. 4, No. 1," 1858, 24 pp. A talk given in New Jersey in celebration of the completion of the first telegraph cable.

6-182 Whitehouse, Edward O.W. **THE ATLANTIC TELEGRAPH: THE RISE, PROGRESS, AND DEVELOPMENT OF ITS ELECTRICAL DEPARTMENT**. London: printed for private circulation by Bradbury and Evans, 1858, 28 pp. "Introduction of gutta-percha, effect of induction, cable troubles" (Weaver, 1-023, I:1433).

6-183 Allan, Thomas. **ALLAN'S SYSTEM OF NATIONAL TELEGRAPHIC COMMUNICATIONS**. London: n.p., 1859, 56 pp. "A series of letters written to different public men in 1858 on sub-oceanic telegraphy" (Weaver, 1-023, I:1435bis).

6-184 [West, Charles]. **THE STORY OF MY LIFE: BY THE SUBMARINE TELEGRAPH**. London: n.p., 1859, 96 pp. "Humorous production containing many curious facts" (Weaver, 1-023, I:1472).

6-185 Window, F[rederick] R. **THE ATLANTIC AND SOUTH ATLANTIC TELEGRAPHS**. London: Smith, Elder, 1859, 32 pp. "Some causes of the failure of the 1858 cable; construction and submergence of the cable" (Weaver, 1-023, I:1474).

6-186 Ansted, D[avid] T. **THE BOTTOM OF THE ATLANTIC AND THE FIRST LAYING OF THE ELECTRIC TELEGRAPH CABLE**. St. Peter Port, Guernsey: F. Le Lievre, 1860, 24 pp. Lecture delivered in November 1858, reprinted from *THE WESTMINSTER REVIEW*.

6-187 Robinson, John. THE ATLANTIC OCEAN TELEGRAPH FROM IRELAND TO NEWFOUNDLAND; OR, THE NORTH ATLANTIC LINE VIA FAROE ISLANDS, ICELAND, GREENLAND AND LABRADOR. London: n.p., 1860, 36 pp. "Difficulties attending the laying and working of an Atlantic cable" (Weaver, 1-023, I:1499).

6-188 Clark, [Josiah] Latimer. EXPERIMENTAL INVESTIGATION OF THE LAWS WHICH GOVERN THE PROPAGATION OF THE ELECTRIC CURRENT IN LONG SUBMARINE CABLES. London: reprinted from Government Report on Submarine Cables, 1861, 48 pp. "Phenomena due to the passage of a current through submarine cables chiefly treated; retardation of signals" (Weaver, 1-023, I:1509).

6-189 Saward, George. DEEP SEA TELEGRAPHS: THEIR PAST HISTORY AND FUTURE PROGRESS. London: *Mechanics Magazine*, 1861, 48 pp.

6-190 Sharpe, Benjamin. A TREATISE ON THE CONSTRUCTION AND SUBMERSION OF DEEP-SEA ELECTRIC TELEGRAPH CABLES. London: n.p., 1861, 16 pp.

6-191 * Preece, Sir William Henry. ON THE MAINTENANCE AND DURABILITY OF SUBMARINE CABLES IN SHALLOW WATERS. WITH AN ABSTRACT OF THE DISCUSSION UPON THE PAPER BY CHARLES MANBY AND JAMES FORREST. London: Institute of Civil Engineers, 1862, 124 pp. Reprinted from ICE's *PROCEEDINGS*, this includes "remarks by Siemens, Latimer Clark, Sir Charles Bright, Willoughby Smith on the durability of gutta percha and the failure of cables" (Weaver, 1-023, I:1496).

6-192 Marcoartu, Arthuro de. TELEGRAPHIC SUBMARINE LINES BETWEEN EUROPE AND AMERICA AND THE ATLANTIC AND PACIFIC. New York: n.p., 1863, 53 pp. Translation from the French with description of existing and projected routes.

6-193 McClenachan, C.T. DETAILED REPORT OF THE SUCCESSFUL LAYING OF THE ATLANTIC TELEGRAPH. New York: Edmund Jones, 1863, 282 pp.

6-194 Piggott, W[illiam] P. ON THE IMPORTANCE OF OCEAN TELEGRAPHY; THE IMPEDIMENTS TO ITS SUCCESS, AND THE WAY TO OBVIATE THEM. London: n.p., 1863, 16 pp.

6-195 THE ATLANTIC TELEGRAPH; ITS HISTORY FROM THE COMMENCEMENT OF THE UNDERTAKING IN 1854, TO THE RETURN OF THE GREAT EASTERN IN 1865. London: n.p., 1865, 117 pp. Photos, maps.

6-196 Rowett, [William]. OCEAN TELEGRAPH CABLE; ITS CONSTRUCTION, THE REGULATION OF ITS SPECIFIC GRAVITY, AND SUBMERSION EXPLAINED. London: n.p., 1865, 122 pp. "Generalities about submarine cables and ocean-depths; preference given to the 'hempen' cable" (Weaver, 1-023, I:1621). With a map and illustrations.

Telegraphy

6-197 ** Russell, William H. **THE ATLANTIC TELEGRAPH**. London: Day & Son, 1865 (reprinted by David & Charles and Naval Institute Press, 1972), 117 pp. An authentic account and a basic source including laying of the first cable from Valentia to Hearts Content in 1858, the disasters of this and following cables, the various attempts, and the role of the *Great Eastern* as a special cable-laying steamship. The plates (tinted lithographs in the original edition) illustrate ships and cable-laying scenes, ship machinery both on deck and within the cable holds, and shore stations. Appendices (no index).

6-198 Van Choate, S.F. **OCEAN TELEGRAPHING: ADAPTATION OF NEW PRINCIPLES FOR THE SUCCESSFUL WORKING OF SUBMARINE CABLES**. Cambridge, MA: Riverside Press, 1865, 41 pp. Folding map.

6-199 THE ATLANTIC TELEGRAPH: ITS HISTORY FROM THE COMMENCEMENT OF THE UNDERTAKING IN 1854 TO THE SAILING OF THE "GREAT EASTERN" IN 1866. London: Bacom, 1866, 116 pp.

6-200 Bright, Sir Charles Tilston. **THE TELEGRAPH TO INDIA AND ITS EXTENSION TO AUSTRALIA AND CHINA**. London: n.p., 1866, 66 pp. "...information on the durability of insulating materials, on the manufacture and failure of cables and other related subjects" (Weaver, 1-023, I:1639).

6-201 *** Field, Henry M. **HISTORY OF THE ATLANTIC TELEGRAPH**. New York: Scribner's, 1866 (reprinted by Books for Libraries, 1972), 364 pp.; 1867 [2nd ed., adding further chapters plus some rewriting], 438 pp.; 1869 [3rd ed., altering some of the appendices], 437 pp.; 1892 and 1893 (4th ed., reprinted by Arno Press "Technology and Society," 1972) 415 pp. A detailed account, by Cyrus Field's brother, from cable attempts of 1850 to the successful cable of 1866. Illustrations (no index). A private-distributed edition (1866, 329 pp.), carrying the story through the 1865 expedition, preceded the editions noted above.

6-202 Macintosh, John. **FACTS AND FIGURES RELATIVE TO SUBMARINE TELEGRAPHY**. London: n.p., 1866, 36 pp. "The commercial features of submarine telegraphy" (Weaver, 1-023, I:1649).

6-203 Thomson, Sir William [Lord Kelvin]. **ATLANTIC TELEGRAPH CABLE**. London: William Brown, 1866, 31 pp. An address before the Royal Society of Edinburgh including many appended documents (speech itself is 11 pp.) on "velocity of settling, angle of immersion, method of grappling for submarine cables, [and] the forces concerned in laying and lifting a deep-sea cable" (Weaver, 1-023, I:1656, 1657).

6-204 Bright, Edward B. **THE ATLANTIC TELEGRAPHS. [A DESCRIPTION OF THE MANUFACTURE, LAYING AND WORKING OF THE CABLES OF 1865 AND 1866; WITH AN EXPLANATION OF THE MEANS EMPLOYED FOR RECOVERING THE LOST CABLE OF 1865]**. Liverpool, England: T. Brakell, 1867, 20 pp. Two plates.

6-205 Griscom, George. **THE ELECTRICAL CABLE; HISTORICAL VIEW OF THE ART OF ELECTROMAGNETIC TELEGRAPHY IN CONNEC-**

TION WITH THE TELEGRAPH CABLE AND ITS INSULATION BY GUTTA PERCHA. Philadelphia: King & Baird, 1867, 40 pp.

6-206 Morat, A.J. de, and J.N. Peirce. **AN EXPOSITION ON THE MOST IMPROVED TELEGRAPH CABLE AND THE THEORIES CONNECTED THEREWITH; WITH TABLES OF COMPARISON AND LISTS OF SUBMARINE CABLES NOW IN USE AND THOSE THAT HAVE FAILED.** Philadelphia: n.p., 1867, 19 pp.

6-207 * Parkinson, J.C. **THE OCEAN TELEGRAPH TO INDIA: A NARRATIVE AND A DIARY.** Edinburgh: William Blackwood, 1870, 328 pp. Contemporary account of the *Great Eastern*'s laying of cable along the Red Sea and across the Arabian Sea to India. Illustrations, maps, diagrams, appendices.

6-208 Blundell, J. Wagstaff. **MANUAL OF SUBMARINE TELEGRAPH COMPANIES.** London: Blundell, 1871; 1872 [2nd ed.].

6-209 Simonton, J.W., and Cyrus Field. **ATLANTIC CABLE MISMANAGEMENT.** New York: privately printed, 1871, 24 pp.

See Hoskiaer, 6-061.

6-210 Saward, George. **THE TRANS-ATLANTIC SUBMARINE TELEGRAPH; A BRIEF NARRATIVE OF THE PRINCIPAL INCIDENTS IN THE HISTORY OF THE ATLANTIC TELEGRAPH COMPANY, COMPILED FROM AUTHENTIC AND OFFICIAL DOCUMENTS BY THE LATE G. SAWARD.** London: privately printed, 1878, 80 pp.

See THE STORY OF CYRUS FIELD, 5-123.

6-211 Smith, Willoughby. **WORKING OF LONG SUBMARINE CABLES.** London: n.p., 1879, 45 pp. Paper read before Society of Telegraph Engineers.

6-212 Field, Cyrus W. **OCEAN TELEGRAPHY: THE 25th ANNIVERSARY OF THE ORGANIZATION OF THE FIRST COMPANY EVER FORMED TO LAY AN OCEAN CABLE.** New York: privately printed, 1879, 64 p.

See Fahie, 6-072.

6-213 Jamieson, A[andrew]. **LAYING AND REPAIRING SUBMARINE TELEGRAPH CABLES.** Glasgow: n.p., 1881, 45 pp. "Popular lecture on submarine cables with map and illustrations" (Weaver, 1-023, I:2236).

6-214 Smith, Willoughby A. **RÉSUMÉ OF THE EARLIER DAYS OF ELECTRIC TELEGRAPHY.** London: Hayman Bros, 1881, 56 pp. Concerns submarine telegraphy. See also author's autobiographical book, 6-216.

6-215 Ballantyne, R.M. **THE BATTERY AND THE BOILER; OR, ADVENTURES IN THE LAYING OF SUBMARINE ELECTRIC CABLES.** Toronto: 1882, 420 pp. Author was an electrician who worked on the 1865 Atlantic cable and the subsequent cables to India. Illustrations.

Telegraphy 149

6-216 ** Smith, Willoughby A. **THE RISE AND EXTENSION OF SUBMARINE TELEGRAPHY**. London: J.S. Virtue, 1891 (reprinted by Arno Press "Telecommunications," 1974), 390 pp. Largely autobiographical work covering the author's career from 1850 to the 1870s. Engravings, tables, appendices (no index).

6-217 Wilkinson, Henry D. **SUBMARINE CABLE-LAYING AND REPAIRING**. London: The Electrician, 1896, 406 pp.; New York: Van Nostrand, 1908 [2nd ed.], 557 pp. Diagrams.

6-218 *** Bright, Charles. **SUBMARINE TELEGRAPHS: THEIR HISTORY, CONSTRUCTION AND WORKING**. London: Crosby Lockwood, 1898 (reprinted by Arno Press "Telecommunications," 1974), 744 pp. Based in part on a revision of a French work (E. Wunschendorff, TRAITE DE TELEGRAPHIE SOUS-MARINE, 1888) this is the epitome of 19th-century cable books—by far the most complete. The first 200 pages focus on history, about 300 deal with construction (insulation, jointing, mechanical protection and strength, and completed cable), and the final 200 with operations (transmission of signals, signaling apparatus, duplex telegraphy, automatic machine transmission, and recent developments—including wireless). An essential book. Engravings, diagrams, maps (several foldout), index.

See Bright, 5-076.

6-219 Bright, Charles. **THE STORY OF THE ATLANTIC CABLE**. London: George Newnes / New York: Appleton, 1903, 220 pp. Evolution of Atlantic telegraphy, making and laying the lines, commercial operation, and cable systems of today. Photos, no index.

6-220 Johnson, George, ed. **THE ALL RED LINE: THE ANNALS AND AIMS OF THE PACIFIC CABLE**. Ottawa: James Hope/ London: Edward Stanford, 1903, 486 pp. Includes a 33-page bibliography.

F. Modern Histories of Submarine Telegraphy and Telephony

(See also chapter 7-F for works focused on submarine telephone cables.)

6-221 * Barnes, C[yril] C. **SUBMARINE TELECOMMUNICATION AND POWER CABLES**. Stevanage, England: Peter Peregrinus for the IEE, "IEE Monograph No. 20," 1977, 206 pp. History of technology involved in cable development for both types of uses. 137 references, appendices, bibliography (in chronological order), index.

6-222 ** Clarke, Arthur C. **VOICE ACROSS THE SEA**. London: Muller; New York: Harper, 1958, 220 pp.; William Luscombe, 1974 [2nd ed.], 228 pp. General account of submarine cables and communications from the early telegraph days. Second edition adds later cable developments and two chapters on satellite communication. Photos, index. See also Clarke, 3-027.

6-223 Clayton, Howard. **ATLANTIC BRIDGEHEAD: THE STORY OF TRANSATLANTIC COMMUNICATIONS**. London: Garnstone Press, 1968, 192 pp. Part 1 of 65 pages deals with the Atlantic telegraph projects while Part 3 (15 pages) discusses Marconi's wireless work—the remainder deals with sea and air transport. Photos, bibliography, index.

6-224 *** Coates, Vary T., and Bernard S. Finn. **A RETROSPECTIVE TECHNOLOGY ASSESSMENT: SUBMARINE TELEGRAPHY—THE TRANSATLANTIC CABLE OF 1866**. San Francisco: San Francisco Press, 1979, 264 pp. Unique and insightful analysis of the many and varied impacts of the first successful transatlantic telegraph cable. Based on careful analysis of contemporary resources. Tables, charts (no index).

6-225 de Cogan, Donard. "Dr. E.O.W. Whitehouse and the 1858 Trans-Atlantic Cable," *HISTORY OF TECHNOLOGY*, 10:1–15 (1985). Argues (based on tests with existing cable segments) that if the cable had been manufactured more carefully, it might well have survived. References.

6-226 de Giuli, Italo. **SUBMARINE TELEGRAPHY**. London: Pitman, 1932, 225 pp. Useful for its survey of 20th-century developments. Photos, diagrams, index.

6-227 * Dibner, Bern. **THE ATLANTIC CABLE**. Norwalk, CT: Burndy Library Publication No. 16, 1959 (reprinted by Blaisdell, 1964), 96 pp. Elegant assessment of the initial telegraph cables. Diagrams, foldout map, notes, bibliography.

6-228 Elwood, Tony. **SHIPS OF THE LINE: A HISTORY OF CABLESHIPS**. London: British Telecommunications, 1986, 30 pp. Details British cable-laying ships over history. Photos, maps, diagrams, bibliography.

6-229 * Finn, Bernard S. **SUBMARINE TELEGRAPHY: THE GRAND VICTORIAN TECHNOLOGY**. London: HMSO (Science Museum), 1973, 48 pp. A brief account by a noted authority. Photos, diagrams, maps, bibliography.

6-230 *** _____, ed. **DEVELOPMENT OF SUBMARINE CABLE COMMUNICATIONS**. New York: Arno Press "Historical Studies in Telecommunications," 1980, 2 vols., ca 800 pp. Invaluable anthology of historical documents, chapters, and articles (the latter usually reproduced in whole) from 1858 to the early 1970s. Vol. 1 reprints more than 30 selections, chiefly from the 19th century. Vol. 2 focuses primarily on the 20th century with another 26 papers. Essential material. Diagrams, photos, maps, notes, references.

6-231 * Garnham, S.A., and Robert L. Hadfield. **THE SUBMARINE CABLE: THE STORY OF THE SUBMARINE TELEGRAPH CABLE FROM ITS INVENTION DOWN TO MODERN TIMES**. London: Sampson, Low, Marston, nd but 1934, 242 pp. Cable history, cables and systems, cable ships, operations and services. Photos, diagrams, notes, index.

6-232 * Garratt, Gerald R.M. **ONE HUNDRED YEARS OF SUBMARINE CABLES**. London: HMSO, 1950, 59 pp. Almost wholly on British achievements, this Science Museum guide covers the early years of land telegraphy and cables, initial Atlantic cables, gutta-percha submarine cables, growth of the long-distance cable network from 1870 to 1920, the British network from 1920 to 1950, development of cables for submarine telephony, and high-speed telegraphy. Photos, maps, bibliography.

6-233 Haglund, H.H. "Atlantic Telegraph Cable Centennial," *WESTERN UNION TECHNICAL REVIEW* (July 1958), pp. 81–102. The ultimately unsuccessful 1858 cable is remembered. Photos, bibliography. Reprinted in Finn, 6-230.

6-234 *** Haigh, K.R. **CABLESHIPS AND SUBMARINE CABLES.** London: Adlard Coles / Washington: United States Underseas Cable Corp., 1968, 418 pp.; London: Standard Cables Ltd., 1978 [rev. ed.], 454 pp. History of the vessels used from 1850 to the present, arranged by company in approximate chronological order. Photos, maps (one folding), index, color chart of house flags. See also Schenck, 6-242.

6-235 Hempstead, Colin A. "The Early Years of Oceanic Telegraphy: Technology, Science, and Politics," *PROCEEDINGS OF THE IEE*, Part A 136:297–305 (November 1989). Focuses on the period 1857–64. Notes, diagrams, references, bibliography.

6-236 _____. "Representations of Transatlantic Telegraphy," *ENGINEERING SCIENCE AND EDUCATION JOURNAL*, 4:S17–S25 (December 1995). In prose, poetry, and pictures in popular magazines—how the early cables were depicted.

6-237 Hunt, Bruce J. "Michael Faraday, Cable Telegraphy, and the Rise of Field Theory," *HISTORY OF TECHNOLOGY*, 13:1–19 (1991).

6-238 Kingsbury, H., and R.A. Goodman. "Methods and Equipment in Cable Telegraphy," *JOURNAL OF THE IEE*, 70:477–521 (May 1932). Diagrams, discussion. Reprinted in Finn, 6-230.

6-239 Lynch, A.C. "Four Pioneer Deep-Sea Cables," *ENGINEERING SCIENCE AND EDUCATION JOURNAL*, 4:S26–S32 (December 1995). The telegraph cables of 1858 and 1861 and the telephone cables of 1956 and 1961 compared and contrasted.

6-240 Merritt, John. **THREE MILES DEEP: THE STORY OF THE TRANS-ATLANTIC CABLES.** London: Hamish Hamilton, 1958, 191 pp.

6-241 Moyer, Claire B. **OCEAN CABLE LORE.** New York: Heath Cote Publishing, 1974, 58 pp. Though brief, this includes useful material on cables, cable ships, cable terminals, and the relation between satellite and cable communications. Photos, glossary, bibliography.

6-242 ** Schenck, Herbert H. **CABLESHIP CHARACTERISTICS.** Washington: Underseas Cable Engineers Inc., 1978. 287 pp. Unique compilation prepared for the U.S. Navy Sea Systems Command which consists of an 80-page general discussion of the cover topic, and a 150-page directory of cable ship data and statistics. Photos, diagrams, tables, appendices, glossary, bibliography, index. See also Haigh, 6-234.

6-243 Scott, R. Bruce. **GENTLEMEN ON IMPERIAL SERVICE: A STORY OF THE TRANS-PACIFIC TELECOMMUNICATIONS CABLE TOLD IN THEIR OWN WORDS BY THOSE WHO SERVED.** Victoria, BC: Sono Nis Press, 1994, 131 pp. Concerns the Pacific Cable Board for which the author worked for four decades. Photos, map, bibliography, index.

6-244 Not used.

6-245 Wood, K.L. "Empire Telegraph Communications," *JOURNAL OF THE IEE*, 84:638–671 (1932). Diagrams, maps. Reprinted in Finn, 6-230.

G. Vintage Telegraphy Collectibles

6-246 French, Tom. INTRODUCTION TO KEY COLLECTING. Maynard, MA: Artifax Books, 1990, 64 pp. Telegraph sending key devices are illustrated and described.

6-247 * Harris, Robert Dalton. THE ELECTRIC MEDIUM: A PATTERN FOR THE EARLY DEVELOPMENT OF THE ELECTRIC TELEGRAPH IN THE UNITED STATES, SIGNIFIED BY A COLLECTION OF TELEGRAPH DELIVERY ENVELOPES. Wynantskill, NY: Author, nd, 149 pp. Offprint from *American Stampless Cover Catalog, Volume III*, this offers valuable insight into dozens of pioneering companies and routes as seen in a collection of remaining telegram delivery envelopes, many of which are pictured. Photos, maps.

6-248 ** Harris, Robert Dalton, and Diane DeBlois. THE TRANSCENDENT CABLE: AN ATLANTIC TELEGRAPH. Schoharie, NY: The Ephemera Society of America, 1994, 80 pp. Fascinating annotated collection of paper materials produced in honor of the five 1857–66 Atlantic telegraph cable expeditions. Each series of voyages is separately covered and provides a very good sense of how contemporary society viewed the technology. References.

6-249 Ingram, D. KEYS, KEYS, KEYS. Hicksville, NY: CQ Communications. Guide to collecting telegraph sending keys.

6-250 Neal, W. Keith. SEARCHING FOR RAILWAY TELEGRAPH INSULATORS. Signal Box Press, 1982. Focuses on British examples.

6-251 Perera, Tom. TELEGRAPH COLLECTOR'S GUIDE. Maynard, MA: Artifax Books, 1998, 80 pp.; 1999 [2nd ed.], 100 pp. Hundreds of photos and supporting text on land-line telegraph equipment, including automatic keys, wireless and spark keys, and foreign apparatus.

Chapter 7

Telephony

The literature on the telephone never achieved the volume of that concerning the telegraph—with the singular exception of regional or city telephone histories. Perhaps all the extended excitement concerning the first wired telecommunications service took the wind out of the sails of technical authors (many of whom wrote on both wired systems in the late 19th century) and their readers.

Paralleling the previous chapter on telegraphy, Section A is an inventory of contemporary works on the telephone for the first three decades of the innovation's development. This is followed by: a brief survey of useful historical works from the 1908–40 period (B); modern histories of American telephony (C); regional American telephone histories (D); foreign systems, divided into subsections on Europe and General, and Britain and the Commonwealth (E); submarine cable telephony (F); video telephones and teleconferencing (G); and guides for collectors of old telephone equipment (H). For related Internet resources, see 15-F.

A. Contemporary Books on Telephony (1877–1907)

(Arranged chronologically, the first three decades of contemporary telephone publications often provide some historical material—as well as valuable engravings of apparatus and installations. Excluded are works devoted to legal or purely business matters. Biographies of Bell, Grey, and others are found in chapter 5-C. Quotations are from cited entries in Weaver, 1-023.)

7-001 *** Dolbear, Amos E. **THE TELEPHONE: AN ACCOUNT OF THE PHENOMENA OF ELECTRICITY, MAGNETISM, AND SOUND, AS INVOLVED IN ITS ACTION, WITH DIRECTIONS FOR MAKING A SPEAKING TELEPHONE**. Boston: Lee and Shepard/New York: Charles

Dillingham/London: Trubner & Co., 1877, 128 pp. First book devoted to the telephone. Early history of electricity and magnetism, theories of electricity, sound, resonance, tone composition, history of early telephones, the author's telephone, and how to make a telephone instrument. Diagrams.

7-002 * ALL ABOUT THE TELEPHONE AND PHONOGRAPH: CONTAINING DESCRIPTIONS OF BELL'S AND DOLBEAR'S TELEPHONES AND EDISON'S PHONOGRAPH; HISTORY OF THE DISCOVER, DETAILS OF CONSTRUCTION AND INTERESTING EXPERIMENTS. London: n.p., 1878, 99 pp. A popular account.

7-003 * Field, Kate, ed. **THE HISTORY OF BELL'S TELEPHONE.** London: Bradbury, Agnew, 1878, 67 pp. The first general description, oddly enough published in London.

7-004 Garner, Samuel. **THE TELEPHONE: ITS HISTORY, CONSTRUCTION, PRINCIPLES AND USES WITH DEFINITE INSTRUCTIONS ON THE MAKING OF TELEPHONES BY WHICH FAILURE IS IMPOSSIBLE AND TO WHICH IS ADDED A CHAPTER ON THE PHONOGRAPH.** London: Brighton, 1878, 32 pp. Author was a science teacher.

7-005 * Gray, Elisha. **EXPERIMENTAL RESEARCHES IN ELECTROHARMONIC TELEGRAPHY AND TELEPHONY, 1867–1878.** New York: Russell Bros., 1878, 96 pp. About the author's work on the transmission of musical tones and other sounds, with a brief autobiography. Reprinted in Shiers, 7-091.

7-006 ** Prescott, George B. **THE SPEAKING TELEPHONE, TALKING PHONO-GRAPH AND OTHER NOVELTIES.** New York: D. Appleton / London: F & E.N. Spon, 1878, 431 pp.; enlarged 2nd ed. published as **THE SPEAKING TELEPHONE, ELECTRIC LIGHT, AND OTHER RECENT INVENTIONS.** 1879, 616 pp. Very detailed survey of the technology of the time, including apparatus and operations. Engravings, index.

7-007 Not used.

7-008 THE TELEPHONE: ITS WONDERS AND LESSONS, A VOICE FROM THE MILLION. London: n.p., 1878, 31 pp. Includes a preface by A.G. Bell.

7-009 Not used.

7-010 ** Du Moncel, Theodore A.L. **THE TELEPHONE, THE MICROPHONE, AND THE PHONOGRAPH.** New York: Harper, 1879, 363 pp. Translated from the French with additions and corrections by the author, this is the first work on the microphone and an early work on the telephone and phonograph. Diagrams.

7-011 Walters, Henry L. **THE PRACTICAL TELEPHONE: A SHORT HISTORICAL SKETCH.** London: publisher not given, 1878. Sources differ: second word in title may be "POLITICAL."

Telephony

7-012 THE TELEPHONE SERVICE IN 1882 IN THE UNITED STATES AND THE WORLD AT LARGE. Worcester, MA: Washburn and Moen Manufacturing Co., 1882, 31 pp.

7-013 Bramwell, Sir Frederick J. **TALKING BY ELECTRICITY: TELEPHONES...** Southport, England: William Milne, 1883, 32 pp. "A lecture on the telephone including the phonograph and photophone" (Weaver, 1-023, I:2350).

7-014 Lockwood, T[homas] D. **PRACTICAL INFORMATION FOR TELEPHONISTS.** New York: W.J. Johnson, 1884, 192 pp. Includes roles of telephone inspectors and repair people in addition to reviewing equipment, especially for a small system.

7-015 ** Prescott, George B. **BELL'S ELECTRIC SPEAKING TELEPHONE: ITS INVENTION, CONSTRUCTION, APPLICATION, MODIFICATION AND HISTORY.** New York: D. Appleton, 1884 (reprinted by Arno Press "Telecommunications," 1972), 526 pp.; revised as **THE ELECTRIC TELEPHONE**, 1900 [2nd ed.], 795 pp. Discusses work of Bell and others with several chapters on Edison's telephone research. Some 330 engravings (several foldouts), increased to 516 in 2nd ed., index.

7-016 ** THE TELEPHONE CASES: CASES ADJUDGED IN THE SUPREME COURT AT OCTOBER TERM, 1887. New York: Banks & Bros. "United States Supreme Court Reports, Vol. 126," 1888, 531 pp. The only official volume of Supreme Court decisions devoted to a single issue (and 10 interrelated cases), this includes much of the technical reasoning and legal findings behind key court decisions concerning Bell's primacy as inventor of the telephone. Diagrams, notes. See also Bell, 7-047.

7-017 * Preece, Sir William H., and Julius Maier. **THE TELEPHONE.** London: Whittaker "The Specialists Series," 1889, 489 pp. Useful for its comparative description of American, British, and European devices and practices. Final chapters on applications. Engravings, tables, references, index.

7-018 ** Poole, Joseph. **THE PRACTICAL TELEPHONE HANDBOOK AND GUIDE TO THE TELEPHONE EXCHANGE.** London: Whittaker, 1891, 288 pp.; 1895 [2nd ed.], 347 pp.; 1906 [3rd ed.], 533 pp.; 1910 [4th ed.], 606 pp.; 1912 [5th ed.], 624 pp.; 1919 [6th ed.], 725 pp.; 1927 [7th ed.], 870 pp. Arranged by type of equipment. Revised by N.V. Knight and W. Prickett as **THE TELEPHONE HANDBOOK.** London: Pitman, 1942, 510 pp. A basic text, especially valuable as its various editions show how British telephone equipment evolved. Engravings, diagrams, appendix, index.

7-019 Hopkins, William J. **TELEPHONE LINES AND THEIR PROPERTIES.** New York: Longmans, Green, 1893, 258 pp.; 1894 [2nd ed.], 272 pp. 1901 [6th ed.], 307 pp. Plates, diagrams.

7-020 Preece, William H., and Arthur J. Stubbs. **A MANUAL OF TELEPHONY.** London: Whittaker / New York: Macmillan, 1893, 508 pp. Hundreds of engravings.

7-021 Not used.

See Houston, 6-083.

7-022 Webb, Herbert Laws. **THE TELEPHONE HAND-BOOK**. Chicago: The Electrician, 1894, 146 pp.; 1902 [2nd ed.], 160 pp.

See Bennett, 7-115.

7-023 Houston, Edwin J., and Arthur E. Kennelly. **THE ELECTRIC TELEPHONE**. New York: Johnstone, 1896, 412 pp.; Electrical World & Engineer, 1902 [2nd ed.] 453 pp. Includes all major types of equipment, including a final chapter on "radiophony" or wireless. Engravings, index.

7-024 Allsop, Frederick C. **TELEPHONES: THEIR CONSTRUCTION AND FITTING**. London and New York: Spon, 1897 [4th ed.], 256 pp.; 1901 [7th ed.], 184 pp.; 1909 [8th ed.], 222 pp. Long subtitle: "A practical treatise on the fitting-up and maintenance of telephones and the auxiliary apparatus." Includes 15 folding plates.

7-025 * **PATENTED TELEPHONY: A REVIEW OF THE PATENTS PERTAINING TO TELEPHONES AND TELEPHONE APPARATUS**. Chicago: American Electrical Engineering Assn., 1897, 102 pp. Issued just after the basic Bell patents had run out (1893–94) as a guide to what remained in force. Includes 18 foldout diagrams from key patents.

7-026 Bell, James, and S. Wilson. **PRACTICAL TELEPHONY**. London: *Electricity*, 1898, 288 pp. Based on articles in *ELECTRICITY*, with much on equipment and operations of the time. Illustrations.

7-027 Byng, M., and F.G. Bell. **A POPULAR GUIDE TO COMMERCIAL AND DOMESTIC TELEPHONY**. Schenectady, NY: General Electric, 1898, 152 pp. Contains 165 illustrations.

See Herbert, 7-134.

7-028 Hopkins, William J. **THE TELEPHONE: OUTLINES OF THE DEVELOPMENT OF TRANSMITTERS AND RECEIVERS**. New York: Longmans, Green, 1898, 83 pp. Includes both granular and electromagnetic instruments. Photos

7-029 Dobbs, A.E. **PRACTICAL FEATURES OF TELEPHONE WORK**. New York: The Electrical Engineer, 1899, 134 pp. Establishing and operating a small exchange system, with illustrated information on all types of equipment and troubleshooting. Illustrated.

7-030 ** Miller, Kempster B. **AMERICAN TELEPHONE PRACTICE**. New York: American Electrician Co., 1899, 458 pp.; 1900 [2nd ed.], 518 pp.; New York: McGraw, 1903 [3rd ed.], 518 pp.; 1905 [4th ed.], 888 pp. A standard operational reference book of the period, valuable historically for its detailed study of then-current operations. Photos, diagrams, tables, index. See also Miller, 7-059.

7-031 Edwards, Stanley R., and A.E. Dobbs. **THE INSPECTOR AND THE TROUBLE MAN**. Chicago: Telephony Publishing, 1900, 106 pp.; 1913 [2nd ed.]

196 pp. Question-and-answer format between an inspector and his student. Second edition adds S. Edwards as primary author. "A complete explanation in plain English of magneto exchanges, line construction, telephone troubles, and the theory of electricity as applied to telephony."

7-032 International Correspondence Schools. **A TREATISE ON TELEPHONY.** Scranton, PA: Colliery Engineering Co., 1900, 4 vols., 1,690 pp. Correspondence school course material including both Bell and independent systems. Diagrams, photos, index.

    VOL. 1: ENGINEERING PRINCIPLES.

    VOL. 2: TELEPHONE EQUIPMENT.

    VOL. 3: EXAMINATIONS.

    VOL. 4: TABLES AND GRAPHS.

7-033 Homans, James E. **ABC OF THE TELEPHONE: A PRACTICAL AND USEFUL TREATISE FOR STUDENTS AND WORKERS IN TELEPHONY, GIVING A REVIEW OF THE DEVELOPMENT OF THE INDUSTRY TO THE PRESENT DATE, AND FULL DESCRIPTIONS OF NUMEROUS VALUABLE INVENTIONS AND APPLIANCES, TOGETHER WITH VERY MANY ILLUSTRATIONS, DIAGRAMS, AND TABLES.** New York: Theo. Audel, 1901, 335 pp.; 1904 [2nd ed.], 349 pp. Especially useful for its details and illustrations of various makes of American equipment. Second edition includes a few pages on wireless. Photos, diagrams, index.

7-034 Baldwin, T.S. **PRACTICAL TELEPHONE HAND BOOK AND GUIDE TO TELEPHONIC EXCHANGE.** Chicago: Drake, 1902, 226 pp.

7-035 Williams, J.A. **MANUAL OF RURAL TELEPHONY.** Cleveland: Manual Publications, 1902, 177 pp. Planning, constructing, and operating a small system.

7-036 TELEPHONY. Chicago (?): International Library of Technology, 1902 [2nd ed.] 662 pp. Designed for correspondence course learning.

7-037 Boyrer, W.C. **TELEPHONY.** Chicago: American School of Correspondence, 1905, 448 pp. Volume 5 of Frank W. Gunsaulus, ed. MODERN ENGINEERING PRACTICE (1902–1908, 12 vols.). Engravings, photos, tables, diagrams, questions, index. Virtually identical to Volume 5 of CYCLOPEDIA OF APPLIED ELECTRICITY from same publisher, though latter has an index.

7-038 Abbott, Arthur V. **TELEPHONY: A MANUAL OF THE DESIGN, CONSTRUCTION AND OPERATION OF TELEPHONE EXCHANGES.** New York: McGraw, 1903–4, 6 vols., 1,473 pp. Hundreds of illustrations, index. Individual volumes:

    VOL. 1: THE LOCATION OF CENTRAL OFFICES. 150 pp.

    VOL. 2: THE CONSTRUCTION OF UNDERGROUND CONDUITS. 175 pp.

158 Chapter 7

VOL. 3: THE CONSTRUCTION OF CABLE PLANT. 153 pp.

VOL. 4: THE CONSTRUCTION OF AERIAL LINES. 259 pp.

VOL. 5: THE SUBSTATION. 465 pp.

VOL. 6: SWITCHBOARDS AND CENTRAL OFFICE. 271 pp.

7-039 Owen, Walter C. TELEPHONE LINES AND METHODS OF CONSTRUCTING THEM OVERHEAD AND UNDERGROUND. London: Whittaker, 1903, 390 pp. Diagrams, tables.

7-040 Webb, Herbert Laws. THE TELEPHONE SERVICE: ITS PAST, ITS PRESENT, AND ITS FUTURE. London: Whittaker, 1904, 118 pp.

7-041 Wilder, George W. TELEPHONE PRINCIPLES & PRACTICE. Madison, WI: Cantwell Printing, 1904, 445 pp.

See U.S. Department of Commerce, 1-069.

7-042 TELEPHONE VOLUME: ICS REFERENCE LIBRARY. Scranton, PA: International Correspondence Schools, 1907, 726 pp.

B. Later Telephony Works (1908–1950)
(This very selective listing of works emphasizes those offering substantial historical intent or value. Arrangement is alphabetical by author.)

7-043 Abbott, Henry H. "Sixty Years of PBX Development," **BELL LABORATORIES RECORD**, 46:8–16 (January 1968). Photos, bibliography of previous papers on PBX in the *RECORD*.

7-044 * Aitken, William. AUTOMATIC TELEPHONE SYSTEMS. London: Benn Bros. and "The Electrician," 1921–24, 3 vols. Very detailed survey of British and other practice, including many large folding diagrams.

VOL. 1: CIRCUITS AND APPARATUS AS USED IN THE PUBLIC SERVICES. 1921, 282 pp. Includes chapters on the Strowger, Siemens, Western Electric, and other systems.

VOL. 2: AUXILIARY SERVICES AND PRIVATE BRANCH EXCHANGES. 1923, 227 pp. Includes a brief supplement to Vol. one.

VOL. 3: LARGE MULTI-OFFICE AUTOMATIC SYSTEMS; SEMI-AUTOMATIC WORKING; MISCELLANEOUS SYSTEMS; LAYOUT AND WIRING; POWER PLANT; TRAFFIC. 1924, 339 pp. Includes pocket of diagrams inside rear cover.

7-045 ** _____. WHO INVENTED THE TELEPHONE? London: Blackie, 1939, 196 pp. A documented attack on Bell's primacy with extensive use of extracts from documents by Elisha Gray and others. An important—and relatively early—example of "revisionist" history in this field. Text references, diagrams, no index.

7-046 Angwin, Sir A. Stanley. "Graham Bell—Pioneer: An Era of Outstanding Developments in World Communications," *JOURNAL OF THE IEE,* 94 (Pt. 1):269–274 (June 1947). Despite biography-type title, this is a recapitulation of historical highlights and general discussion of telephone progress to 1943. Reprinted in Shiers, 7-091.

See Baldwin, 7-128.

7-047 *** Bell, Alexander G. **THE BELL TELEPHONE: THE DEPOSITION OF ALEXANDER GRAHAM BELL IN THE SUIT BROUGHT BY THE UNITED STATES TO ANNUL THE BELL PATENTS.** Boston: American Bell Telephone Co., 1908 (reprinted by Arno Press, "Telecommunications," 1974), 469 pp. The inventor's 1892 deposition in question and answer format—perhaps his definitive statement of what happened and when. See also THE TELEPHONE CASES, 7-016.

7-048 * Casson, Herbert N. **THE HISTORY OF THE TELEPHONE.** Chicago: McClurg, 1910 (reprinted by Books for Libraries, 1971), 315 pp. A popular and nontechnical account, one of the first, still useful for early details. Photos, index.

7-049 Gray, George H. "The Evolution of Wire Transmission," *ELECTRICAL COMMUNICATION,* 20:235–245 (1942). Review of telephone transmission, carrier and coaxial cable, with comments on speech quality, system reliability, and economics. Photos, charts, bibliography. Reprinted in Shiers, 7-091.

7-050 Gust, Friedrich Wilhelm. **THE BLIND TELEPHONIST: TECHNOLOGY AND ORGANIZATION OF A VOCATION FOR THE BLIND.** Munich: Siemens & Halske, 1957, 142 pp. Possibly unique item, translated from the German, outlining and illustrating history, equipment and approaches, vocational requirements, and examples of employing blind people in different parts of the telephone industry. Photos.

7-051 * Jewett, Frank B. "The Telephone Switchboard—Fifty Years of History," *BELL TELEPHONE QUARTERLY,* 7:149–165 (July 1928). Review from 1878 as to types and apparatus. Photos. Reprinted in Shiers, 7-091.

7-052 _____. "A Quarter Century of Transcontinental Telephony," *ELECTRICAL ENGINEERING,* 59:3–11 (January 1940). Photos, diagrams, maps. Reprinted in Shiers, 7-091. Traces the story from the first services in 1915.

7-053 *** Kingsbury, John E. **THE TELEPHONE AND TELEPHONE EXCHANGES: THEIR INVENTION AND DEVELOPMENT.** London: Longmans, Green, 1915 (reprinted by Arno Press "Technology and Society," 1972), 558 pp. An important and detailed technical history, with emphasis on British and European developments. Photos, diagrams, tables, references, appendices, index.

7-054 Langdon, William C. "Myths of Telephone History," *BELL TELEPHONE QUARTERLY*, 12:123–140 (April 1933). Deals primarily with those contesting Bell's primacy, based on AT&T archives. Reprinted in Shiers, 7-091.

7-055 Lawson, R. "A History of Automatic Telephony," *POST OFFICE ELECTRICAL ENGINEERS JOURNAL*, 5:192–207 (1912/13). Diagrams. Reprinted in Shiers, 7-091.

7-056 Not used.

7-057 McMeal, Harry B. **THE STORY OF INDEPENDENT TELEPHONY.** City not given: Independent Pioneer Telephone Assn., 1934, 289 pp. The "non-Bell" portion of the industry is described, chiefly from a business and service point of view. See also Pleasance, 7-088.

7-058 McMeen, Samuel G. **TELEPHONY: A COMPLETE AND DETAILED EXPOSITION OF THE THEORY AND PRACTICE OF THE TELEPHONE ART.** Chicago: American Technical Society, 1912, 1922 [rev. ed.], 950 pp. A "general historical and descriptive survey" with many photos and diagrams.

7-059 * Miller, Kempster. **TELEPHONE THEORY & PRACTICE.** New York: McGraw-Hill, 1930–33, 3 vols. A standard handbook, something of an update to the author's pioneering work (7-030). Useful for its portrayal of period practices and equipment. Photos, diagrams, index.

    VOL. 1: THEORY & ELEMENTS. 1930, 486 pp.

    VOL. 2: MANUAL SWITCHING AND SUBSTATION EQUIPMENT. 1933, 439 pp.

    VOL. 3: AUTOMATIC SWITCHING & AUXILIARY EQUIPMENT. 1933, 494 pp.

7-060 Moudin, R. **THE STROWGER AUTOMATIC TELEPHONE EXCHANGE.** London: E. & F.N. Spon / New York: Spon & Chamberlin, 1919, 186 pp. One of the earlier treatments of automatic telephony. Photos, diagrams (many foldouts), index.

7-061 *** Rhodes, Frederick L. **BEGINNINGS OF TELEPHONY.** New York: Harper, 1929 (reprinted by Arno Press "History of Telecommunications," 1974), 261 pp. A thorough study of progress in the United States up to the mid-1890s. The first 75 pp. concern Bell and his patents while the remainder deals with microphones, overhead lines, cables, loaded lines, switchboards, station apparatus, and long-distance lines. Much information on companies, people, patents and litigation, mechanical and electrical details, technical improvements, installations and operations, with many quotes and extracts from a variety of sources. Photos, diagrams, appendices (including lists of important lawsuits and patents), references, index.

7-062 Romnes, Haakon I. "Advances in Communications," *ELECTRICAL ENGINEERING*, 78:481–492 (May 1959). Progress in telephone technology over previous 25 years by a man who became AT&T's chairman.

7-063 ** Shaw, Thomas. "The Conquest of Distance by Wire Telephony," *BELL SYSTEM TECHNICAL JOURNAL*, 23:337–421 (October 1944). Story of transmission development from the early days of loading to the wide use of thermionic repeaters, concentrating on the Bell System. Organizational tables, notes, references, appendices. Reprinted in Shiers, 7-091.

7-064 Smith, Arthur Bessey. **THE EARLY HISTORY OF THE AUTOMATIC TELEPHONE.** Dublin, CA: Telephone Historical Institute, 1996, 86 pp. Collects and reprints a series of articles from about 1908 from an unidentified periodical which trace the largely American story, with a few mentions of developments elsewhere. Photos, diagrams.

7-065 * Smith, Arthur Bessey, and Wilson Lee Campbell. **AUTOMATIC TELEPHONY: A COMPREHENSIVE TREATISE ON AUTOMATIC AND SEMI-AUTOMATIC SYSTEMS.** New York: McGraw-Hill, 1914, 1921 [2nd ed.], 430 pp. Defines itself as the first book devoted to the topic. Covers both independent company and Western Electric apparatus. Photos, diagrams, index.

7-066 Taylor, Lloyd W. "The Untold Story of the Telephone," *AMERICAN JOURNAL OF PHYSICS*, 5:243–251 (December 1937). Review of little-known experiments, apparatus, and documents concerning the early work of Bell and Gray, particularly the latter's receivers. Photos, references. Reprinted in Shiers, 7-091.

7-067 Theiss, J. Bernhard, and Guy A. Joy. **TOLL TELEPHONE PRACTICE.** New York: Van Nostrand, 1912, 418 pp. Twenty-two chapters include background, operation in rural areas, toll switchboards and related topics. Photos, diagrams, index.

See U.S. Federal Communications Commission, 4-108.

7-068 * Watson, Thomas A. **THE BIRTH AND BABYHOOD OF THE TELEPHONE.** New York: AT&T, 1913, 1938 (and many later reprints), 47 pp. Bell's assistant looks back to the pioneering experiments in the 1870s. Initially given as an address before the Telephone Pioneers in 1913. Photos, maps. See also the same author's "How Bell Invented the Telephone," *PROCEEDINGS OF THE AIEE*, 34:1503–1513 (August 1915—reprinted in Shiers, 7-091); and *ELECTRICAL ENGINEERING*, 66:232–236 (March 1947) for variants.

C. Modern Histories of U.S. Telephony

7-069 * Boettinger, H.M. **THE TELEPHONE BOOK: BELL, WATSON, VAIL AND AMERICAN LIFE, 1876–1983.** Croton-on-Hudson, NY: Riverwood Publishers, 1977, 191 pp.; New York: Stearn Publishers, 1983 [2nd ed.], 220 pp. Lavish pictorial history (much in color) of the men (especially Bell and his assistant, plus early AT&T chief Theodore Vail) and appliances of telephony, especially those of AT&T. Photos (including foldouts).

7-070 Brittain, James E. "The Introduction of the Loading Coil: George A. Campbell and Michael I. Pupin," *TECHNOLOGY AND CULTURE*, 11:36–57 (January 1970). Extensively referenced study of an initial long-distance device.

7-071 Not used.

See Brooks, 4-097.

7-072 Brouwer, J.F. "Half a Century of 'Electrification' in Telephony Systems," *PHILLIPS TECHNICAL REVIEW*, 42:361–373 (September 1986).

See Chapuis, 7-116, 7-117.

7-073 * Coe, Lewis. **THE TELEPHONE AND ITS SEVERAL INVENTORS: A HISTORY**. Jefferson, NC: McFarland, 1995, 230 pp., photos, diagrams, appendices relating to telephone patents, bibliography, index. Parallels the author's earlier book on the telegraph (6-104) and updates, and to a degree refutes, Aitken's prewar study (7-045).

7-074 Collins, Arthur, and Robert Pederson. **A TIME FOR INNOVATION**. Dallas: Merle Collins Foundation, 1973, 184 pp. Argues for immediate introduction of digital technology into the telephone network.

7-075 "Diamond Jubilee: 1909–1984, Seventh-Fifth Anniversary," *TELEPHONE ENGINEER & MANAGEMENT*, 88:11:88–303 (June 1, 1984). More than a dozen articles emphasizing the independent telephone industry including foldout chart of equipment changes and trends.

7-076 Fischer, Claude S. "The Revolution in Rural Telephony, 1900–1920," *JOURNAL OF SOCIAL HISTORY*, 21:5-26 (Fall 1987). See Fischer, 7-079.

7-077 _____. "Technology's Retreat: The Decline of Rural Telephony in the United States, 1920-1940," *SOCIAL SCIENCE HISTORY*, 11:295–327 (Fall 1987). See Fischer, 7-079.

7-078 Fischer, Claude S., and Glenn R. Carroll. "Telephone and Automobile Diffusion in the United States, 1902–1937," *AMERICAN JOURNAL OF SOCIOLOGY*, 93:1153-1178 (March 1988). See Fischer, 7-079.

7-079 ** _____. **AMERICA CALLING: A SOCIAL HISTORY OF THE TELEPHONE TO 1940**. Berkeley: University of California Press, 1992, 424 pp. Important analysis of how the technology impacted American life before World War II, told through analysis of contemporary literature and studies of several California communities. The articles above are largely subsumed in this valuable study. Photos, charts, notes, bibliography, appendices, index.

7-080 Friedman, Herb. "The Blue Book and Ma Bell," *RADIO-ELECTRONICS*, 58:11:49–52 (November 1987). Diagrams.

7-081 Gabel, David. "Competition in a Network Industry: The Telephone Industry, 1894–1910," *JOURNAL OF ECONOMIC HISTORY*, 54:543–572 (September 1994). Discusses the period after Bell's patents entered the public domain.

Telephony

7-082 Green, Venus. "Goodbye Central: Automation and the Decline of 'Personal Service' in the Bell System, 1878–1921," *TECHNOLOGY AND CULTURE*, 36:912–949 (October 1995). Early development of automatic dialing.

7-083 Jordan, D.W. "The Adoption of Self-Inductance by Telephony, 1886–1889," *ANNALS OF SCIENCE*, 39:433–461 (September 1982). Innovation of an important technology in voice transmission.

7-084 Langdale, John V. "The Growth of Long-Distance Telephony in the Bell System: 1875–1907," *JOURNAL OF HISTORICAL GEOGRAPHY*, 4:145–159 (1978). Expansion of distance covered as technology improved. Maps, references.

7-085 Leon, George D. "Who Really Invented the Telephone?" *RADIO-ELECTRONICS*, 55:1:47–50 (January 1984). Yet another exploration of the old topic.

7-086 Lipartito, Kenneth J. "Component Innovation: The Case of Automatic Telephone Switching, 1891–1920," *INDUSTRIAL AND CORPORATE CHANGE*, 3:325–357 (1994).

7-087 * Mueller, Milton "The Switchboard Problem: Scale, Signaling, and Organization in Manual Telephone Switching, 1877–1897," *TECHNOLOGY AND CULTURE*, 30:534–560 (July 1989). An authority on early telephone development focuses on the central role of switchboard development in telephone expansion.

7-088 Pleasance, Charles A. **THE SPIRIT OF INDEPENDENT TELEPHONY**. Johnson City, TN: Independent Telephony Books, 1989, 304 pp. The only current overall history of the many non-Bell local exchanges. Notes, bibliography, index. See also McMeal, 7-057.

7-089 ** Pool, Ithiel de Sola, ed. **THE SOCIAL IMPACT OF THE TELEPHONE**. Cambridge, MA: MIT Press "Bicentennial Studies," 1977, 502 pp. Twenty-one original papers from a two-day seminar at MIT (see also THE TELEPHONE'S FIRST CENTURY, 7-094), these range over the instrument's technology and its cultural impact over its first century. Charts, notes, index.

7-090 *_____. **FORECASTING THE TELEPHONE: A RETROSPECTIVE TECHNOLOGY ASSESSMENT OF THE TELEPHONE**. Norwood, NJ: Ablex "Communication and Information Science," 1983, 162 pp. Reviews and assesses forecasts of the telephone's future and impact made from 1876 through 1940. References, index.

7-091 *** Shiers, George, ed. **THE TELEPHONE: AN HISTORICAL ANTHOLOGY**. New York: Arno Press "Historical Studies in Telecommunications," 1977, ca 600 pp. Invaluable collection reprints all or part of 23 historically important papers, many noted elsewhere in this chapter. Photos, diagrams, references.

7-092 "Special Issue: Telephone Centenary," *TELECOMMUNICATION JOURNAL*, 43:111:166–244 (March 1976). Illustrated survey of changes in the industry. Photos, tables, charts.

7-093 *TELECOM HISTORY.* Dublin, CA: Telephone History Institute, 1994, two issues, 116 and 164 pp. An attempt to develop an historical quarterly devoted to telephone equipment and operations, only these two issues appeared. Included are quite detailed and well-illustrated studies of telephone exchange service, and the development of the automatic telephone from 1889 to 1918.

7-094 THE TELEPHONE'S FIRST CENTURY—AND BEYOND. New York: Crowell, 1977, 119 pp. An AT&T-produced commemorative volume based on a two-day seminar at MIT, this includes five talks and summary papers. See also Pool (7-089) for remainder of the sessions.

7-095 ** Wasserman, Neil H. **FROM INVENTION TO INNOVATION: LONG-DISTANCE TELEPHONE TRANSMISSION AT THE TURN OF THE CENTURY.** Baltimore: Johns Hopkins University Press "AT&T Series in Telephone History," 1985, 160 pp. Study of the application of loading coils in long-distance voice transmission. Photos, diagrams, notes, appendices, index. For the rest of this series, see Garnet, 4-101; Lipartito, 7-106; and Smith, 4-106.

D. U.S. Regional Telephony Histories

(It seems virtually every local or regional telephone company has issued some commemoration of its own history. Most focus on people and events of course, but provide some useful insights into technology as well. What follows is a sample.)

7-096 Bendix, Ben. **SERVING HAWAII: THE FIRST 100 YEARS: THE 100th ANNIVERSARY OF THE HAWAIIAN TELEPHONE COMPANY.** Honolulu: Hawaiian Telephone Company, 1983, 88 pp. Largely pictorial review of what became a GTE subsidiary. See also Simonds, 7-111.

7-097 Boylan, John T. **TALES OF THE "NINETIES."** Cleveland, OH: Lakeside Press, 1952, 137 pp. The Cleveland Telephone Company's early years.

7-097a Chanecka, Stephen R. **HISTORY OF THE ROSEVILLE TELEPHONE COMPANY.** Roseville, CA: Roseville Telephone Company, 1997, 427 pp. Long history of one of the larger California independent exchanges. Photos, index.

7-098 Clement, Edwin A. **THE NORTH CAROLINA TELEPHONE STORY: THE FIRST NINETY-EIGHT YEARS.** Raleigh: North Carolina Independent Telephone Association, 1978, 223 pp. The independent (non-Bell System) operators.

7-099 Cooper, Dennis R. **THE PEOPLE MACHINE: AN ILLUSTRATED HISTORY OF THE TELEPHONE ON THE CENTRAL WEST COAST OF FLORIDA.** Tampa: General Telephone of Florida, 1971, 301 pp. Titles sums it up. Photos, chronology, index.

7-100 Cromwell, Joseph H. **THE C&P STORY: SERVICE IN ACTION— WASHINGTON, D.C.** Washington: C&P Telephone, 1981, 260 pp. One of four parallel volumes (the others cover Maryland, Virginia, and West Virginia). Photos.

7-101 THE FLIGHT OF SPEECH: A STORY OF FREE ENTERPRISE, 1897–1947, ISSUED ON THE OCCASION OF THE 50th ANNIVERSARY

OF THE PENNSYLVANIA TELEPHONE CORPORATION. Erie, PA: Pennsylvania Telephone Corp., 1947, 107 pp. One of the independent companies. Photos, diagrams, maps, chronology.

7-101a Griffith, Tom, and the Iowa Telephone Association. **LINES BETWEEN THE RIVERS: A HISTORY OF TELEPHONY IN IOWA.** West Des Moines, IA: Iowa Telephone Assn., 1991, 599 pp. The independent companies are detailed.

7-102 Hackenburg, Herbert J. **MUTTERING MACHINES TO LASER BEAMS: A HISTORY OF MOUNTAIN BELL.** Denver: Mountain Bell, 1986, 400 pp. A lavishly illustrated (much in color) history of what is now a U.S. West subsidiary. Tables, index.

7-102a Hall, Jerry F., ed. **HELLO TEXAS: A HISTORY OF TELEPHONY IN THE LONE STAR STATE.** Austin: Texas Telephone Association, 1990, 231 pp. Overall history of member companies, both Bell and independent. Photos.

7-103 Haskin, Eleanor, ed. **INDEPENDENT TELEPHONY IN NEW ENGLAND: A HISTORY 1876-1976.** Weare, NH: Independent Telephone Pioneer Assn, 1976, 364 pp. Typescript volume with a company-by-company historical survey. Photos.

7-104 Howe, F.L. **THIS GREAT CONTRIVANCE: THE FIRST HUNDRED YEARS OF THE TELEPHONE IN ROCHESTER.** Rochester, NY: Rochester Telephone Co., 1979, 124 pp. Another of the independent companies. Photos, index.

7-105 Jepson, Thomas C. "The Telegraph Comes to Colorado: A New Technology and Its Consequences," *ESSAYS AND MONOGRAPHS IN COLORADO HISTORY,* 7:1-25 (1987). Technology on the frontier.

7-106 ** Lipartito, Kenneth. **THE BELL SYSTEM AND REGIONAL BUSINESS: THE TELEPHONE IN THE SOUTH, 1877-1920.** Baltimore: Johns Hopkins University Press "AT&T Series in Telephone History," 1989, 283 pp. Based on AT&T archival sources, this study explores how the southern U.S. "caught up" with the rest of the country in terms of telephone availability and application. Tables, photos, notes, index. For rest of this series, see Garnet, 4-101; Smith, 4-106; and Wasserman, 7-095.

7-107 Masters, R.S., et al. **AN HISTORICAL REVIEW OF THE EAST BAY EXCHANGE: OAKLAND, BERKELEY, ALAMEDA, PIEDMONT, ALBANY, EMERYVILLE, SAN LEANDRO...** San Francisco: Pacific Telephone, 1827, 119 pp. History of the East Bay, California, companies. Photos, foldout map.

7-107a Missouri Telephone Association. **THE SHOW-ME STATE STORY: A HISTORY OF TELEPHONY IN MISSOURI.** Jefferson City, MO: Missouri Telephone Association, 1989, 401 pp. The independent companies. Photos, index.

7-108 Park, David G. Jr. **GOOD CONNECTIONS: A CENTURY OF SERVICE BY THE MEN & WOMEN OF SOUTHWESTERN BELL.** St. Louis, MO: Southwestern Bell, 1984, 300 pp. Photos, chronology, appendices, index.

166   *Chapter 7*

7-109 Rippey, James C. **GOODBYE, CENTRAL: HELLO, WORLD: A CENTENNIAL HISTORY OF NORTHWESTERN BELL**. Omaha, NE: Northwestern Bell, 1975, 344 pp. Issued by Telephone Pioneers and covering operations in Iowa, Minnesota, Nebraska, North and South Dakota. Photos, index.

7-110 Shearer, Stephen R. **HOOSIER CONNECTIONS: THE HISTORY OF THE INDIANA TELEPHONE INDUSTRY AND THE INDIANA TELEPHONE ASSOCIATION**. Indianapolis: Indiana Telephone Association, 1992, 424 pp. The first 120 pages are historical by period; much of the remainder concerns the association and more recent developments. Photos, maps, glossary.

7-111 Simonds, William A. **THE HAWAIIAN TELEPHONE STORY**. Honolulu: Hawaiian Telephone Co., 1958, 114 pp. Another of the independent firms. Photos. See also Bendix, 7-096.

7-112 Tyler, Victor Morris. **A SHORT REVIEW OF CONNECTICUT TELEPHONY, 1878 TO 1907**. New Haven, CT: Author, 1957, 125 pp. Reviews the history of Southern New England Telephone (SNET). Photo, appendices.

7-113 *** Walsh, J. Leigh. **CONNECTICUT PIONEERS IN TELEPHONY**. New Haven, CT: Telephone Pioneers of America, 1950, 444 pp. One of the best of the breed (it is based on considerable research) this is a TPA historical publication detailing telephone development in the state from its inception in the 1870s through the terrible floods of the late 1930s. Photos, notes, appendices, index.

7-114 ** Wilson, Clark M., compiler and editor. **TRACING THE TELEPHONE IN WESTERN MASSACHUSETTS, 1877–1930**. Springfield, MA: Author, 1958, 430 pp. Details expansion of the instrument and industry in various Massachusetts towns. Photos.

E. Foreign Telephony Systems

(The items cited here deal largely with foreign systems—especially those of Canada and Britain—though they may include some comment on American telephony as well. See related but more general histories in chapter 3-E.)

Europe and General

7-115 ** Bennett, A.R. **THE TELEPHONE SYSTEMS OF THE CONTINENT OF EUROPE**. London: Longmans, Green, 1895 (reprinted by Arno Press "Telecommunications," 1974), 436 pp. Twenty-six chapters arranged by country or province. Important work illustrating the initial development of telephone service over a large area. Engravings, maps, diagrams, tables. See also Webb, 7-126.

7-116 *** Chapuis, Robert J. **100 YEARS OF TELEPHONE SWITCHING (1878–1978), PART 1: MANUAL AND ELECTROMECHANICAL SWITCHING (1878–1960s)**. Amsterdam: North-Holland Publishing "Studies in Telecommunication, Vol. 1," 1982, 482 pp. Definitive study of European and North American developments. Photos, diagrams, charts, tables, references, index, list of acronyms.

Telephony

7-117 *** Chapuis, Robert J., and Amos E. Joel Jr. **ELECTRONICS, COMPUTERS AND TELEPHONE SWITCHING: A BOOK OF TECHNOLOGICAL HISTORY AS VOLUME 2: 1960-1985 OF "100 YEARS OF TELEPHONE SWITCHING."** Amsterdam: North-Holland Publishing "Studies in Telecommunication, Vol. 13," 1990, 595 pp. Continues item above with about (despite the subtitle) 100 pages covering period before 1960. Photos, diagrams, charts, tables, references, index. See also Joel, 7-120 and 7-121.

7-118 Not used.

7-119 "Development and Evolution of the Telephone Service," *TELECOMMUNICATION JOURNAL,* 50:679-682 (December 1983). The telephone in Argentina. For a parallel work on telegraphy, see Hodge, 6-131.

7-120 Joel, Amos E. Jr., ed. **ELECTRONIC SWITCHING: CENTRAL OFFICE SYSTEMS OF THE WORLD.** New York: IEEE Selected Reprint Series, 1976, 279 pp. Includes papers on the U.S., Canada, Britain, France, Sweden, Japan, the Netherlands, and Germany. Photos, diagrams, references.

7-121 _____. **ELECTRONIC SWITCHING: DIGITAL CENTRAL OFFICE SYSTEMS OF THE WORLD.** New York: IEEE Selected Reprint Series, 1981, 268 pp. Papers on the U.S. and Canada, France, Italy, the Netherlands, West Germany, East Germany, Belgium, Britain, Japan, and Sweden. Photos, diagrams, references.

7-122 Kragh, Helge. "The Krarup Cable: Invention and Early Development," *TECHNOLOGY AND CULTURE,* 35:129-157 (January 1994). Telephone cable "loading" over a two-decade period at the turn of the century.

7-123 Technical Staff of the Copenhagen Telephone Company, trans. by Alexander C. Jarvis. **THE DEVELOPMENT OF TELEPHONIC COMMUNICATION IN COPENHAGEN, 1881-1931.** Copenhagen: Danmarks Naturvidenskabelige Samfund GEC, 1932, 163 pp. Originally part of a commemorative volume on the first half century of Danish telephones. Photos, map, plans.

7-124 "Tivador Puskas—A Forerunner," *TELECOMMUNICATION JOURNAL,* 54:307-310 (May 1987). The Budapest-based pioneer of sending information and entertainment by means of telephone wires.

7-125 Uges, H.J. **BEYOND THE DIAL: DEVELOPMENT OF THE NETHERLANDS TELEPHONE SYSTEM.** The Hague: Netherlands Postal and Telecommunications Services, 1953, 189 pp. From a series of articles published in a Dutch journal, 1947-48. Photos, diagrams.

7-126 Webb, Herbert Laws. **THE DEVELOPMENT OF THE TELEPHONE IN EUROPE.** London: Electrical Press, 1911 (reprinted by Arno Press "Telecommunications," 1974), 78 pp. Based on 1910 articles reprinted from *ELECTRICAL INDUSTRIES,* this brief volume reviews both British and European telephone systems and practices. See also Bennett, 7-115.

7-127 * Young, Peter. **PERSON TO PERSON: THE INTERNATIONAL IMPACT OF THE TELEPHONE.** Cambridge, England: Granta Editions, 1991,

295 pp. An informal journalistic history of the instrument and its varied effects around the world, arranged in approximate chronological fashion. Published with some support from British Telecom. Photos, index.

Britain and Commonwealth

7-128 *** Baldwin, Francis G.C. **THE HISTORY OF THE TELEPHONE IN THE UNITED KINGDOM.** London: Chapman & Hall/New York: Van Nostrand, 1925 (reprinted by Chapman & Hall, 1938), 728 pp. Lavish and definitive treatment of the subject. First 120 pages deal with Bell's telephone and its introduction into England and early telephone companies in London and the provinces to 1886. Considerable focus on National Telephone Company operations prior to its nationalization in 1912. There are no formal references, but names, dates, patent numbers, and other particulars from journals, books, and other sources, along with extracts, are plentiful in the text. Photos, charts, maps, tables, appendices, index.

7-129 Benson, Don. **WIRE SONG: AN ILLUSTRATED HISTORY OF THE TELEPHONE IN BRITISH COLUMBIA, 1880–1930.** New Westminster, BC: Westminster Publishing, 1991, 63 pp. Photos, bibliography, index.

7-130 Cashman, Tony. **SINGING WIRES: THE TELEPHONE IN ALBERTA.** Edmonton: Alberta Government Telephones Commission, 1972, 496 pp. Well-illustrated provincial history of the telephone with color photos of vintage equipment.

7-131 Earl, R.A.J. **THE DEVELOPMENT OF THE TELEPHONE IN OXFORD, 1877–1977.** Oxford, England: British Telecom Museum, 1983 [2nd ed.], 158 pp. Useful example of the varied trends in British telephony as seen in a single city (there are many similar books and booklets for other British communities). Photos, glossary, index.

See EVENTS IN TELECOMMUNICATIONS HISTORY, 1-051.

7-132 THE FIRST CENTURY OF SERVICE: BELL 1880–1980. Montreal: Bell Canada, 1980, 60 pp. Company promotional piece offers useful brief survey. Photos, maps.

7-133 Foreman-Peck, James. "The Development and Diffusion of Telephone Technology in Britain, 1900–1940," *TRANSACTIONS OF THE NEWCOMEN SOCIETY,* 63:165–180 (1991/92).

7-134 ** Herbert, Thomas Ernest. **THE TELEPHONE SYSTEM OF THE BRITISH POST OFFICE: A PRACTICAL HANDBOOK.** London: Page, 1898; London and New York: Whittaker, 1901 [2nd ed.], 218 pp.; 1904 [3rd. ed.], 218 pp.; Pitman, 1923 [4th ed.], 868 pp. A standard descriptive work. References after most chapters. Revised, with William S. Proctor, as **TELEPHONY: A DETAILED EXPOSITION OF THE TELEPHONE SYSTEM OF THE BRITISH POST OFFICE.** London: Pitman "The Specialists' Series."

VOL. 1: MANUAL SWITCHING SYSTEMS AND LINE PLANT. (1932; 1934 [2nd ed.]), 1,212 pp.

VOL. 2: AUTOMATIC TELEPHONY. (1938), 749 pp.; and **SUPPLEMENT** (1940)

Further revised by J. Atkinson (Pitman, 1948–50, 1966), under the same main title:

VOL. 1: GENERAL PRINCIPLES AND MANUAL EXCHANGE SYSTEMS. 513 pp.

VOL. 2: AUTOMATIC EXCHANGE SYSTEMS. 871 pp. Photos, diagrams, references, index.

For Herbert's parallel work on the telegraph, see 6-159.

7-135 * Johannessen, Neil, ed. **"RING UP BRITAIN": THE EARLY YEARS OF THE TELEPHONE IN THE UNITED KINGDOM.** London: British Telecommunications, 1991. 60 pp. Reprints three articles from the magazine of the National Telephone Company (which in 1912 was taken over by the government) plus part of a post office history that, together, detail the rise of British telegraphy and telephony to about 1910. Photos, tables.

7-136 Melick, Donald M., and Robert J. Walker. "Bell Canada: A Case Study of Dynamics of New Technologies in a Changing Environment," *JOURNAL OF ENGINEERING AND TECHNOLOGY MANAGEMENT,* 9:339–353 (December 1992).

7-137 Morris, Robert C. **BETWEEN THE LINES: A PERSONAL HISTORY OF THE BRITISH PUBLIC TELEPHONE AND TELECOMMUNICATIONS SERVICE, 1870–1990.** London: Just Write Publishing, 1994, 126 pp. Informal narrative in form of a memoir. Photos, bibliography.

7-138 Occomore, Dave. **"NUMBER PLEASE!" A HISTORY OF THE EARLY LONDON TELEPHONE EXCHANGES FROM 1880 TO 1912.** Romford, England: Ian Henry Publications, 1995, 92 pp. Takes the story up to the takeover of the private national Telephone Company by the General Post Office. Photos, map, notes.

7-139 * Ogle, E.B. **LONG DISTANCE PLEASE: THE STORY OF THE TRANS CANADA TELEPHONE SYSTEM.** Toronto: Collins, 1979, 300 pp. Relates the development and operation of the transcontinental network by a consortium of ten companies (which was renamed Stentor in the 1990s). Chronology, glossary, bibliography, index.

7-140 Patten, William. **PIONEERING THE TELEPHONE IN CANADA... CONTRIBUTIONS TO THE HISTORY OF THE TELEPHONE BASED ON ORIGINAL NOTES AND REMINISCENCES, INCLUDING A HITHERTO UNPUBLISHED ACCOUNT OF THE TELEPHONE BY A. MELVILLE BELL.** Montreal: privately printed, 1926, 139 pp. Photos.

7-141 *** Anniversary issues of *THE POST OFFICE ELECTRICAL ENGINEERS JOURNAL.* Two issues contain numerous well-illustrated technical articles detailing British telegraph, telephone, and radio developments. Photos, diagrams, maps, references. Specifically:

"Jubilee Number Commemorating the 50th Anniversary of Foundation of the Institution of Post Office Electrical Engineers, 1906–1956" 49:3:150–274 (October 1956), includes 14 papers ranging over a half century of development.

"IPOEE 75th Anniversary, 1906–1981" 74:3:150–302 (October 1981), includes 26 papers most of which focus on the 1956–81 period.

7-142 Robertson, John H. THE STORY OF THE TELEPHONE: A HISTORY OF THE TELECOMMUNICATIONS INDUSTRY OF BRITAIN. London: Pitman, 1947 (reprinted by Scientific Book Club, 1948), 299 pp. A general nontechnical account by a political journalist. No list of contents or chapter headings, no illustrations, no references. Some names, dates, places, titles, and extracts in the text. Index.

7-143 Scowen, Frank. "The Automatic Telephone Exchange," *TRANSACTIONS OF THE NEWCOMEN SOCIETY*, 47:35–46 (1974/75 and 1975/76). Concerns developments in Britain from 1912 on.

7-144 Surtees, Lawrence. WIRE WARS: THE CANADIAN FIGHT FOR COMPETITION IN TELECOMMUNICATIONS. Scarborough, ON: Prentice-Hall Canada, 1994, 420 pp. Largely a discussion of changing telephone industry and governmental policies over the past two decades, with some reference to technology. Bibliography, chronology, index.

F. Submarine Telephony
(See also chapter 6-F for earlier works and those combining telegraph and telephone cables.)

7-145 Affel, H.A., et al. "The New Key West–Havana Carrier Telephone Cable," *BELL SYSTEM TECHNICAL JOURNAL*, 11:197–212 (January 1932). Reviews a decade of development.

7-146 Buckley, O.E. "The Future of Transoceanic Telephony," *BELL SYSTEM TECHNICAL JOURNAL* 21:1–19 (January 1942); also in *JOURNAL OF THE IEE*, 89:1:454–461 (1942). Early Bell Labs thinking on cable telephony options.

7-147 FIRST TRANSATLANTIC TELEPHONE CABLE. London: Submarine Cables Ltd, 1957, 39 pp. Describes TAT-1. Illustrated.

7-148 Kelly, Mervin J., et al. "A Transatlantic Telephone Cable," *COMMUNICATION AND ELECTRONICS*, 17:124–139 (March 1955). Four authors, two with Bell labs and two with the British Post Office, related plans and technical details for what became TAT-1, followed by discussion. Diagrams, notes. A shorter version appeared in *ELECTRICAL ENGINEERING* 74:192–197 (March 1955). Reprinted in Shiers, 7-091.

7-149 Malcolm, H.W. THE THEORY OF THE SUBMARINE TELEGRAPH AND TELEPHONE CABLE. London: Benn Bros., 1917, 565 pp. One of the first books to include submarine *telephone* cables. Photos, diagrams, index.

7-150 Nichols, Russell T. **SUBMARINE TELEPHONE CABLES AND INTERNATIONAL COMMUNICATION.** Santa Monica, CA: Rand Corp. RM-3472-RC, 1963, 35 pp. The major services and some sense of their use. Tables.

7-151 * O'Meara, Major W.A.J. "Submarine Cables for Long-Distance Telephone Circuits," *JOURNAL OF THE IEE* (December 15, 1910), pp. 309–356. Focuses on cables from Britain to France, Ireland, and the Isle of Wight. Photos, maps, diagrams, tables. Reprinted in Finn, 6-230.

7-152 *** Schenck, Herbert H., and Leo Waldick. **1990 WORLD'S SUBMARINE TELEPHONE CABLE SYSTEMS.** Washington: Government Printing Office (NTIA Contract Report 91-42), 1991 [3rd ed.], 513 pp. Massive compilation, first published in 1980, of cables past and present with, for each one, a full page of data and statistics and a map showing route. The first 32 pages are a concise history of ocean cables, followed by a discussion of fiber optic cables, and listings of systems in being and planned. Photos, diagrams, maps, glossary, bibliography.

7-153 ** Shimura, Seiichi, ed. **INTERNATIONAL SUBMARINE CABLE SYSTEMS.** Tokyo: KDD Engineering and Consulting, 1984, 509 pp. Historical retrospect, basic design technique, project planning, system engineering and procurement, construction and installation, maintenance, and optical fibre systems. Diagrams, photos, tables, maps, references, appendices, index.

7-154 Solomon, Louis. **VOICEWAY TO THE ORIENT: FIRST U.S.–JAPAN TELEPHONE CABLE.** New York: McGraw-Hill, 1964, 64 pp. While designed for younger readers, this is a useful synopsis. Photos, index.

7-155 "Symposium on the Transatlantic Telephone Cable, 1957" *PROCEEDINGS OF THE IEE,* 104 (Part B, Supp. 4): 1–125 (1957). Discusses the development and installation of TAT-1. Maps, diagrams, tables, references.

7-156 * "Transatlantic Communications... ," *BELL SYSTEM TECHNICAL JOURNAL,* 36:1–326 (January 1957). First of several issues that year devoted to details of the planning, design, and construction of TAT-1. Photos, diagrams, maps, notes.

7-157 Tucker, D. G. "The First Cross-Channel Telephone Cable: The London-Paris Telephone Links of 1891," *TRANSACTIONS OF THE NEWCOMEN SOCIETY,* 47:116–132 (1974/76). Inception of submarine telephone cables.

G. Video Telephony and Teleconferencing
(For three decades, AT&T [and more briefly other firms] tried to effectively innovate video telephone service under such brand names as "Picturephone," or "VideoPhone," but failed for reasons discussed in these citations.)

7-158 Carson, D.N. "The Evolution of Picturephone Service," *BELL LABORATORIES RECORD,* 46:282–291 (October 1968). Photos. Reprinted in Shiers, 7-091.

7-159 ** Dickson, Edward M. with Raymond Bowers. **THE VIDEO TELEPHONE: IMPACT OF A NEW ERA IN TELECOMMUNICATIONS—A PRELIMINARY TECHNOLOGY ASSESSMENT.** New York: Praeger, 1974,

242 pp. Cornell University research underwritten by the National Science Foundation, this is one of the better studies of the difficulties in launching a new service. Tables, chapter references, glossary (no index).

7-160 Elton, Martin C.J. "Visual Communication Systems: Trials and Experiences," *PROCEEDINGS OF THE IEEE,* 73:700–705 (April 1985). Details more than two decades of work. Photos, references.

7-161 Falk, Howard. "Picturephone and Beyond," *IEEE SPECTRUM,* (November 1973), 10:45–49. A brief survey. Photos.

7-162 ** Noll, A. Michael. "Anatomy of a Failure: Picturephone Revisited," *TELECOMMUNICATIONS POLICY,* 16: 307–316 (May–June 1992). A former Bell Labs researcher traces what went wrong. References.

7-163 "Picturephone," *BELL LABORATORIES RECORD,* 47:5:134–193 (May–June 1969) is a special issue devoted to the system's technology which had been developed in the Labs. Photos.

7-164 * "The Picturephone System," *BELL SYSTEM TECHNICAL JOURNAL,* 50:219–709 (February 1971) is a special issue of 20 technical papers devoted to all aspects of the system. This remains the single best resource on the original technology as publically demonstrated. Photos, diagrams, references.

H. Vintage Telephony

7-165 Clark, Paul, and Guy Ryecart. **THE PHONE—AN APPRECIATION**. San Diego, CA: Laurel Glen "Design Icons," 1998, 32 pp. Brief illustrated album of photos depicting telephone design over the years.

7-166 * Dooner, Kate E. **TELEPHONES ANTIQUE TO MODERN**. West Chester, PA: Schiffer, 1992. 175 pp. See next entry.

7-167 _____. **TELEPHONE COLLECTING: SEVEN DECADES OF DESIGN**. Atglen, PA: Schiffer, 1993, 128 pp. With 7-166, these two color-illustrated albums provide a good sense of the growing market in old telephones and related objects, here with many photos and guides for collectors.

7-168 Emerson, Andrew. **OLD TELEPHONES**. Aylesbury, England: Shire Books "Album 161," 1986, 32 pp. Brief survey of British telephone design including telephone call boxes. Photos.

7-169 _____. **ELECTRONIC CLASSICS: COLLECTING, RESTORATION, AND REPAIR**. London: Newnes, 1998, 413 pp. Includes telephones.

7-170 Knappen, Ron, and Mary. **HISTORY AND IDENTIFICATION OF OLD TELEPHONES: COMMENTS, EDITORIALS, ORGANIZATION, SALES**. Galesville, WI: R. Knappen, 1978, 2 vols., 822 pp. A scrapbook of old catalogue and advertising material. Photos, drawings, index.

Telephony

7-171 Langdon, William C. "The American Telephone Historical Collection," and "The Growth of the Historical Collection," *BELL TELEPHONE QUARTERLY* (January 1924, April 1925). Two brief surveys of the company's own equipment archive.

7-172 Luff, Peter P. "The Electronic Telephone," *SCIENTIFIC AMERICAN* (March 1978), 238:3:58–64. Traces development of the instrument to the integrated ciruits then in the offing. Photos, diagrams.

7-173 * Martin, William H. "Seventy-Five Years of the Telephone: An Evolution in Technology," *BELL SYSTEM TECHNICAL JOURNAL*, 30:215–238 (April 1951). Experiments, theory, design, and measurements related to telephone instruments. Photos, bibliography. Reprinted in Shiers, 7-091.

7-174 ** Meyer, Ralph O. **OLD TIME TELEPHONES! TECHNOLOGY, RESTORATION AND REPAIR.** New York: TAB/McGraw-Hill, 1995, 304 pp. Useful historical material and perhaps the best how-to information. Covers development of components, telephone instruments, electrical circuits, and restoration and repair. Photos, diagrams, index.

7-175 * Montjoy, Richard. **100 YEARS OF BELL TELEPHONES.** West Chester, PA: Schiffer, 1995, 176 pp. Very detailed catalog of (especially) early wooden telephones and related devices with supporting text. Designed for collectors.

7-176 ** Povey, P.J., and R.A.J. Earl. **VINTAGE TELEPHONES OF THE WORLD.** London: Peter Peregrinus with Science Museum "IEE History of Technology 8" [should be 9], 1988, 202 pp. A British volume useful for its details on European telephone instruments in what is otherwise a narrative history of the instrument. Photos, references, index.

7-177 Soresini, Franco. **TELEFONI/TELEPHONE SETS.** Milan: BE-MA Editrice, 1989, 143 pp. Small album of color photos with bilingual (Italian and English) captions.

7-178 Stamp, Gavin. **TELEPHONE BOXES.** London: Chatto & Windus "Chatto Curiosities of the British Street," 1989, 106 pp. All you ever wanted to know about the famous red telephone booths now being replaced with gray modern fixtures. Photos, bibliography, index.

7-179 Wolff, Lawrence. **DESK TELEPHONES OF THE BELL SYSTEM.** Burbank, CA: CompuGraphix & Printing, 1993.

7-180 Yves, Arden. **TELEPHONE CARDS.** Aylesbury, England: Shire Publications, 1994, 32 pp. Brief guide to collecting phonecards, some of which feature historical telephone illustrations. Photos, bibliography.

Chapter 8

Electromagnetic Waves

Entries in this chapter cover early experiments, discoveries, and theories concerning electromagnetic radiation, the evolution of antennas, propagation theory, and spectrum management. Kindred subjects included deal with theories of the ether, resonance, interference, transmission lines, and modulation. For later applications of this information to wireless, see radio entries in chapter 9. For more information on specific inventors noted here, including Maxwell, Hertz, and Marconi, see chapter 5-C.

The chapter begins with (A) modern survey histories and moves on to (B) observations and experiments up to about 1900, (C) developments after 1900, (D) microwaves, and (E) spectrum management.

A. Modern Histories

8-001 *** Aitken, Hugh G.J. SYNTONY AND SPARK: THE ORIGINS OF RADIO. New York: Wiley-Interscience (reprinted by Princeton University Press), 1976, 447 pp. Elegantly written assessment of the inception of wireless as seen primarily through the innovative work of Clerk Maxwell and Hertz in the 19th century, to developments in the early 1900s by Lodge and Marconi, each discussed in a chapter. Continued by Aitken, 9-090. Line drawings, notes, index.

See Garratt, 5-022.

8-002 ** Gluckman, Albert Gerard. THE INVENTION AND EVOLUTION OF THE ELECTROTECHNOLOGY TO TRANSMIT ELECTRICAL SIGNALS WITHOUT WIRES: AN ANNOTATED BIBLIOGRAPHY OF 17th, 18th-, AND 19th-CENTURY EXPERIMENTAL STUDIES OF ELECTROSTATIC INDUCTION, SPARK-GAP AND LIGHTNING DISCHARGES,

Chapter 8

MAGNETIC INDUCTION, OSCILLATING CIRCUITS, RESONANCE, AND ELECTROMAGNETIC WAVE PROPAGATION. Washington: Washington Academy of Sciences, 1993, 239 pp.; 1996 [2nd ed.], 256 pp. Useful for information on historical basic research on wireless communications. Describes 311 books, articles, and reports. Arranged in *reverse* chronological order from 1922 back to 1664 "in order to project the appearance of looking backward into time from the viewpoint of electromagnetic science at the turn of the 20th century." Includes chronology of "evolutionary milestones," name index.

See O'Hara and Pritcha, 5-031.

See Hunt, 5-025.

8-003 Süsskind, Charles. "Observations of Electromagnetic Wave Radiation before Hertz," *ISIS*, 55:32–42 (March 1964). Galvani, Henry, Edison, Thomson and Houston, S.P. Thompson, Dolbear, and Hughes are discussed. References.

8-004 Whittaker, Edmund T. **A HISTORY OF THE THEORIES OF AETHER AND ELECTRICITY.** London: Longmans Green, 1910; reprinted by Thomas Nelson and Philosophical Library, 1951. A second volume was first published in 1953. Both reprinted 1958 and (by Harper) 1961 and again in 1987 (by Tomash/American Institute of Physics as Volume 7 [in two parts] of "The History of Modern Physics, 1800–1950"). Highly technical.

> **VOL. 1: THE CLASSICAL THEORIES.** (434 pp.). The original volume covers from the earliest times to 1900, including Maxwell and those that followed. References, index.

> **VOL. 2: THE MODERN THEORIES, 1900–1926.** (319 pp.). Includes subject index for this volume only. References, index.

B. Observations and Experiments to 1900

(Included here are the major English-language papers of Maxwell and Hertz, arranged in an approximate chronological order, reporting theories and experiments. For lesser papers, see Shiers [9-080], pp. 149–152. See also the closely related material cited in chapter 9-A, and the biographical material on the inventors in 5-C.)

8-005 Larmor, L. "The Origins of Clerk Maxwell's Electric Ideas, As Described in Familiar Letters to W. Thomson," *PROCEEDINGS OF THE CAMBRIDGE PHILOSOPHICAL SOCIETY*, 32:695–750 (1936). See also Campbell and Garnett, 5-169 and Harmon, 5-175.

8-006 *** Maxwell, James C. "A Dynamical Theory of the Electromagnetic Field," *PROCEEDINGS OF THE ROYAL SOCIETY*, 23:531–536 (1864); *PHILOSOPHICAL TRANSACTIONS*, 155:459–512 (1865); *PHILOSOPHICAL MAGAZINE*, 29: 152–157 (1865). One of the monumental papers of the 19th century. The first part is a nonmathematical explanation of the theory that links electromagnetic radiations ("and other radiations if any") with light. The rest of the paper develops the general equations of the electromagnetic field. For more on Maxwell, see 5-169 through 5-183.

8-007 Poincaré, H., and Frederick K. Vreeland. **MAXWELL'S THEORY AND WIRELESS TELEGRAPHY.** New York: McGraw, 1904, 255 pp. The first 107 pages are Ponicaré's paper (as translated by Vreeland); the remainder is Vreeland's application of the theory to early wireless. Diagrams, notes, index.

8-008 de Tunzelmann, G.W. "Hertz's Researches on Electrical Oscillations," **ANNUAL REPORT OF THE BOARD OF REGENTS OF THE SMITHSONIAN INSTITUTION, 1889.** Washington: Government Printing Office, 1890, pp. 145–203. Reprinted from an 1898 series of articles in *THE ELECTRICIAN.* Contemporary discussion of Hertz's research. Diagrams, notes. See also Blanchard, 5-139 and Buchwald, 5-140.

See Hertz, 5-141

8-009 Not used.

8-010 *** Lodge, Oliver J. **THE WORK OF HERTZ AND SOME OF HIS SUCCESSORS.** London: "The Electrician" Printing and Publishing, 1894, 42 pp.; 1898 [2nd ed.], 58 pp.; retitled SIGNALLING THROUGH ["ACROSS" on title page] SPACE WITHOUT WIRES: BEING A DESCRIPTION OF THE WORK OF HERTZ AND HIS SUCCESSORS, 1900 [3rd ed.], 133 pp. (Reprinted by Arno Press, 1974); 1908 [4th ed.], 154 pp. Based on a lecture before the Royal Institution, June 1, 1894 (first 42 pages). Besides being a memorial to Heinrich Hertz, the volume also triggered interest in the initial experiments on wireless telegraphy (see 9-A). With the third edition, later chapters focused on coherer (detector) developments, and photoelectric researches. Diagrams, notes and references.

8-011 Hyndman, Hugh H.F. **RADIATION: AN ELEMENTARY TREATISE ON ELECTROMAGNETIC RADIATION AND ON RÖNTGEN AND CATHODE RAYS.** London: Swan Sonnenschein/New York: Macmillan, 1898, 307 pp. This contemporary account of the exciting work of the 1890s omits mention of the practical experiments of Lodge, Marconi, and others concerning wireless communication. The first 152 pages are largely devoted to the ether, ethereal vibrations, and early experiments and theories on electromagnetic waves. Vacuum tube phenomena, cathode rays and effects are treated on pp. 154–204. The remaining text concerns Röntgen or X-rays. Illustrations, tables, notes and references.

8-012 MacDonald, H.M. **ELECTRIC WAVES: BEING AN ADMANS PRIZE ESSAY IN THE UNIVERSITY OF CAMBRIDGE.** Cambridge, England: Cambridge University Press, 1902, 200 pp. On Faraday's and Maxwell's theories and the characteristics of propagation, radiation, and transmission media.

C. Developments after 1901

(Arranged alphabetically by author. See also IRE 50th Anniversary issue of its PROCEEDINGS, 3-036.)

8-013 Armstrong, Edwin H. "Frequency Modulation—1922 and 1948," *RADIO-CRAFT,* 19:20:55 (June 1948). Refutes Carson's 1922 paper (8-015). Photos. See also 9-130, 9-131.

8-014 Benton, Mildred C., comp. **SINGLE-SIDEBAND IN COMMUNICATION SYSTEMS: A BIBLIOGRAPHY.** Washington: U.S. Naval Research Laboratory, Office of Technical Services, Bibliography No. 9," 1956, 99 pp. Nearly 500 entries from 1920 on, including papers, books, reports, with author and subject indexes.

8-015 Carson, John R. "Notes on the Theory of Modulation," *PROCEEDINGS OF THE IRE,* 10:57–64 (February 1922); reprinted in same journal 51:893–896 (June 1963). The infamous mathematical analysis rejecting the value of frequency modulation which was later refuted by Armstrong (see 8-013). Notes.

8-016 Deloraine, E. Maurice, and Alec. H. Reeves. "The 25th Anniversary of Pulse Code Modulation," *IEEE SPECTRUM,* 2:56–63 (May 1965). Historical background and the past, present, and future. Photos, charts, references.

8-017 Greenwood, Thomas L. "The Radio Spectrum below 550 kHz," *IEEE SPECTRUM,* 4:121–123 (March 1967). This brief survey includes much historical matter.

8-018 Gunther, Frank A. "Tropospheric Scatter Communications: Past, Present, and Future," *IEEE SPECTRUM,* 3:79–100 (September 1966). Systems and companies including a triple-fold chart. Photo, maps, diagrams, references.

8-019 IRE Committee on Wave Propagation. "Tropospheric Propagation: A Selected Guide to the Literature," *PROCEEDINGS OF THE IRE,* 41:588–594 (May 1953). Brief survey arranged into 12 sections. Notes.

8-020 Lodge, Oliver J. **THE ETHER OF SPACE.** New York: Harper, 1909, 168 pp. A foremost supporter of the ether theory, the author describes his own experiments and provides some historical background from the time of Isaac Newton. Illustrations, text references, no index.

8-021 _____. **ETHER AND REALITY.** London: Hodder & Stoughton, 1925, 179 pp. "A series of discussions on the many functions of the either of space."

8-022 Marconi, Guglielmo. "On Methods Whereby the Radiation of Electric Waves may be Mainly Confined to Certain Directions...."*PROCEEDINGS OF THE ROYAL SOCIETY,* Series A, Vol. 77, No. A518: 413–421 (April 30, 1906). A paper presented for Marconi by J. Ambrose Fleming in March 1906.

8-023 National Bureau of Standards. "Bibliography of Radio Wave Phenomena and Measurement of Radio Field Intensity," *PROCEEDINGS OF THE IRE,* 19:1034–1089 (June 1931). Comprehensive list of 620 citations from 1900 arranged chronologically under 14 headings with a list of journals and author index.

8-024 Oswald, Arthur A. "Early History of Single-Sideband Transmission," *PROCEEDINGS OF THE IRE,* 44:1676–1679 (December 1956). Some 58 references beginning in 1907.

8-025 Smith-Rose, Reginald L. "Fifty Years Research in Radio Wave Propagation," *WIRELESS WORLD,* 67:203–207 (April 1961). Diagrams.

Electromagnetic Waves 179

D. Microwaves
(For microwave tubes, see chapter 11-E.)

8-026 Adams, Stephen. "Microwave Instrumentation: An Historical Perspective," *IEEE TRANSACTIONS ON MICROWAVE THEORY & TECHNOLOGY*, MTT-32:9:1157–1161 (September 1984). References.

8-027 Beverage, Harold H., et al. "The Wave Antenna," *TRANSACTIONS OF THE AIEE*, 42:215–266 (February 1923). Photos, diagrams, tables, notes, bibliography.

8-028 Bryant, John H. "Coaxial Transmission Lines, Related Two-Conductor Transmission Lines, Connectors, and Components: A U.S. Historical Perspective," *IEEE TRANSACTIONS ON MICROWAVE THEORY & TECHNOLOGY*, MTT-32:9:970–983 (September 1984). Useful chronological survey of major developments. Photos, references.

8-029 Button, Kenneth J. "Historical Sketch of Ferrites and Their Microwave Applications," *MICROWAVE JOURNAL*, 3:73–79 (March 1960). Graphs, bibliography.

8-030 Cantelon, Philip L. "The Origins of Microwave Telephone—Waves of Change," *TECHNOLOGY AND CULTURE*, 36:560–582 (July 1995). A useful survey which is closely related to the appendix in his history of MCI (4-147).

8-031 Kemp, J. "Waveguides in Electrical Communication," *JOURNAL OF THE IEE*, 90 (Pt. III):90–114 (September 1943). Historical review. References and patents.

8-032 Marconi, Guglielmo. "Micro Radio Waves," *ELECTRICIAN*, 110:3–6 (January 6, 1933). Recent experiments with very short waves, with illustrations of transmitting equipment, circuits, and wave patterns. Part of an address delivered before Royal Institution in December 1932. Photos, diagrams.

8-033 Meinel, Holger H. "Communication Applications of Millimeter Waves: History, Present Status, and Future Trends," *IEEE TRANSACTIONS ON MICROWAVE THEORY & TECHNOLOGY*, MTT 43:1639–1653 (July 1995).

8-034 Ramsay, John F. "Microwave Antenna and Waveguide Techniques before 1900." *PROCEEDINGS OF THE IRE*, 46:405–415 (February 1958). A survey of quasi-optical experiments and devices from 1888 to 1900. Photos, diagrams, notes.

8-035 Schweitzer, Ellis. "40 Years of Waveguides: A Glimpse at History," *BELL LABORATORIES RECORD*, 48:72–79 (March 1970). Pictorial essay with 18 photos and supporting text.

8-036 Steneck, Nicholas H. **THE MICROWAVE DEBATE**. Cambridge, MA: MIT Press, 1984, 279 pp. Historical study on health effects of microwave radiation from 1930 into the 1980s. Bibliographic note, index.

8-037 Wiltse, J.C. "History of Millimeter and Sub-Millimeter Waves," *IEEE TRANSACTIONS ON MICROWAVE THEORY & TECHNOLOGY*, MTT-32:1118–1127 (September 1984).

E. Spectrum Management

8-038 Aitken, Hugh G.J. "Allocating the Spectrum: The Origins of Radio Regulation," *TECHNOLOGY AND CULTURE*, 35:686–716 (October 1994). Discusses the decisions made prior to the 1927 formation of the Federal Radio Commission.

8-039 Glatzer, Hal. WHO OWNS THE RAINBOW? CONSERVING THE RADIO SPECTRUM. Indianapolis, IN: Howard W. Sams, 1984, 302 pp. Useful introduction to spectrum principles, issues, and some history. Diagrams, notes, index.

8-040 Hale, W.K. FREQUENCY ASSIGNMENT METHODOLOGY: AN ANNOTATED BIBLIOGRAPHY. Washington: Government Printing Office (NTIA Special Publication 80-10), 1980, 42 pp. About 60 citations in sections on general, engineering applications, algorithms, complexity of frequency assignment problems, and both upper and lower bounds.

8-041 Jackson, Charles. "The Allocation of the Radio Spectrum," *SCIENTIFIC AMERICAN*, 242:2:32–37 (February 1980). Very good introduction, with a color chart of U.S. allocations.

8-042 ** Joint Technical Advisory Committee [of the Institute of Radio Engineers and the Radio-Television Manufacturers Association]. RADIO SPECTRUM CONSERVATION: A PROGRAM OF CONSERVATION BASED ON PRESENT USES AND FUTURE NEEDS. New York: McGraw-Hill, 1952, 221 pp. See next item.

8-043 ** Joint Technical Advisory Committee [of the Institute of Electrical and Electronics Engineers and the Electronic Industries Association]. RADIO SPECTRUM UTILIZATION: A PROGRAM FOR THE ADMINISTRATION OF THE RADIO SPECTRUM. New York: IEEE, 1964, 272 pp. These two reports (see also item above) include considerable historical material on spectrum propagation research and management. Tables, diagrams, notes, bibliography, index.

8-044 * Kittross, John M. TELEVISION FREQUENCY ALLOCATION POLICY IN THE UNITED STATES. New York: Arno Press, 1979, 600 pp. Primarily a facsimile reprint of a 1960 dissertation, but including updated notes and a related article. Notes, tables, bibliography.

8-045 Kobb, Bennett Z. SPECTRUM GUIDE: RADIO FREQUENCY ALLOCATIONS IN THE UNITED STATES, 30Mhz–300GHz. Falls Church, VA: New Signals Press, 1995, 311pp. A descriptive "snap-shot" of American allocations, arranged by frequency, and offering some background.

8-046 *** Levin, Harvey J. THE INVISIBLE RESOURCE: USE AND REGULATION OF THE RADIO SPECTRUM. Baltimore: Johns Hopkins University Press, 1971, 431 pp. Largely an economic study of technical basics, with some history, this remains one of the finest books on the subject. Tables, notes, glossary, spectrum chart, index.

8-047 Pestre, Dominique. "Studies of the Ionosphere and Forecasts for Radiocommunications: Physicists and Engineers, the Military, and National Laboratories in

France (and Germany) after 1945," *HISTORY AND TECHNOLOGY,* 13:183–205 (1997).

8-048 Scholtz, Robert A. "The Origins of Spread—Spectrum Communications," *IEEE TRANSACTIONS ON COMMUNICATIONS,* COM-30:822–854 (May 1982).

8-049 Slotten, Hugh Richard. "Radio Engineers, the Federal Radio Commission and the Social Shaping of Broadcast Technology: Creating 'Radio Paradise,'" *TECHNOLOGY AND CULTURE,* 36:950–986 (October 1995). The role of spectrum allocation in early radio broadcasting. References.

8-050 _____. "Rainbow in the Sky: FM Radio, Technical Superiority, and Regulatory Decision-Making," *TECHNOLOGY AND CULTURE,* 37:686–720 (October 1996). Revisionist review of Edwin Armstrong's early work and the innovation of FM broadcasting.

8-051 Television Allocations Study Organization. "Report and Papers." *PROCEEDINGS OF THE IRE,* 48:991–1154 (June 1960). Special issue with results of the two-and-a-half-year industry study of television spectrum allocations, TASO being especially concerned with the UHF propagation problem. Photos, diagrams, tables, references.

8-052 U.S. Congress, Congressional Budget Office. **WHERE DO WE GO FROM HERE? THE FCC AUCTIONS AND THE FUTURE OF RADIO SPECTRUM MANAGEMENT.** Washington: Government Printing Office, 1997, 80 pp. One of the better resources on why auctions have been instituted, their outcome, and problems. Tables, charts.

8-053 U.S. Congress, Office of Technology Assessment. **RADIO FREQUENCY USE AND MANAGEMENT: IMPACTS FROM THE WORLD ADMINISTRATIVE RADIO CONFERENCE OF 1979.** Washington: Government Printing Office, 1982, 163 pp. Surveys the changing requirements, influences and motivations among nations in spectrum allocations, U.S. and international management of the process, alternatives. Tables, charts, glossary, bibliography

8-054 _____. **THE 1992 WORLD ADMINISTRATIVE RADIO CONFERENCE: ISSUES FOR U.S. INTERNATIONAL SPECTRUM POLICY—BACKGROUND PAPER.** Washington: Government Printing Office, 1991, 134 pp. Good survey of technologies and services, the international context for spectrum policy, etc. Figures, tables, notes, appendices, index.

8-055 _____. **THE 1992 WORLD ADMINISTRATIVE RADIO CONFERENCE: TECHNOLOGY AND POLICY IMPLICATIONS.** Washington: Government Printing Office, 1993, 190 pp. Discusses implications for U.S. radio technology. Tables, charts, appendices, index.

8-056 * U.S. Department of Commerce, National Telecommunications and Information Administration. **U.S. SPECTRUM MANAGEMENT POLICY: AGENDA FOR THE FUTURE.** Washington: Government Printing Office, 1991, 200 pp. A good status report which focuses on traditional block allocations and alternatives to it. Notes, appendices.

Chapter 8

8-057 _____. **SPECTRUM REALLOCATION FINAL REPORT.** Washington: Government Printing Office (NTIA Special Publication 95-32), 1995, 250 pp. The key document in an ongoing, multiyear process of shifting roughly 200 MHz of frequencies from government-only use to shared and commercial use. Tables, charts, maps.

8-058 _____. **REPORT OF THE INTERDEPARTMENT RADIO ADVISORY COMMITTEE.** Washington: NTIA, biennial. Each report covers the activities of the previous six months in coordination of government use of spectrum.

8-059 U.S. Department of Commerce, Telecommunication Science Panel of the Commerce Technical Advisory Board. **ELECTROMAGNETIC SPECTRUM UTILIZATION—THE SILENT CRISIS.** Washington: Government Printing Office, October 1966, 80 pp. Broad assessment of problems and prospects of over-utilized spectrum and what might be done about the issue. Charts, references, appendices.

8-060 * U.S. Executive Office of the President, Office of Telecommunications Policy. **THE RADIO FREQUENCY SPECTRUM: UNITED STATES USE AND MANAGEMENT.** Washington: Government Printing Office, 1975, 250 pp. Handbook of then-current practice with brief history. Useful details on major users and issues. Tables, charts.

8-061 U.S. Federal Communications Commission. **COMPATIBILITY FOR UHF TELEVISION: FINAL REPORT.** Washington: FCC, September 1980, 275 pp. Final conclusions of a multiyear staff study on how to bring UHF television stations to more of a comparable status with VHF outlets as far as station coverage and signal strength is concerned. Ironically, the goal was finally attained through cable television carriage of TV station signals. Tables, notes.

8-062 Webbink, Douglas. "The Value of the Frequency Spectrum Allocated to Specific Uses," *IEEE TRANSACTIONS ON ELECTROMAGNETIC COMPATIBILITY,* EMC-19:3:343–351 (August 1977). An economist assesses options. See next entry.

8-063 _____. **FREQUENCY SPECTRUM DEREGULATION ALTERNATIVES.** Washington: FCC Office of Plans and Policy Working Paper 2, October 1980. 52 pp. A forward-looking analysis of market options (such as auctions) other than traditional block allocations. Notes, appendices.

Chapter 9

Radio

Early wireless or radio history is found in experiments with and observations of electromagnetic waves (see chapter 8). For the lives and work of radio's many key inventors, see chapter 5. The 1920s witnessed the rise of radio as a major industry centered on electron-tube technology which is covered in chapter 11. Popular books and articles, as well as technical material, appeared in growing numbers after 1900 and those of substantial historical interest are included here.

As with the chapters on telegraphy and telephony, this chapter begins with Section A, a listing of the earliest (in this case the first 15 years) of English-language works on wireless (to 1908). Section B cites historical and modern material more selectively from 1909 to 1925. Section C describes modern survey histories. Section D notes material on radio and broadcasting after 1925 as radio became a large industry (subdivided into sections on general, foreign, and high frequency and shortwave). Section E includes works on specific devices (such as detectors) and patents. Section F provides an indicator of the revival of interest in old radio in more recent times. For related Internet resources, see 15-G.

A. Contemporary Works on Wireless (1892–1908)

(Parallel to similar sections on the telegraph [6-A] and telephone [7-A], this check list records the initial 15 years of English-language wireless works—in this case including articles—listed here chronologically and then by author within each year. See also the closely-related material cited in chapter 8-C. Only the first decade and a half is covered in this case, as the vast majority of titles by that point and later were how-to guides with little historical content.)

9-001 Crookes, William. "Some Possibilities of Electricity," **FORTNIGHTLY REVIEW,** 51:173–181 (February 1892). This famous paper contains a prediction of the means for radio telegraphy: transmitter, receiver, directive antenna, circuit tuning, and selectivity.

See Hertz, 5-141.

See Lodge, 8-010.

9-002 ** Kerr, Richard. **WIRELESS TELEGRAPHY POPULARLY EXPLAINED**. London: Seeley/New York: Scribner's, 1898, 111 pp. ; 1899 [2nd ed.]; 1900 [3rd ed.], 116 pp.; 1901 [5th ed.], 116 pp.; 1902 [6th ed.]; 1903 [7th ed.], 120 pp. The first English-language book devoted to wireless. Eight brief chapters on the background science, induction experiments and Hertzian waves. (As was then typical with British works, several of these "editions" appear in fact to be merely reprintings.) Photos, diagrams, no index.

9-003 Thompson, Silvanus P. "Telegraphy across Space," *JOURNAL OF THE ROYAL SOCIETY OF ARTS,* 46:453–460 (1898); reprinted in **SMITHSONIAN INSTITUTION ANNUAL REPORT, 1898**. Washington: Government Printing Office, 1899, pp. 235–247. A brief sketch of communication without wires by the conduction, induction, and electric wave methods.

9-004 *** Fahie, John J. **A HISTORY OF WIRELESS TELEGRAPHY 1838–1899, INCLUDING SOME BARE-WIRE PROPOSALS FOR SUBAQUEOUS TELEGRAPHS**. London: Blackwood/New York: Dodd, Mead, 1899, 325 pp.; 1901 [2nd ed., reprinted by Arno Press "History of Broadcasting," 1971], 348 pp. Very important historical record. Divided into three parts: the possible (through the work of Mahlon Loomis in 1872); the practicable (through to the mid-1890s); and the practical (Preece, Smith, and Marconi methods), plus several appendices. Diagrams, notes, index.

9-005 * Marconi, Guglielmo. "Wireless Telegraphy," *PROCEEDINGS OF THE ROYAL INSTITUTION* 16:247–256 (1899); *JOURNAL OF THE IEE,* 28:273–297, 300–316 (1899). A survey of his work up to early 1899. Reprinted in Eastwood, 9-101.

9-006 * Bottone, Stanley R. **WIRELESS TELEGRAPHY AND HERTZIAN WAVES**. London: Whittaker, 1900, 116 pp.; 1901 [2nd ed.]; 1905 [3rd ed.], 127 pp.; 1910 [4th ed.], 136 pp.; 1912 [5th ed.], 136 pp. Pitman, 1919 [6th ed.], 133 pp. A simple account with some construction details for amateurs. Useful historical material to p. 60. A pioneering practical book in the field. Diagrams, index.

9-007 * Marconi, Guglielmo. "Signals across the Atlantic," *ELECTRICAL WORLD* 38:1023–1025 (December 21, 1901). His own report on the historic Morse code "S" transmissions. See also Coulson, 9-052.

9-008 Tunzelmann, George W. von **WIRELESS TELEGRAPHY: A POPULAR EXPOSITION**. London: The Office of Knowledge, 1901, 104 pp.; 1902 [2nd ed.], 104 pp. Illustrations. From a 1900 series of articles in *KNOWLEDGE*.

9-009 Marconi, Guglielmo. "The Progress of Electric Space Telegraphy," *PROCEEDINGS OF THE ROYAL INSTITUTION,* 17:195–210 (1902); also *ELECTRICIAN* 49:388–392 (June 27, 1902). Survey of his work from 1900. Diagrams. Reprinted in Eastwood, 9-101.

9-010 Trevert, Edward [pseud.]. **THE ABC OF WIRELESS TELEGRAPHY: A PLAIN TREATISE ON HERTZIAN WAVE SIGNALING, EMBRACING**

THEORY, METHODS OF OPERATION, AND HOW TO BUILD VARIOUS PIECES OF THE APPARATUS EMPLOYED. Lynn, MA: Bubier Publishing, 1902, 116 pp.; 1904 [2nd ed.], 116 pp.; 1911 [3rd ed.]. Includes a 24 page list of station call letters in the final edition.

9-011 * Blaine, Robert Gordon. AETHERIC OR WIRELESS TELEGRAPHY. London: Biggs and Sons, 1903, 232 pp.; 1906, 1909. Valuable comparison of different national systems at this early date. Diagrams, photos, notes, index.

9-012 Ernst, Maurice, comp. WIRELESS TELEGRAPHY AND TELEPHONY. London: S. Rentell, 1903, 31 pp. Practical guide. Illustrations.

9-013 Fleming, John A. "Hertzian Wave Wireless Telegraphy," *POPULAR SCIENCE MONTHLY* (June 1903+), 125 pp. A series of seven articles covering theory and the history of both transmitters and detectors. Diagrams.

9-014 Hudgins, Lt. J[ohn] M.. INSTRUCTIONS FOR THE USE OF WIRELESS-TELEGRAPH APPARATUS. Washington: Government Printing Office (U.S. Navy, Bureau of Equipment), 1903, 29 pp. Appears to be first U.S. Navy radio publication. Diagrams. See also Robison, 9-035.

9-015 THE MARCONI WIRELESS TELEGRAPH SYSTEM. New York: Munroe & Munroe, 1903, 22 pp. Brief description.

9-016 Maver, William Jr. "Wireless Telegraphy—Its Past and Present Status and Its Prospects," SMITHSONIAN INSTITUTION ANNUAL REPORT FOR 1902. Washington: Government Printing Office, 1903, pp. 261–274. A summary of events from the early 1890s. Photos, diagrams.

9-017 ** Sewall, Charles Henry. WIRELESS TELEGRAPHY: ITS ORIGIN, DEVELOPMENT, INVENTIONS, AND APPARATUS. New York: van Nostrand / London: Crosby, Lockwood, 1903, 229 pp. Overview of history, principles, theory, apparatus, and practical operation. Diagrams, photos, appendices, index.

9-018 * DEVELOPMENT OF WIRELESS TELEGRAPHY. New York: American DeForest Wireless Telegraph Co., nd, but 1904, 26 pp. Interesting narrative booklet prepared for sale (for ten cents!) at the St. Louis World's Fair. Useful example of early promotion aimed at consumers. Photos.

9-019 INTERNATIONAL CONFERENCE ON WIRELESS TELEGRAPHY: PRELIMINARY CONFERENCE AT BERLIN, AUGUST 1903. London: HMSO, 1904, 61 pp. Translation by G.R. Nielson of the official report of the first international conference on wireless.

9-020 Johnson, A[rthur] T[homas] M[etcalf]. ELECTRIC FLASHES; OR, THE SYSTEMS OF WIRELESS TELEGRAPHY... London: R.A. Everett, 1904, 174 pp.

9-021 Maver, William Jr. "Progress in Wireless Telegraphy," SMITHSONIAN INSTITUTION ANNUAL REPORT FOR 1903. Washington: Government Printing Office, 1904, pp. 275–280. On recent work, especially with coherers and other detectors. Diagrams.

Chapter 9

9-022 _____. **MAVER'S WIRELESS TELEGRAPHY: THEORY AND PRACTICE.** New York: Maver Publishing, 1904, 199 pp.; and as **MAVER'S WIRELESS TELEGRAPHY AND TELEPHONY: A HANDBOOK.** 1910 [4th ed.], 316 pp. Basic text.

9-023 Moore, Edmund B. **WIRE AND WIRELESS TELEGRAPHY.** Springfield, VT: Reporter Publishing Co, 1904, 38 pp. Describes the history of the telegraph from its birth to the time of publication. Compares various systems of wireless and considers role of wireless at sea. Diagrams.

See Poincaré and Vreeland, 8-007.

9-024 SIGNALING: MARCONI WIRELESS TELEGRAPHY EXPLAINED. Glasgow: n.p., 1904, 107 pp.

9-025 * Story, Alfred T. **THE STORY OF WIRELESS TELEGRAPHY.** London: Newnes/New York: Appleton, 1904, 215 pp.; London: Hodder & Stoughton, 1912 (reprinted in 1971), 225 pp. Conduction and induction experiments, experimental work of Hertz and Lodge and their contemporaries, Marconi, and other pioneering work. Diagrams, index.

9-026 WIRELESS TELEGRAPHY: REPORT OF THE INTER-DEPARTMENTAL BOARD APPOINTED BY THE PRESIDENT TO CONSIDER THE ENTIRE QUESTION OF WIRELESS TELEGRAPHY IN THE SERVICE OF THE NATIONAL GOVERNMENT. Washington: Government Printing Office, 1904, 25 pp. The "Roosevelt Board" offered a number of recommendations in its 11-page report, supplemented by several appendices. Reprinted in Kittross, 4-031.

9-027 ** Collins, A. Frederick. **WIRELESS TELEGRAPHY: ITS HISTORY, THEORY AND PRACTICE.** New York: McGraw, 1905, 299 pp. A brief historical introduction is provided in 13 of 20 chapters. A thorough coverage of contemporary theory and practice with chapters on oscillators, induction coils, interruptors, detectors, transmitters, receptors, subsidiary apparatus, aerials and earths, syntonization, and wireless telegraphy. Diagrams, photos, notes, index of names.

9-028 Marconi, Guglielmo. "Recent Advances in Wireless Telegraphy," **SMITHSONIAN INSTITUTION ANNUAL REPORT FOR 1905.** Washington: Government Printing Office, 1906, pp. 131–145. Based on a Royal Institution paper delivered March 3, 1905. Illustrations. Reprinted in Eastwood, 9-101.

9-029 Collins, A. Frederick. **MANUAL OF WIRELESS TELEGRAPHY AND TELEPHONY.** New York: John Wiley, 1906, 1909 [2nd ed.], 250 pp.; 1913 [3rd ed.], 300 pp. Theory, commercial apparatus, transmitters, receivers, wiring diagrams, adjustments and operations, different systems, antennas and components. Photos, glossary, annotated list of books on wireless telegraphy, index.

9-030 _____. **WIRELESS TELEGRAPHY; INSTRUCTION PAPER.** Chicago: American School of Correspondence of Armour Institute of Technology, 1906, 50 pp.

Radio

9-031 Eichhorn, Gustave. **WIRELESS TELEGRAPHY**. London: Charles Grifin/Philadelphia: J.B. Lippincott, 1906, 116 pp. Mathematical approach to the Telefunken system. Diagrams.

9-032 *** Fleming, John A[mbrose]. **THE PRINCIPLES OF ELECTRIC WAVE TELEGRAPHY AND TELEPHONY**. London: Longmans, Green, 1906–1919, four eds. listed below. A foremost treatise and a valuable record of early development, though historical material is found primarily in the first two editions. All are valuable for their discussion of contemporary methods and thinking. Photos, line drawings, tables, index.

> **1906** (title does not include AND TELEPHONY, 671 pp.): Chap. 7 (pp. 419–463) discusses the inception of wireless to 1905 with an emphasis on the work of Marconi, Fleming, and Lodge, while chap. 8 (pp. 465–544) goes on to discuss development with progress by all workers after 1897. Chapters include extensive reference and detailed patent information. Four appendices include the text of the 1904 British law and a bibliography of 32 books and 44 original papers and lectures.

> **1910** (Title changed to add AND TELEPHONY; 2nd ed., 906 pp.): Chap. 7 (pp. 511–635) details the evolution of electric wave telegraphy through 1909. Chap. 10 (pp. 844–865) covers radiotelephony. Appendices add the Berlin 1906 protocol and service regulations. Bibliography has 45 entries. Chapter 7 reprinted in Shiers, 9-080.

> **1916** (3rd ed., 911 pp.): To accommodate new developments, most historical material was removed. Chapter 8 (pp. 720–791) expanded to include military and aeronautical applications of radiotelegraphy. Appendices and bibliography from earlier editions also deleted.

> **1919** (4th ed., 707 pp.): Deletes most historical material and all appendices to allow full inclusion of contemporary developments. Annotations include many references to patents.

9-033 Kennelly, Arthur E. **WIRELESS TELEGRAPHY AND TELEPHONY: AN ELEMENTARY TREATISE**. New York: Moffat, Yard "Present Day Primers," 1906, 211 pp.; 1909 [2nd ed.], 279 pp.; 1910; 1915; 1917. Photos, diagrams, index.

9-034 Mazzotto, Domenico, Prof., trans. by S.R. Bottone. **WIRELESS TELEGRAPHY AND TELEPHONY**. London and New York: Whittaker, 1906, 416 pp. Diagrams, photos.

9-035 Robison, Cmdr. S[amuel] S. **MANUAL OF WIRELESS TELEGRAPHY (RADIO) FOR THE USE OF NAVAL ELECTRICIANS**. Annapolis: U.S. Naval Institute/London: S. Rentell, 1906, 144 pp.; 1909 [2nd ed.], 129 pp.; 1911 [2nd rev. ed.], 212 pp.; 1913 [3rd ed.], 241 pp.; 1920, 307 pp. Later editions included revisions and additions by Capt. D.W. Todd and Lt-Cmdr. S.C. Hooper. Diagrams, tables, appendices, index. See also Hudgins, 9-014.

9-036 St. John, Thomas M. **WIRELESS TELEGRAPHY FOR AMATEURS AND STUDENTS: CONTAINING THEORETICAL AND PRACTICAL INFORMATION, TOGETHER WITH COMPLETE DIRECTIONS FOR PERFORMING**

NUMEROUS EXPERIMENTS ON WIRELESS TELEGRAPHY WITH SIMPLE HOME-MADE APPARATUS. New York: Author, 1906, 171 pp.

9-037 Tissot, C.P. RESONANCE OF ANTENNA SYSTEMS IN WIRELESS TELEGRAPHY. 1906.

9-038 White, William J. WIRELESS TELEGRAPHY AND TELEPHONY. London: Pitman "Shilling Scientific Series," 1906, 173 pp.; 1912 [2nd ed.], 202 pp.

9-039 * Erskine-Murray, James. A HANDBOOK OF WIRELESS TELEGRAPHY: ITS THEORY AND PRACTICE, FOR THE USE OF ELECTRICAL ENGINEERS, STUDENTS, AND OPERATORS. London: Crosby, Lockwood/New York: van Nostrand, 1907, 322 pp.; 1909 [2nd ed.], 370 pp.; 1911 [3rd ed.]; 1913 [4th ed.]; 1914 [5th ed.], 442 pp.; 1915 [6th ed.], 442 pp; 1918 [7th ed.]. A growing volume that became a standard textbook and reference volume. It offers historical material in an early chapter, plus later chapters comparing then-contemporary wireless systems with details on apparatus and trials. Photos, line drawings, charts, tables, index.

9-040 Howgrave-Graham, Robert P. WIRELESS TELEGRAPHY FOR AMATEURS. London: Percival Marshall, 1907, 160 pp.; 1909; 1912; 176 pp.; 1914.

9-041 Fleming, John Ambrose. AN ELEMENTARY MANUAL OF RADIO-TELEGRAPHY AND RADIOTELEPHONY FOR STUDENTS AND OPERATORS. London: Longmans, Green 1908, 340 pp.; 1911 [2nd ed.]; 1916 [3rd ed.], 360 pp. Includes an account of early history. Diagrams.

9-042 Massie, Walter W., and Charles R. Underhill. WIRELESS TELEGRAPHY AND TELEPHONY POPULARLY EXPLAINED. New York: D. van Nostrand, 1908, 76 pp. Includes a chapter contributed by Nikola Tesla.

9-043 Monckton, C.C.F. RADIO-TELEGRAPHY. London: Archibald Constable/New York: van Nostrand "Westminster Series," 1908, 289 pp. Photos, diagrams.

9-044 Ruhmer, Ernst, trans. by James Erskine-Murray. WIRELESS TELEPHONY IN THEORY AND PRACTICE. London: Crosby, Lockwood/New York: van Nostrand, 1908, 237 pp. Diagrams, bibliography, index.

B. Wireless and Radio (1909–1925)
(These largely contemporary works for the 15 years from 1909 to 1925 are cited more selectively than in the previous section. Arrangement is by author.)

9-045 Alexanderson, E.F.W. "Trans-Oceanic Radio Communication," *PROCEEDINGS OF THE IRE,* 8:263–285 (1920). The GE engineer describes his alternator in design and application.

9-046 * Ashley, Charles G., and Charles B. Hayward. WIRELESS TELEGRAPHY AND WIRELESS TELEPHONY: AN UNDERSTANDABLE PRESENTATION OF THE SCIENCE OF WIRELESS TRANSMISSION OF INTELLIGENCE. Chicago: American Technical Society, 1909, 1911, 1916, 141 pp. Includes one of the earliest discussions of wireless telegraphy in aeronautics. Diagrams, photos, index.

9-047 * Beare, Robert W. **MODERN WIRELESS: THE CONSTRUCTION AND CONTROL OF ALL TYPES OF RADIO SETS SIMPLY EXPLAINED...** London: Virtue, nd but 1925, 3 vols., approx 250 pp. each. Broad and quite detailed survey of British practices in manufacturing and operation. Diagrams, photos.

9-048 * Brittain, James E. "The Alexanderson Transoceanic Radio System," *PROCEEDINGS OF THE IEEE*, 72:625–633 (May 1984). The rise and application of the alternator. Photos, map, diagrams. See also the author's detailed biography of the inventor, 5-038.

9-049 Bucher, Elmer E. **PRACTICAL WIRELESS TELEGRAPHY: A COMPLETE TEXT BOOK FOR STUDENTS OF RADIO COMMUNICATION.** New York: Wireless Press, 1917, 322 pp.; 1921 [2nd ed.], 336 pp. A widely used textbook with useful historical material. Photos, diagrams, charts, index.

9-050 Coleman, Arthur, and James B. Harrietts. **RADIO STATIONS OF THE WORLD.** New York: n.p., 1916, 197 pp. A tabular listing, by country.

9-051 Collins, Francis A. **THE WIRELESS MAN: HIS WORK & ADVENTURES ON LAND & SEA.** New York: Century, 1912, 251 pp. Aimed at younger readers, this narrative surveys real-life wireless operator experiences. Photos, no index.

9-052 Coulson, Thomas. "Radio's Memorable Anniversary," *JOURNAL OF THE FRANKLIN INSTITUTE*, 253:287–292 (April 1952). On the 50th anniversary of transatlantic radio. See also, Marconi, 9-007.

9-053 Coursey, Philip R. "The Methods Employed for the Wireless Communication of Speech," *WIRELESS WORLD*, 4:47–52, 90–96, 216–223, 305–313, 385–387 (April–August 1916). Useful five-part technical paper with historical treatment: arc, spar, vacuum tube and alternator systems, modulation, receivers. Photos, diagrams, references. Reprinted in Shiers, 9-080.

9-054 _____. **TELEPHONY WITHOUT WIRES.** London: Wireless Press, 1919, 414 pp. A comprehensive coverage of techniques, devices, and developments, especially from 1905 to 1918. Photos, diagrams, references, index.

See de Forest, 11-016.

9-055 * de Forest, Lee. "Recent Developments in the Work of the Federal Telegraph Company," *PROCEEDINGS OF THE IRE*, 1:37–57 (January 1913); reprinted in *PROCEEDINGS OF THE IEEE*, 51:426–433 (March 1963). Description of the company installations and operations including Poulsen arc generators and other transmitters, service performance, frequency-shift keying and selective fading. Photos, diagrams. Reprinted in Shiers, 9-080.

9-056 * Dowsett, H.M. **WIRELESS TELEPHONY AND BROADCASTING.** London: Gresham Publishing, 1925, 2 vols., 210 and 233 pp. A technical exposition and a broad survey of the radio art in Britain up to the mid-1920s with several historical chapters. Photos, diagrams, maps, notes, biographical notes, index.

9-057 Eccles, William H. **WIRELESS TELEGRAPHY AND TELEPHONY: A HANDBOOK OF FORMULAE, DATA AND INFORMATION**. London: Benn Brothers "The Electrician Series," / New York: van Nostrand, 1918 [2nd ed.], 514 pp. There is much historical material including names, dates and other particulars, especially patents. Photos, tables, diagrams, glossary, index

9-058 Elwell, Charles F. **THE POULSON ARC GENERATOR**. London: Ernest Benn/New York: van Nostrand, 1923, 192 pp. The major source on this device by its major American promoter. Some history. Photos, diagrams, bibliography, index.

9-059 _____. "Radio: Its Past, Present, and Future," *JOURNAL OF THE ROYAL SOCIETY OF THE ARTS*, 74:757–769 (1925). Brief survey.

9-060 * Fessenden, Reginald A. "Wireless Telephony," *PROCEEDINGS OF THE AIEE*, 27:558–627 (July 1908); reprinted in **SMITHSONIAN INSTITUTION ANNUAL REPORT FOR 1908**. Washington: Government Printing Office, 1909, pp. 161–195. An important early paper. Brief history of wireless signaling and wireless telephony, description of methods, apparatus, circuits and operation, discussion of possibilities. Photos, diagrams, references. Reprinted in Shiers 9-080.

See Fleming, 11-019.

9-061 Fleming, John A. **THE WONDERS OF WIRELESS TELEGRAPHY EXPLAINED IN SIMPLE TERMS FOR THE NON-TECHNICAL READER**. London: Society for Promoting Christian Knowledge, 1913, 280 pp. and later editions to 1923. An explanation for nontechnical readers includes history and some early circuits.

9-062 _____. "Radiotelegraphy: A Retrospect of Twenty Years," *ELECTRICIAN*, 77:831–836 (September 15, 1916). Photos, diagrams.

9-063 * Gibson, Charles R. **WIRELESS TELEGRAPHY AND TELEPHONY WITHOUT WIRES**. London: Seeley, Service/Philadelphia: J.B. Lippincott, 1914, 156 pp. Revised, with William B. Cole, as **WIRELESS OF TO-DAY: DESCRIBING THE GROWTH OF WIRELESS TELEGRAPHY & TELEPHONY FROM THEIR INCEPTION TO THE PRESENT DAY**. London: Seeley, Service "Science of To-Day Series," 1924, 318 pp., which retains the historical material and expands upon it. A popular historical treatment. Photos, diagrams, index.

9-064 * Goldsmith, Alfred N. **RADIO TELEPHONY**. New York: Wireless Press, 1918, 247 pp. The first comprehensive text of the period with numerous circuit diagrams and pictures of apparatus. Generators, modulation, antennas, and reception. Photos, diagrams, notes, index.

9-065 ** Jensen, Peter R. **EARLY RADIO: IN MARCONI'S FOOTSTEPS, 1894 TO 1920**. Kenthurst, Australia: Kangaroo Press, 1994, 176 pp. Three parts: a traveler's history of radio which reviews "then and now" at Marconi sites in Italy, Ireland, and England; the replication projects showing how to rebuild early wireless devices; and appendices (key documents). Photos, maps, diagrams, references, index.

9-066 Lescarboura, Austin C. **RADIO FOR EVERYBODY**. New York: Scientific American Publishing, 1922, 334 pp.; 1923 [revised ed], 353 pp.; 1924 [rewritten ed.], 361 pp. Widely used basic guide with discussion of radio's uses and equipment as well as good photos and diagrams. Index.

9-067 Lodge, Oliver J. "The Origin or Basis of Wireless Communication," *NATURE*, 111:328–332 (March 10, 1923). See also Lodge, 5-148 and 8-010.

9-068 * Low, A.M. **WIRELESS POSSIBILITIES**. New York: E.P. Dutton "To-day and To-morrow Series," 1924, 77 pp. Interesting prognostications on the changing roles of sound, radio television, and wireless at war. No index.

9-069 MARCONI AND HIS SOUTH WELLFLEET WIRELESS. Chatham, MA: The Chatham Press "Cape Cod History Guide Vol. 1," 1969, 30 pp. See also Whately, 9-086.

9-070 *** Marconi Company. **THE YEAR-BOOK OF WIRELESS TELEGRAPHY AND TELEPHONY**. London: The Wireless Press, 1913–25, annual (13 volumes published). Invaluable annual record of development, of immense value for company, key personnel, events, patents, and legal specifics of this era. Regular features in all volumes include: calendar, historical record to that point, current version of the International Radio Convention, lists of land and ship wireless stations, glossary, tables of useful data, patents issued in the previous year, biographical and obituary notices, list of books and periodicals published in different countries, directory of wireless societies, laws and regulations in various countries, etc. About 100 pages contain special articles by noted authorities. Volumes through 1920 include photographs of wireless installations and key personnel. Advertisements for services, shipping companies, equipment, and supplies usually occupy about 100 pages of front and end matter (less in later volumes). Through 1921, volumes include a foldout color map of world wireless stations tipped into the book near the end (now often missing from remaining copies as it was intended for wall use). In 1922 this was changed to a map showing distances from London. The 1923–25 volumes include a color atlas section in the main body of the volume. Each issue is a mine of contemporary information despite the predominance of Marconi Company emphasis. More specifically the special articles in each volume include:

> **1913** (564 pp.): Electrical measurements in wireless telegraphy, wireless time signals, the wireless direction finder, distress signaling, wireless telegraphy and the mercantile marine, the Marconi system, principles of wireless telegraphy explained by mechanical analogies, syntony, the technical situation of radiotelephony, the wireless telegraphy transmitter, wireless telegraphy for military purposes, facts and theories of long-distance signaling, and methods of producing continuous waves.
>
> **1914** (745 pp.): Waves and wave motion, function of the atmosphere in transmission, measurement of the strength of wireless signals, problems of wireless telephony, application of wireless telegraphy to meteorology, wireless

time signals and longitudes, international time and weather signals, radiotelegraphic investigations.

1915 (804 pp.): Function of the earth in radiotelegraphy, radiotelephony, international radiotelegraphic research during 1914, wireless and the war at sea, influence of wireless telegraphy on modern strategy, long-distance services, wireless newspapers at sea, some applications of radiotelegraphy, application of wireless telegraphy to meteorology, wireless telegraphy in survey.

1916 (880 pp.): Intelligence in naval warfare, photoelectric phenomena, the Allies' strategy in 1915, capacitance, inductance and wavelengths of antenna, wireless waves in World War I, the progress of radiotelephony in the U.S. during 1915, measurement of signal intensity, the problems of interference, report of the committee on standardization.

1917 (940 pp.): Fleming valve and de Forest Audion (U.S. decision), heroic wireless operators, the electric arc as a generator of persistent electric oscillations, the wireless drama, ionic valves, inductance capacity and natural frequency of aerials, the Heaviside layer, wireless to the rescue, some features of the long-distance stations of the American Marconi company, achievements of wireless telegraphy.

1918 (1,158 pp.): Waves in water, air, earth, and aether; some recent examples of devotion to duty displayed by ship's telegraphists, on the energy transmission in wireless telegraphy, how wireless telegraphy has been affected by America's entrance into the World War, the magnetic behavior of iron in alternating fields of radio frequency, wireless possibilities, a record of lifesaving at sea.

1919 (1,160 pp.): Maxwell's electromagnetic theory of light, maps for radiotelegraphy and aeronautics, a review of the methods and progress of radiotelephony, determination of electrical and acoustic characteristics of telephone receivers, and some radio-frequency phenomena.

1920 (1,148 pp.): Technical progress in France, Germany, Great Britain, Holland, Italy, Japan, Norway, and the U.S.; the progress of wireless telephony, wireless telephony and its application to aircraft, and direction finding.

1921 (1,355 pp.). Technical progress of radiotelegraphy in Canada, China, France, Germany, Italy, Japan, Netherlands, Norway, Spain, United Kingdom, and the U.S. during 1920; some outstanding problems of radio reception-interference, the radio compass, historical landmarks in wireless invention, valve amplifiers in shipboard use, aviation section.

1922 (1,477 pp.): Technical progress in Australia, Canada, China, France, Germany, Britian, Holland, Italy, Japan, New Zealand, Norway, Spain, and the U.S. during 1921; the recording of wireless signals, the Earthing resistance of antennae, the birth and early history of long distance wireless telegraphy, the rectification effect in its relation to the composition and structure of crystals.

1923 (about 1,500 pp.—volume is paginated in many separate sections): Technical development in Australia, Canada, China, France, Germany, Britain,

Holland, Italy, Japan, New Zealand, and the U.S.; recent progress in automatic reception, aerials, valve design and manufacture, multiple aerial arrangements, five papers on radio and aviation; papers on radio direction finding.

1924 (910 pp.): Direction-finding (two articles), scientific signals (meteorological, time, hydrographic, and general), recent experiments on atmospherics, dull emitter valves, the ideal empire chain, progress in wireless telegraphy in the British mercantile marine, and some polar diagrams of reception for systems of special aerials.

1925 (889 pp.): Some recent work at the National Physical Laboratory in connection with wireless research, broadcasting and loud speakers, allocation of wavelengths to prevent interference, atmospherics, and the design of a broadcasting station.

9-071 ** Marconi, Guglielmo. "Wireless Telegraphic Communication," in **NOBEL LECTURES, PHYSICS 1901–1921**. Amsterdam: Elsevier, 1967, pp. 196–222. Nobel Prize lecture, December 11, 1909. Diagrams, maps, photos, references. Reprinted in Shiers, 9-080.

9-072 _____. "Radiotelegraphy," **SMITHSONIAN INSTITUTION ANNUAL REPORT FOR 1911**. Washington: Government Printing Office, 1912, pp. 117-131. Based on Royal Institution paper delivered June 2, 1911. Photos, diagrams. Reprinted in Eastwood, 9-101; and Shiers, 9-080.

9-073 * _____. "Radio Telegraphy," *PROCEEDINGS OF THE IRE*, 10:215–238 (August 1922). A pioneer paper. Joint AIEE/IRE lecture in New York, June 20, 1922. Survey of early long-distance communications, recent developments in multiple-tube transmitters, receiver design, propagation observations, point-to-point transmission with reflectors on shortwaves (1–20 meters).

9-074 ** Marconi International Marine Co. **MARCONI WIRELESS TELEGRAPHY: A SHORT HISTORY OF ITS INVENTION, EVOLUTION, DEVELOPMENT**. London: Marconi, 1909, 102 pp.; 1911, 1920. Focuses, as the title suggests, on Marconi achievements. Photos.

9-075 * Marriott, Robert H. "United States Radio Development," *PROCEEDINGS OF THE IRE* , 5:179–197 (1917). Includes two useful historical charts. Reprinted in Shiers, 9-080.

9-076 Morse, A.H. **RADIO: BEAM AND BROADCAST—ITS STORY AND PATENTS**. London: Ernest Benn, 1925, 192 pp. Based on a series in *RADIO NEWS* (May–September 1925). Photos, illustrations, patent abstracts, bibliography, notes, index.

9-077 Phillips, V.J. "Without the Valve: Some Alternatives to Thermionic Devices," *PROCEEDINGS OF THE IEE*, Part A 136:313–320 (November 1989). Includes the Alexanderson Alternator, coherers, and other detectors. Diagrams, photo, references.

9-078 Pickworth, George. "Germany's Imperial Wireless System," *ELECTRONICS WORLD AND WIRELESS WORLD,* 99:427–432 (May 1993). The long-wave worldwide system prior to World War I is described showing that the German system was at least as impressive as Marconi's chain of stations.

9-079 Pierce, George W. **PRINCIPLES OF WIRELESS TELEGRAPHY.** New York: McGraw-Hill, 1910, 350 pp. Technical discussion with considerable historical material. Photos, diagrams, tables, notes, index.

9-080 *** Shiers, George, ed. **THE DEVELOPMENT OF WIRELESS TO 1920.** New York: Arno Press "Historical Studies in Telecommunications," 1977, ca 500 pp. This anthology reprints 20 pioneering technical and historical papers tracing developments from the late 19th century, many by the inventors themselves including Fleming, de Forest, Fessenden, Marconi, Carl Braun, Armstrong, Elwell, and Alfred Goldsmith. Most are listed separately here. Photos, diagrams, notes.

9-081 * Swierstra, N. Tj. "The Birth of Broadcasting," *EBU REVIEW* (1969), pp. 10-15. Argues case for a Dutch station which began regular transmissions in 1919, preceding KDKA by a year. Photos.

9-082 Taussig, Charles William. **THE BOOK OF RADIO.** New York: Appleton, 1922, 447 pp. Roughly half of this is a typical 1920s guide for amateurs, but the remainder includes useful information on radio stations for the government, in Europe, Radio Central, etc. Photos, diagrams, appendices, index.

9-083 U.S. Department of Commerce and Labor, Bureau of Navigation (later Radio Bureau). **RADIO STATIONS OF THE UNITED STATES.** Washington: Government Printing Office, 1913–1934, annual. Tabulation of land stations with details on transmitter, power, call letters, etc. Tables, notes.

See U.S. Federal Trade Commission, 4-072.

9-084 Verrill, A.H. **HARPER'S WIRELESS BOOK: HOW TO USE WIRELESS ELECTRICITY IN TELEGRAPHING, TELEPHONING, AND THE TRANSMISSION OF POWER.** New York: Harper "Practical Books for Boys," 1913, 185 pp. Diagrams, photos, index.

9-085 Wander, Tim. **2MT WRITTLE: THE BIRTH OF BRITISH BROADCASTING.** Stowmarket, England: Capella Publications, 1988, 178 pp. The first British broadcast station was operated by Marconi in 1922 in Writtle, just outside of Chelmsford. Photos, glossary, appendices, index.

9-086 Whately, Michael E. **MARCONI WIRELESS ON CAPE COD: SOUTH WELLFLEET, MASSACHUSETTS, 1901-1917.** n.p., 1987, 32 pp. Two booklets (see also 9-069) cover similar ground with similar photographs.

9-087 Yates, Raymond F., and Louis G. Pacent. **THE COMPLETE RADIO BOOK.** New York: Century, 1922, 330 pp. Useful for being about half devoted to various applications of radio. Photos, maps, diagrams, brief collection of biographies, index.

9-088 Zenneck, Jonathan, trans. by A.E. Seelig. **WIRELESS TELEGRAPHY.** New York: McGraw-Hill, 1915, 443 pp. Based on the German editions of 1909 and

1913. A first rate engineering text with a good deal of historical material. Photos, diagrams, tables, bibliography, index.

C. Survey Histories
(The material that follows is largely modern but includes a few earlier works that attempted to be comprehensive.)

9-089 ** "A Century of Wireless," *EBU REVIEW,* 263:2–96 (Spring 1995). Special issue on the centennial of early Marconi work, with several illustrated articles each on developments in the studio, transmitter, receiver, international frequency planning, and key pioneers. Photos, tables, bibliographies.

See Aitken, 8-001.

9-090 *** Aitken, Hugh G.J. **THE CONTINUOUS WAVE: TECHNOLOGY AND AMERICAN RADIO, 1900–1932.** Princeton, NJ: Princeton University Press, 1985, 588 pp. Continues the story of wireless from author's earlier book (8-001), focusing on American developments: Fessenden and the alternator, Elwell and the arc transmitter, de Forest and his Audion, radio and cables and the national interest, the development of RCA (three chapters), and the expansion of the business based on tube technology in the 1920s. Line drawings, notes, index.

9-091 Archer, Gleason L. **HISTORY OF RADIO TO 1926.** New York: American Historical Society, 1938 (reprinted by Arno Press "History of Broadcasting," 1971), 421 pp. While largely concerned with business aspects, this first volume of two (the other, BIG BUSINESS AND RADIO, 1939, includes little on technology) includes considerable pre-broadcast technical background and context, especially patent and related corporate rivalries. Photos, notes, appendices, index.

9-092 ** **THE AWA REVIEW.** Bloomfield, NY: Antique Wireless Association, 1986–date, approximately annual (11 vols. published through 1998). Excellent original research on all aspects of early wireless and radio. Each volume includes six to eight often quite extensive illustrated papers by AWA members which focus on the work of a key inventor, an important company, trends in receiver development and design, and the like. Photos, diagrams, references. Selected contents of individual issues include:

> **VOL. 1: 1986** (123 pp.): Farnsworth's television contribution, Atwater Kent's early radio development, Armstrong's 1934-35 FM tests, and John Stone Stone on Nicola Tesla.
>
> **VOL. 2: 1987** (135 pp.): DeForest's many radio companies from 1907–1920, an appraisal of spark gap engineering, a century of telegraph key development, and the first 50 years of Philco.
>
> **VOL. 3: 1988** (136 pp.): Program transmission and early radio networks, foreign and military telegraph keys, and Alexanderson Alternators.
>
> **VOL. 4: 1989** (156 pp.): Early GE development of vacuum tubes, the feminine touch in telecommunications, a decade of electroacoustic reproduction (1920–30), preserving early television history, and early microphone history.

VOL. 5: 1990 (191 pp.): General Electric and RCA, America's signals intelligence in World War I, radio intelligence in World War II, San Francisco's network broadcast centers of the 1930s, and key U.S. radio patents.

VOL. 6: 1991 (184 pp.): Early years of wireless in Hawaii, early television in Britain, who invented the superheterodyne, the military communications explosion in World War I, and more U.S. radio patents.

VOL. 7: 1992 (186 pp.): Early electronic tube development, Marconi's first Poldhu transmitter, a brief history of the valve audio amplifier, and evolution of broadcast transmitters.

VOL. 8: 1993 (154 pp.): How to recognize rare tubes and why they are rare, a new bibliography on Fessenden, the Federal telephone company, a history of National Electric Supply Co., and a glimpse at old-time transmitter development. [Note: no 1994 volume appeared.]

VOL. 9: 1995 (251 pp.): Nearly 100 pages review the life and apparatus of Marconi, the international contest for radar, the story of the Magnetron, and eight years of amateur licensing.

VOL. 10: 1996 (261 pp.): Major paper on Atwater Kent, the race for development of the radiotelephone to 1920, the 75th anniversary of RCA's "Radio Central" and the Collins Radio Company.

VOL. 11: 1998 (236 pp.): E.H. Scott Radio Labs, early vacuum tube radio research at Western Electric, and the national Company's coil-catacomb radios.

VOL. 12: 1999 (320 pp.) Issue is devoted to "The Atwater Kent Radios" by Ralph C. Williams, and includes coverage of all radios manufactured by the Philadelphia firm.

9-093   Not used.

See Blake, 9-144.

9-094 * BROADCASTING NETWORK SERVICE. New York: AT&T Long Lines Department, 1934, 53 pp. The terrestrial telephone line links described. Photos, maps (two foldout), diagrams.

9-095 Clarkson, R.P. **THE HYSTERICAL BACKGROUND OF RADIO.** New York: J.H. Sears, 1927, 257 pp. A light and nontechnical survey of electrical progress discussing inventors, experimenters, and organized research. Index.

9-096 Coe, Lewis. **WIRELESS RADIO: A BRIEF HISTORY.** Jefferson, NC: McFarland, 1996, 192 pp. Parallels the author's earlier histories of telegraphy (6-104) and telephony (7-073) and provides useful concise survey of the work of many inventors, bringing the story up to date. Photos, glossary, appendices, select bibliography, index.

9-097 Constable, A. **EARLY WIRELESS.** London: Midas "Collector's Library," 1980, 160 pp. Traces the history of wireless and radio communication.

9-098 * Dalton, W. M. **THE STORY OF RADIO**. London: Adam Hilger, 1975, 3 vols. Intended as an eight-volume set relating the full history of electronic media, only the first three, taking the story to about 1930, appeared. Useful for its British focus on the trend from wireless to radio broadcasting. Photos, line drawings, index. Specifically:

> **VOL. 1: HOW RADIO BEGAN.** (150 pp.) Takes the story to about 1920 in chapters on magnetism and electricity, electrical engineering, wireless-telegraphy, and the thermionic valve.
>
> **VOL. 2: EVERYONE AN AMATEUR.** (157 pp.) Focuses on the early 1920s with chapters on amateur wireless, the BBC, home radio constructors, and the rediscovery of shortwaves.
>
> **VOL. 3: THE WORLD STARTS TO LISTEN.** (154 pp.) Covers 1926 to 1930 in chapters on receiver measurements, the BBC, compromise in design (of receivers), and the allied fields of telephony, movies, and early television.

9-099 *** Douglas, Susan J. **INVENTING AMERICAN BROADCASTING, 1899–1922**. Baltimore: Johns Hopkins University Press, 1987, 363 pp. Readable yet scholarly analysis of the combination of technological innovation, institutional development, and both visions and business realities that led to the radio broadcasting business in the early 1920s. Chapters focus on Marconi as inventor-hero, the inventors' struggles for technical distinction, wireless telegraphy in the navy, the ups and downs of wireless as a business, the important role of amateur operators prior to World War I, initial radio regulation, the rise of military and corporate control, and the social construction of broadcasting. Photos, notes, index.

9-100 Dunlap, Orrin E. **THE STORY OF RADIO**. New York: Dial Press, 1927, 226 pp.; 1935 [2nd ed.], 326 pp. A useful journalistic account for general readers. Photos, index.

9-101 *** Eastwood, Sir Eric, ed. **WIRELESS TELEGRAPHY**. New York: John Wiley "Royal Institution's Library of Science," 1974, 391 pp. "Brings together [23] lectures presented at The Royal Institution during the years 1857 to 1932 which describe important advances in the science and practice of electrical communication." Many are listed separately here. The first six papers focus on aspects of telegraphy and telephony. Lecturers on wireless include Preece, Marconi (six examples are included here), Fleming (four examples), and W.H. Eccles. Photos, diagrams, notes, index.

9-102 Eccles, William H. **WIRELESS**. London: Thornton Butterworth, 1933, 256 pp. Semitechnical and popular survey from the experiments of Hertz into the 1930s. Useful study by a recognized authority includes a few patent references. Photos, bibliography, index.

9-103 Espenschied, Lloyd. "The Origins and Development of Radiotelephony," *PROCEEDINGS OF THE IRE*, 25:1101–1123 (September 1937). A unified account devoted primarily to American contributions from 1912 on. Photos, bibliography.

9-104 * **INTERNATIONAL CONFERENCE ON 100 YEARS OF RADIO**. London: IEE "Conference Publication No. 411," 1995, 260 pp. Includes 42 papers from a September 1995 meeting held by the IEE, the British Vintage Wireless Society, and the International Union of Radio Science, in London.

9-105 Jackson, G.A. "The Early History of Radio Interference," *IERE JOURNAL*, 57:244–250 (December 1987).

9-106 Kennedy, T.R. "From Coherer to Spacistor," *RADIO-ELECTRONICS*, 29:44–59 (April 1958). A half century of Gernsback's radio and allied publications with highlights of events in radio progress. Photos, list of Gernsback publications, 1908–1958.

9-107 Leinwoll, Stanley. **FROM SPARK TO SATELLITE: A HISTORY OF RADIO COMMUNICATION.** New York: Charles Scribner's, 1979, 242 pp. A wide-ranging popular history of radio into the 1970s, emphasizing the role of key inventors and developments in the expanding roles of the medium. Photos, drawings, references, index.

9-108 *** Maclaurin, W. Rupert. **INVENTION AND INNOVATION IN THE RADIO INDUSTRY.** New York: Macmillan, 1949 (reprinted, Arno Press "History of Broadcasting," 1971), 304 pp. Thorough treatment with supporting data and much critical analysis of the process and nature of radio inventions. A study of struggles, litigation, progress, and failure of both individual inventors and industrial organizations. Includes chapters on FM and television. Photos, diagrams, notes, tables, appendices (one details radio patent litigation), bibliography, index.

9-109 ** McNicol, Donald. **RADIO'S CONQUEST OF SPACE: THE EXPERIMENTAL RISE IN RADIO COMMUNICATION.** New York: Murray Hill Books, 1946 (reprinted by Arno Press, "Telecommunications," 1974), 374 pp. A semitechnical record of radio progress from the mid-19th century which is one of the best surveys of inventive achievement—both readable and reliable. Photos, diagrams, notes, index.

9-110 Muller-Fischer, Erwin. **CHRONOLOGICAL TABLE OF THE HISTORY OF RADIO ENGINEERING.** Berlin: Telefunken G.M.B.H., 1962. Not seen.

9-111 O' Day, William T. **RADIO COMMUNICATION: ITS HISTORY AND DEVELOPMENT.** London: HMSO (Science Museum), 1934 (reissued 1949), 95 pp. An authoritative survey based on objects in the Science Museum collections. Photos, references.

9-112 Schubert, Paul. **THE ELECTRIC WORD: THE RISE OF RADIO.** New York: Macmillan, 1928 (reprinted by Arno Press, "History of Broadcasting," 1971), 311 pp. A readable narrative of radio developments that touches lightly on technical matters. Its 12 chapters appear in three parts: the era of maritime adoption, the era of military use, and the era of popular use (broadcasting). Unfortunately, it lacks documentation or index.

9-113 * Sivowitch, Elliot N. "A Technological Survey of Broadcasting's Pre-History," *JOURNAL OF BROADCASTING*, 15:1–20 (Winter 1970–71). Highlights of experimental work from invention of the telephone to the advent of broadcasting. Includes work of Stubblefield, Edison, Dolbear, Thomson, Stone, Fessenden, and de Forest. References.

9-114 * Sturmey, S.G. **THE ECONOMIC DEVELOPMENT OF RADIO**. London: Gerald Duckworth, 1958, 284 pp. A thorough study of inventions, innovations, companies, and group activities in the British radio industry including marine and long-distance communications, broadcasting, and development of equipment and systems. Much of historical interest. Tables, notes, references, index.

9-115 Vyvyan, Richard N. **WIRELESS OVER THIRTY YEARS**. London: George Routledge, 1933 (reprinted by EP Publishing as **MARCONI AND WIRELESS**, 1974), 256 pp. A general and nontechnical account with personal reminiscences by a close associate of Marconi. Early history, transatlantic wireless, high-power stations, histories of Imperial Wireless, beam development, wireless during World War I, commercial development, broadcasting, and the work of the British Post Office. Photos, diagrams, tables, no index.

9-116 Wedlake, G.E.C. **SOS: THE STORY OF RADIO COMMUNICATION**. Newton Abbot, England: David & Charles/New York: Crane Russak, 1973, 240 pp. Popular and informal history of radio from its inception through World War II, with a special focus on the era around World War I. Photos, line drawings, chronology, bibliography, index.

D. Radio and Broadcasting after 1925

(The literature on radio broadcasting, which began in the early 1920s, is huge—the material included here focuses on the medium's technical background. See also chapter 10-E for stereo radio and 13-F for digital radio.)

General

9-117 ** Hammond, John H., and Ellison S. Purington. "A History of Some Foundations of Modern Radio-Electronic Technology," *PROCEEDINGS OF THE IRE* 45:1191–1208 (September 1957). This important paper reveals early work done at Hammond's laboratory, before and during World War I, most concerning remote control by radio. Photos, references. See also a rebuttal and commentary in L. Espenschied, "Discussion of 'A history....'," in the same journal 47:1253–1268 (July 1959). References. Taken together, these two papers contain highly personalized accounts of the background of some key radio research up to about 1930.

9-118 Horn, C.W. "Ten Years of Broadcasting," *PROCEEDINGS OF THE IRE*, 19:356–376 (March 1931). Brief review of the beginnings of broadcasting and subsequent development of receivers, transmitters, and networking. Photos, diagrams.

9-119 Landry, Robert J. **THIS FASCINATING RADIO BUSINESS**. Indianapolis: Bobbs-Merrill, 1946, 343 pp. Broad popular survey of the radio industry just before television entered the scene, written by the editor of the *Variety*, trade paper, and including numerous references to history and technical factors. Photos, index.

9-120 Sterling, George E. **THE RADIO MANUAL**. New York: van Nostrand, 1928, 666 pp.; 1929 [2nd ed.], 798 pp.; 1938 [3rd ed.], 1,120 pp.; 1950 [4th ed.], 888 pp. Author became an FCC commissioner. This standard text aimed at amateurs includes all aspects of radio-broadcast, public safety, aviation, and marine use, with television added in final edition. Photos, diagrams, index.

9-121 Terman, Frederick E. **RADIO ENGINEERING**. New York: McGraw-Hill, 1932, 688 pp.; 1937 [2nd ed.], 807 pp.; 1947 [3rd ed.], 969 pp.; 1955 [4th ed.], 1,074 pp. Standard textbook treatment designed for engineers with each edition showing changes in industry equipment and practice. Photos, diagrams, index.

Foreign

9-122 Amos, D.J. **THE STORY OF THE COMMONWEALTH WIRELESS SERVICE**. Adelaide, Australia: E.J. McAlister, 1936.

See BBC, 4-186 and 4-187.

9-123 Huth, Arno. **RADIO TODAY: THE PRESENT STATE OF BROADCASTING**. Geneva, Switzerland: Geneva Research Centre, Studies XII:6, July 1942 (reprinted by Arno Press "International Communication and Propaganda," 1972), 160 pp. Updates and summarizes the author's longer French-language study (1937) surveying the status of radio in all areas. Notes, tables, selected bibliography, no index.

See Macleod, 5-165.

9-124 Mott, Robert L. **RADIO SOUND EFFECTS: WHO DID IT, AND HOW, IN THE ERA OF LIVE BROADCASTING**. Jefferson, NC: MacFarland, 1993, 303 pp. Written by a radio sound-effects authority, this unique history reviews the development of sound effects in radio drama and comedy programs. Photos, diagrams, index.

9-125 Muscio, Winston T. **AUSTRALIAN RADIO: THE TECHNICAL STORY, 1923–1983**. Kenthurst: Kangaroo Press, 1984, 243 pp. Photos, bibliography, index.

9-126 Rosenthal, Eric. **YOU HAVE BEEN LISTENING...THE EARLY HISTORY OF RADIO IN SOUTH AFRICA**. Cape Town: Purnell, 1974, 165 pp. Published to mark the 50th anniversary of broadcasting in South Africa; the first four chapters cover early wireless development and amateur operations. Photos, index.

See NHK, 4-223 and 4-224.

See Pawley, 4-196.

9-127 Schwoch, James. **THE AMERICAN RADIO INDUSTRY AND ITS LATIN AMERICAN ACTIVITIES, 1900–1939**. Champaign: University of Illinois Press, 1990, 184 pp. American wireless companies, growth of broadcasting, the rise of a military-industrial complex, and international conferences on radio. Notes, bibliography, index.

9-128 Shepherd, Frank M. **TECHNICAL DEVELOPMENT OF BROADCASTING IN ASIA-PACIFIC, 1964–1984**. Kuala Lumpur, Malaysia: Asia Pacific Broadcasting Union, 1984, 324 pp. Includes overall trends and country-by-country assessment. Photos, maps, diagrams, no index.

9-129 Zimmerman, Arthur Eric. **IN THE SHADOW OF THE SHIELD: THE DEVELOPMENT OF WIRELESS TELEGRAPHY AND RADIO BROADCASTING IN KINGSTON AND AT QUEEN'S UNIVERSITY, 1902–1957**.

Kingston, ON: Queen's University, 1991, 658 pp. Detailed study with extensive quotations from contemporary papers and documents. Photos, diagrams, index.

FM, Short Wave, and High Frequency Radio
(See also chapter 3-F for international radio services, 4-D for the ARRL amateur radio organization, and 8-D for microwaves.)

9-130 ** Armstrong, Edwin H. "A Method of Reducing Disturbances in Radio Signaling by a System of Frequency Modulation," *PROCEEDINGS OF THE IRE,* 24:689–740 (May 1936); reprinted in *PROCEEDINGS OF THE IEEE,* 72:1041–1062 (August 1984). The pioneering paper on FM radio. Diagrams, photos, notes.

9-131 _____. "Evolution of Frequency Modulation," *ELECTRICAL ENGINEERING,* 59:485–493 (December 1940). Based on two lectures. Photos, references.

9-132 _____. "Some Recent Developments in the Multiplexed Transmission of Frequency Modulated Broadcast Signals," *PROCEEDINGS OF THE RADIO CLUB OF AMERICA,* 30:3–13 (October 1953). Historical survey from 1906 is followed by detailed discussion of the system problem, the receiver problem, and field tests. Photos, diagrams, notes.

9-133 ** Berg, Jerome S. **ON THE SHORT WAVES, 1923–1945: BROADCAST LISTENING IN THE PIONEER DAYS OF RADIO.** Jefferson, NC: McFarland, 1999, 272 pp. Three main sections trace the rise of the medium in the 1920s, short wave broadcasting and listening in the 1930s ("DXing"—the emphasis here), and wartime use of short wave services, with considerable discussion of equipment, QSL cards, short wave publications (an excellent historical survey), and station operations. Photos, notes, reading list, index.

9-133a **FREQUENCY MODULATION.** Princeton, NJ: RCA Review, January 1948, 515 pp. Dubbed as "Volume I," this includes RCA research papers and summaries from 1936 to 1947 on general topics (5), transmission (17), reception (11), and miscellaneous topics (12), plus a bibliography of 76 further papers from the same period. Photos, diagrams, references.

See Headrick, 3-089.

9-134 Kelsey, Elizabeth, comp. **TRAIL BLAZERS TO RADIONICS AND REFERENCE GUIDE TO ULTRA HIGH FREQUENCIES.** Chicago: Zenith Radio Corp., 1943, 85 pp. Two parts: the first 29 pages offer 45 short biographies from Pythagoras to Farnsworth with references; the second of 56 pages lists some 726 articles and books in 21 categories.

9-135 * Leutz, Charles R., and Robert B. Gable. **SHORT WAVES.** Altoona, PA: C.R. Leutz, 1930, 384 pp. Appears to be the first English-language book on the subject. First chapter is historical, while others review propagation, commercial and ship to shore radio, directional antennas, aircraft radio, broadcast receivers, medical applications, amateur radio, and television. Photos, diagrams, tables, index.

9-136 RADIO AT ULTRA-HIGH FREQUENCIES. Princeton, NJ: RCA Review, 1940, 1949, 2 vols. Papers by RCA engineers. Photos, diagrams, references. As follows:

> **VOL. I:** (1940, 456 pp.). Includes 23 papers from 1930–39 on transmission, propagation, relaying, measurement, reception, and television—plus abstracts of several more.
>
> **VOL. II: (1940-1947).** (1948, 485 pp.). Includes more than 20 papers and many summaries on antennas and transmission lines, propagation, reception, radio relays, microwaves, measurements and components, and aids to navigation. Bibliography of papers by RCA engineers.

9-137 Sleeper, Milton B., ed. **FM RADIO HANDBOOK: 1946 EDITION.** Great Barrington, MA: FM Company, 1946, 174 pp. Includes some historical material as well as contemporary sense of the industry. Photos, diagrams, tables.

9-138 THE STORY OF THE FIRST TRANS-ATLANTIC SHORT WAVE MESSAGE. New York: Radio Club of America, 1950, 78 pp. Commemorates 1921 IBCG transmissions, and reprints contemporary reports in *QST* and other press stories. Photos.

9-139 Wenstrom, William H. "Historical Review of Ultra-Short-Wave Progress," *PROCEEDINGS OF THE IRE,* 20:95–112 (January 1932). Primarily on the evolution of oscillator circuits; spark, regenerative, Barkhausen-Kurz, magnetrons, etc. Diagrams, references.

E. Circuits and Patents

(Includes patents and patent battles, but also specific devices, especially detectors. See also chapters 10 and 11.)

9-140 * Armstrong, Edwin H. "A New Method for the Reception of Weak Signals at Short Wave Lengths," *PROCEEDINGS, RADIO CLUB OF AMERICA* (December 1919), pp. 1–6. Pioneer paper on the superheterodyne circuit. Diagrams.

9-141 _____. "The Story of the Super-Heterodyne," *RADIO BROADCAST,* 5:198–207 (July 1924). Photos, diagrams. See next entry.

9-142 * _____. "The Super-Heterodyne—Its Origin, Development and Some Recent Improvements," *PROCEEDINGS OF THE IRE,* 12:539–552 (October 1924). As seen by its inventor.

9-143 * _____. "A Study of the Operating Characteristics of the Ratio Detector and Its Place in Radio History," *PROCEEDINGS, RADIO CLUB OF AMERICA,* 25:3-20 (November 1948). This important paper by the inventor of wideband FM critically appraises limiting circuits and covers much history. Diagrams, graphs, photos, tables, appendices.

9-144 *** Blake, George G. **HISTORY OF RADIO TELEGRAPHY AND TELEPHONY.** London: Radio Press, 1926; Chapman & Hall, 1928 (reprinted by Arno Press "Telecommunications," 1974), 425 pp. Valuable reference in the form of entries rather than a formally organized history, with special emphasis and reference to key patents on components, circuits, and systems. Includes early electrical signaling, early wireless, electromagnetic wave experiments, coherers, magnetic detectors, electrolytic and crystal detectors, early radio telephony and telegraphy, various kinds of transmitters, microphones, facsimile, thermionic valves and circuits. Photos, diagrams, notes, reference list of 1,125 items, index.

9-145   Branly, Eduoard. "Variations of Conductivity under Electrical Influence," *ELECTRICIAN*, 27:221–223, 448–449 (June 26, and August 21, 1891). A pioneer paper on the coherer.

9-146 * "The Centennial of the Semiconductor Diode Detector," *PROCEEDINGS OF THE IEEE*, 86:218–285 (January 1998). Seven historical papers largely devoted to the work of Sir J.C. Bose who developed the "self-recovering coherer" detector Marconi used to receive a signal across the Atlantic in 1901. Photos, diagrams, references.

9-147   Harrison, Arthur P. Jr. "Single-Control Turning: An Analysis of an Invention," *TECHNOLOGY AND CULTURE*, 20:296–321 (April 1979). Factors involved in development of the device after about 1925—specifically demand from a growing radio broadcast business. Photos, references.

9-148   Harrison, Charles W. "Some Aspects of the Genesis of Radio Engineering," *IEEE ANTENNAS AND PROPAGATION MAGAZINE*, 35:29–33 (December 1993). Early components.

9-149 *** Kraeuter, David W. **RADIO AND TELEVISION PIONEERS: A PATENT BIBLIOGRAPHY**. Metuchen, NJ: Scarecrow Press, 1992, 329 pp. Lists each of the patents of some 40 radio and television inventors—some 3,000 in all. Inventors are listed in alphabetical order. For each is given the name, patent number, date, and a citation to the Patent Office's *Official Gazette*. List of patent depository libraries, index.

9-150 *** _____. **BRITISH RADIO AND TELEVISION PIONEERS: A PATENT BIBLIOGRAPHY**. Metuchen, NJ: Scarecrow Press, 1993, 217 pp. Twenty-nine inventors (not all of them British—such as Bell and Marconi) are included totaling some 1,100 patents. For each the author provides title, number, and date. Index.

9-150a * _____. **RADIO AND ELECTRONICS PIONEERS: A PATENT BIBLIOGRAPHY**. Ann Arbor, MI: University Microfilms Research Abstracts, 1994, 309 pp. Continues the two volumes above (9-148, 9-149) with another 34 inventors in the same format. Index.

9-150b ** _____. **INDEX TO RADIO AND ELECTRONICS PATENTS: A KEYWORD INDEX TO THE TITLES OF OVER 6,400 U.S. AND BRITISH PATENTS ISSUED TO 100 INVENTORS, WITH AN EMPHASIS ON RADIO AND TELEVISION**. Ann Arbor, MI: University Microfilms Research Abstracts, 1995, 512 pp. Covers the preceding three volumes (9-148 through 9-150a), providing a 10,000-entry index to patents from the 1830–1980 period.

9-150c ** _____. **NUMERICAL INDEX TO RADIO AND ELECTRONICS PATENTS**. Ann Arbor, MI: University Microfilms Research Abstracts, 1997, 323 pp.

Covers the first three volumes of this series (9-148 through 9-150a) in patent number order.

9-151 Lodge, Oliver J. "The History of the Coherer Principle," *ELECTRICIAN*, 40:87–91 (November 12, 1897). The English innovator relates a story he was intimately involved with.

See Maclaurin, 9-108.

See McNicol, 9-109.

9-152 O'Dell, T.H. "Marconi's Magnetic Detector: 20th-Century Technique Despite 19th-Century Normal Science," *PHYSIS: REVISTA INTERNAZIONALE DI STORIA DELLA SCIENZA*, 25:525–548 (1983). Notes that while users of the device were pleased, scientists were negative though they did not test it. Includes comparison to work of Rutherford and Poulsen. References.

9-153 Pickworth, George. "The Spark That Gave Radio to the World," *ELECTRONICS WORLD AND WIRELESS WORLD*, 99:937–942 (November 1993). Rise of the spark gap transmitter.

9-154 _____. "Coherer-Based Radio," *ELECTRONICS WORLD AND WIRELESS WORLD*, 100:563–567 (July 1994). Continues paper above.

9-155 * Phillips, V[ivian] J. **EARLY RADIO WAVE DETECTORS**. London: Peter Peregrinus and the Science Museum "IEE History of Technology 1," 1980, 223 pp. Chapters on spark-gap detectors, coherers, electrolytic detectors, magnetic detectors, thin-film and capillary detectors, thermal detectors, tickers and tone-wheels, and other means of detecting wireless signals. Photos, diagrams, references, index.

9-156 Thrower, K.R. "Evolution of Circuit Design for AM Broadcast Receivers," *IERE JOURNAL*, 56:325–341 (October–December 1986).

F. Vintage Radios

(There is a vast and growing literature for radio receiver collectors—this is but an indication of what is now available.)

9-157 Alth, Max. **COLLECTING OLD RADIOS AND CRYSTAL SETS**. Des Moines, IA: Wallace-Homestead, 1977, 76 pp. Photos, index.

9-158 Bryant, John H., and Harold N. Cones. **THE ZENITH TRANS-OCEANIC: THE ROYALTY OF RADIOS**. Atglen, PA: Schiffer, 1995, 160 pp. A collector's guide to the high-end AM-FM-SW receivers of the 1940s. Photos (many in color), maps. See also Cone's overall history of the firm, 4-182.

9-159 Bunis, Marty, and Sue Bunis. **THE COLLECTOR'S GUIDE TO ANTIQUE RADIOS**. Paducah, KY: Collector Books, 1991, 175 pp.; 1992 [2nd ed.] 215 pp., 1995 [3rd ed.], 278 pp.; 1997 [4th ed.], 248 pp. Largely a price guide but useful for tracing developments.

Radio

9-160 _____. **COLLECTOR'S GUIDE TO TRANSISTOR RADIOS.** Paducah, KY: Collector Books, 1994, 254 pp.; 1996 [2nd. ed.], 320 pp. Identification and price guide of some 2,000 models, also useful for tracing overall developments.

9-161 Bussey, Gordon. **VINTAGE CRYSTAL SETS 1922–1927.** London: IPC Electrical-Electronic Press/Wireless World, nd., but ca 1976. 128 pp. Well-illustrated guide to this particular type of radio receiver. (See also the author's brief booklet on one of the more important British manufacturers: **THE STORY OF PYE WIRELESS** published by that company in 1979 and reprinted in 1981, 16 pp.)

9-162 Constable, Anthony. **EARLY WIRELESS.** Tunbridge Wells, England: Midas Books "Collectors Library," 1980, 160 pp. Photos, index.

9-163 Dachis, Chuck. **RADIOS BY HALLICRAFTERS.** Atglen, PA: Schiffer, 1996, 220 pp.; 1999 [2nd ed. "revised and expanded"]. The company was a primary maker of shortwave and other kit radios and related equipment, virtually all of which is described here with a collector's price guide. Photos (considerable color), bibliography, index.

9-164 Not used.

9-165 Hawes, Robert. **RADIO ART.** London: Greenwood Books, 1978, 96 pp. Wonderfully nostalgic romp through British wireless and radio receivers—and related objects—to about 1945. Photos, bibliography.

9-166 Not used.

9-167 Not used.

9-168 *Hawes, Robert. **RADIO! RADIO!** Bampton, England; Sunrise Press, 1986, 244 pp.; 1996 [3rd ed.], 320 pp. Comprehensive study in text and photos from Victorian wireless to radios of the 1960s. Photos, diagrams, index.

9-169 _____. **AUDIO! AUDIO! THE BRITISH HI-FI SPOTTER'S DIRECTORY OF CLASSIC AUDIO AMPLIFIERS AND CONTROL UNITS.** Bampton, England: Sunrise Press, 1995, 95 pp. Useful guide to early British high fidelity equipment. Photos.

9-170 Lane, David, and Robert Lane. **TRANSISTOR RADIOS: A COLLECTOR'S ENCYCLOPEDIA AND PRICE GUIDE.** Radnor, PA: Wallace-Homestead, 1994, 170 pp. Photos, bibliography.

9-171 Moore, Raymond S. **COMMUNICATIONS RECEIVERS: 1932–1981.** Key Largo, FL: RSM Communications, [4th ed.], 136 pp. Covers the tube-based receiver output of 66 manufacturers. Photos.

9-172 _____. **TRANSMITTERS, EXCITERS, AND POWER AMPLIFIERS, 1930–1980.** Key Largo, FL: RSM Communications, 144 pp. Focuses on U.S.-made equipment. Photos.

9-173 Osterman, Fred. **SHORT-WAVE RECEIVERS PAST AND PRESENT: COMMUNICATIONS RECEIVERS, 1942–1997.** Reynoldsburg, OH: Universal Radio, 1987, 1997 [2nd ed.], 351 pp.; 1998 [3rd ed.], 473 pp. Includes 770 receivers from 98 manufacturers both foreign and domestic. Photos.

9-174 Ramirez, Ron, with Michael Prosise. **PHILCO RADIO: A PICTORIAL HISTORY OF THE WORLD'S MOST POPULAR RADIOS, 1928–1942.** Atglen, PA: Schiffer, 1993, 190 pp. Lavish album picturing the massive output of this company, with brief narrative of company history. Photos (most color), tables, appendices, bibliography, index.

9-175 Rutland, David. **BEHIND THE FRONT PANEL: THE DESIGN AND DEVELOPMENT OF 1920s RADIOS.** Philomath, OR: Wren, 1994, 158 pp. Intended for radio collectors who rebuild old sets, this includes useful historical information on manufacturing methods. Photos, diagrams, glossary, references, index.

9-176 *** Schiffer, Michael Brian. **THE PORTABLE RADIO IN AMERICAN LIFE.** Tucson: University of Arizona Press, 1991, 260 pp. Definitive history of the "small" radio from the 1920s to date, exploring the search for miniaturization from tubes to transistors and the varied social impacts of portables as seen by an archeologist. Photos, bibliography, index.

9-177 _____. "Cultural Imperatives and Product Development: The Case of the Shirt Pocket Radio," *TECHNOLOGY AND CULTURE,* 34:98–113 (January 1993). See also author's related book immediately above.

9-178 * Sievers, Maurice L. **CRYSTAL CLEAR: VINTAGE AMERICAN CRYSTAL SETS, CRYSTAL DETECTORS, AND CRYSTALS.** Vestal, NY: Vestal Press, 1991 (reprinted by Sonoran Publications, 1995), 292 pp.; **VOLUME 2,** Chandler, AZ: Sonoran Publications, 1995, 252 pp. Featuring hundreds of crystal sets and detectors (570 in vol. 1 and 172 more in vol. 2), this two-volume set is the definitive collector's guide to the type. Photos, diagrams, index.

9-179 Smith, Norman. **TRANSISTOR RADIOS, 1954-1968.** Atglen, PA: Schiffer, 1998, 160 pp. Includes more than 1,000 models in photos and descriptions.

9-180 Wood, Scott. **EVOLUTION OF THE RADIO.** Gas City, IN: L-W Books Sales, 1991 (Vol. 1), 1993 (Vol. 2). Photos.

Chapter 10

Electroacoustics and Recording

This chapter differs from others in this volume in that it veers away from a strict definition of telecommunications to a closely allied field. The introduction of the telephone in the mid-1870s marks the beginning of practical electroacoustics—the application of the science of electricity to the art of acoustics. The music recording industry, radio, the motion picture industry, and later television and cable, have all applied the science to specific needs.

These entries appear in sections on (A) general surveys of electroacoustics and the development of sound reproduction devices in general; (B) selected works on the phonograph; (C) electronic/magnetic audio recording; (D) selected works on motion picture sound recording; (E) stereo sound and recording; (F) video recording; and (G) consumer video disc and cassette recorders. All of these are vital to full application of radio and television services. For digital audio, see 13-F; for related Internet sites, see 15-H.

A. General Surveys
(Includes overall histories of electroacoustics and microphones, loudspeakers and other sound reproduction devices.)

10-001 ** Bauer, B.B. "A Century of Microphones," *PROCEEDINGS OF THE IRE*, 50:719–729 (May 1962). Reprinted in *JOURNAL OF THE AUDIO ENGINEERING SOCIETY*, 35:246–258 (April 1987). Good brief survey of technological trends. Diagrams, bibliography.

10-002 * Bergmann, Ludwig, trans. by H.S. Hatfield. *ULTRASONICS AND THEIR SCIENTIFIC AND TECHNICAL APPLICATION*. London: G. Bell, 1938, 264 pp. A standard work. Includes 575 references from 1847 on.

10-003 Boyle, Robert W. "Ultrasonics," *SCIENCE PROGRESS,* 23:75–105 (July 1928). Report on research and development in Britain during and after World War I.

10-004 Frederick, Halsey A. "The Development of the Microphone," *JOURNAL OF THE ACOUSTICAL SOCIETY OF AMERICA,* 3: Supp to July 1931, 25 pp.; also *BELL TELEPHONE QUARTERLY,* 10:1674–1688 (July 1931). One of the earlier historical treatments.

10-005 Gander, Mark R. "Fifty Years of Loudspeaker Developments As Viewed through the Perspective of the Audio Engineering Society," *JOURNAL OF THE AUDIO ENGINEERING SOCIETY,* 46:43–58 (January–February 1998).

10-006 * Hoover, Cynthia A. **MUSIC MACHINES—AMERICAN STYLE.** Washington: Smithsonian Institution Press, 1971, 139 pp. Based on a museum exhibition, this illustrates 19th-century music machines, the phonograph, acoustic and electric recordings, the impact of radio, movies and music, the jukebox, high fidelity, and electronic instruments. Bibliography, photos.

10-007 ** Hunt, Frederick V. **ELECTROACOUSTICS: THE ANALYSIS OF TRANSDUCTION AND ITS HISTORICAL BACKGROUND.** Cambridge, MA: Harvard University Press "Harvard Monographs in Applied Science No. 5," 1954, 260 pp. A thorough survey of history in chapter one (to p. 91) with extensive references including leading papers and patents concerning telephones, microphones, loudspeakers, and systems of particular interest to audio specialists. Charts, tables, notes, appendices, index.

10-008 * Jordan, Robert Oakes, and James Cunningham. **THE SOUND OF HIGH FIDELITY.** Chicago: Windsor Press/Popular Mechanics, 1958, 208 pp. Excellent photos and two-color diagrams (and a brief historical chapter) makes this a useful snapshot of hi-fi devices of the era.

10-009 Lipshitz, Stanley P. "Dawn of the Digital Age," *JOURNAL OF THE AUDIO ENGINEERING SOCIETY,* 46:37–42 (January–February 1998). Initial marketing of digital home recording devices.

10-010 Massa, Frank. "Some Personal Recollections of Early Experiences on the New Frontier of Electroacoustics during the Late 1920s and Early 1930s," *JOURNAL OF THE ACCOUSTICAL SOCIETY OF AMERICA,* 77:1296–1302 (April 1985). The author describes developments in which he participated.

10-011 * Olson, Harry F. **ACOUSTICAL ENGINEERING.** Princeton, NJ: van Nostrand, 1957 (republished by Professional Audio Journals, 1971), 718 pp. A basic work, based in part on the author's ELEMENTS OF ACOUSTICAL ENGINEERING (van Nostrand, 1947 [2nd ed.]). Covers all aspects of theory and practice and includes a survey of means for communication of information. Photos, diagrams, notes, index.

10-012 Paul, Floyd A. "A Decade of Electroacoustic Reproduction (1920–1930), *THE A.W.A. REVIEW,* 4:84–98 (1989). The peak of pre-electronic recording is reviewed. Photos.

10-013 * Rice, Chester W., and Edward W. Kellogg. "Notes on the Development of a New Type of Hornless Loud Speaker," *TRANSACTIONS OF THE AIEE,* 44:461–

Electroacoustics and Recording 209

475 (April 1925). A basic paper showing the transition to electrical sound reproduction. Photos, graphs, diagrams, notes. See also discussion of the paper on pp. 475–480.

10-014 Sivowitch, Eliot N. "Musical Broadcasting in the 19th Century," *AUDIO*, 51:19–23 (June 1967). On experiments with telephone installations used to provide public concerts from the mid-1870s. Photos, diagrams.

10-015 Vigoureux, Paul. **ULTRASONICS**. London: Chapman and Hall, 1950, 163 pp. General survey. Photos, diagrams, tables, bibliography (328 entries from 1894 on).

B. Phonographs

10-016 * Bachman, William S., et al. "Disk Recording and Reproduction," *PROCEEDINGS OF THE IRE*, 50:738–744 (May 1962). History and highlights of development from 1878. References.

10-017 Batten, Joe. **JOE BATTEN'S BOOK: THE STORY OF SOUND RECORDING; BEING THE MEMOIRS OF JOE BATTEN...** London: Rockliffe, 1956, 201 pp. The author (1885–1955) was an English music figure. Photos.

10-018 Botone, S[elimo] R[omeo]. **TALKING MACHINES AND RECORDS; A HANDBOOK FOR THOSE WHO USE THEM**. London: Pitman, 1904, 91 pp. Includes instructions on making a phonograph. Photos, diagrams.

10-019 * Chew, V.K. **TALKING MACHINES: 1877–1914**. London: HMSO for the Science Museum, 1967, 81 pp.; 1981 [2nd ed., without subtitle], 80 pp. Useful, brief survey from the invention of the phonograph to the turn of the century with discussion of American, British, and European firms and developments. Photos, bibliography, index.

10-020 Clements, Henry B. **GRAMOPHONES AND PHONOGRAPHS: THEIR CONSTRUCTION, MANAGEMENT AND REPAIR**. London: Cassell, 1912. Example of relatively early how-to manual.

See Du Moncel, 7-010.

10-021 Edison, Thomas A. **THE PHONOGRAPH AND ITS FUTURE; AND THE AUROPHONE AND ITS FUTURE....** Toronto: Rose-Belford Publishing "Religio-Science Series No. 4." 1878. See next entry.

10-022 _____. **INVENTOR'S HANDBOOK OF THE PHONOGRAPH**. Newark, NJ: Ward & Tichenor, 1889. With 10-021, two very early booklets attributed to the inventor (but likely written by others), explaining how the phonograph worked. See also Garbit 10-026.

10-023 Fabrizio, Timothy C., and George F. Paul. **ANTIQUE PHONOGRAPH GADGETS, GIZMOS AND GIMMICKS**. Atglen, PA: Schiffer, 1999, 253 pp. A well-illustrated collector's guide (including current prices). Photos, bibliography.

10-024 Frayne, John G. "A History of Disc Recording," *JOURNAL OF THE AUDIO ENGINEERING SOCIETY*, 33:263–266 (April 1985). Brief survey of analog methods.

Chapter 10

10-025 Frow, George L., and Albert F. Sefl. **THE EDISON CYLINDER PHONOGRAPHS**. Sevenoaks, England: Authors, 1978, 207 pp.; and by G.L. Frow alone as THE EDISON CYLINDER PHONOGRAPH COMPANION, 1994 [2nd ed.], 583 pp. Collections guide to all Edison phonographs with illustrations and descriptions of each model, 1877–1929. Photos, diagrams, bibliography.

10-026 Garbit, Frederick J. **THE PHONOGRAPH AND ITS INVENTOR, THOMAS ALVA EDISON**. Boston: Gunn, Bliss, 1878, 15 pp. "...being a description of the invention and a memoir of its inventor."

10-027 * Gelatt, Roland. **THE FABULOUS PHONOGRAPH 1877–1977**. New York: Macmillan, 1955, 320 pp.; 1965 [1st rev. ed.], 336 pp.; 1977 [2nd rev. ed.], 349 pp. A standard popular narrative from cylinders to stereo recording. Largely about companies, promoters, legal and business battles, growth of the industry, and artists and their associates in the music field. Photos, chronology, index.

10-028 Gillett, W. **THE PHONOGRAPH AND HOW TO CONSTRUCT IT**. London: E. and F.N. Spon, 1892 (reprinted 1978), 86pp. A guide for home equipment builders of tinfoil or wax cylinder machines.

10-029 Harvith, John, and Susan Edwards Harvith, eds. **EDISON, MUSICIANS, AND THE PHONOGRAPH: A CENTURY IN RETROSPECT**. New York: Greenwood "Contributions to the Study of Music and Dance No. 11," 1987, 461 pp. Largely based on interviews with recording artists, thus offering a user perspective of the technology. Notes, bibliography, index.

10-030 Hitchcock, H. Wiley, ed. **THE PHONOGRAPH AND OUR MUSICAL LIFE**. New York: Brooklyn College of CUNY "Institute for Studies in American Music Monograph 14," 1977, 91 pp. Proceedings of a December 1977 conference to note the centennial of recorded sound.

10-031 Kellogg, Edward W. "Electrical Reproduction from Phonograph Records," *JOURNAL OF THE AIEE*, 46:903–911 (February 1927). The transition from acoustical to electrical methods of recording. Photos, charts, diagrams, bibliography.

10-032 ** Koenigsberg, Allen. **THE PATENT HISTORY OF THE PHONOGRAPH, 1877–1912**. Brooklyn, NY: AMS Press, 1990, 87 pp. Self-published guide by a recognized authority to some 2,118 patents by 1,013 inventors. Patents are indexed by number, chronology, and inventor. Drawings. See also Meador, 10-037.

10-033 Mackenzie, Compton, ed. **THE GRAMOPHONE: ITS PAST; ITS PRESENT; ITS FUTURE**. London: Musical Association of London "Proceedings at Leeds, No. 51," 1924–25. A quite early history emphasizing developments in Britain.

10-034 * Marco, Guy A., and Frank Andrews, eds. **ENCYCLOPEDIA OF RECORDED SOUND IN THE UNITED STATES: 1870–1970**. New York: Garland "Reference Library of the Humanities 963," 1993, 1,000 pp. Covers the machines, companies, and artists, with some commentary and references.

10-035 Marty, Daniel, trans. by Douglas Tubbs. **THE ILLUSTRATED HISTORY OF PHONOGRAPHS**. New York: Dorset Press, 1981, 190 pp. Handsome album of photos (many in color) of vintage machines, both American and European. Glossary, bibliography.

10-036 McKendrick, John G. **WAVES OF SOUND AND SPEECH AS REVEALED BY THE PHONOGRAPH.** London: Macmillan, 1897. Early use of the phonograph as a research device.

10-037 Meador, John **AMERICAN PHONOGRAPH DESIGN PATENTS, 1897–1916.** n.p., 314 pp. See also Koenigsberg, 10-032.

10-038 Millard, A[ndre] J. **AMERICA ON RECORD: A HISTORY OF RECORDED SOUND.** New York: Cambridge University Press, 1995, 413 pp. Photos, discography, bibliographies, index. A social history with some reference to technology.

10-039 Mitchell, Ogilvie. **THE TALKING MACHINE INDUSTRY.** London: Pitman "Common Commodities and Industries," 1922, 120 pp. A survey of the British phonograph industry on the eve of the electronics revolution.

10-040 Newville, Leslie J. "Development of the Phonograph at Alexander Graham Bell's Volta Laboratory," *U.S. NATIONAL MUSEUM BULLETIN 218*, Paper 5, pp. 70–79 (1959). Early work with recorded sound.

10-041 ** "The Phonograph and Sound Recording after One-Hundred Years: Centennial Issue," *JOURNAL OF THE AUDIO ENGINEERING SOCIETY,* 25:656–990 (October–November 1977). Important collection of 33 original papers in sections on historical (Edison, Berliner, and others), technology (players and recordings), the industry (in the U.S. and overseas), innovations (including stereo and tape recording), and those who create. Photos, diagrams, references.

10-042 PHONOGRAPHS AND GRAMOPHONES. Edinburgh: Royal Scottish Museum, 1977. Proceedings of a symposium on the 100th anniversary of recorded sound.

See Prescott, 7-006.

10-043 *** Read, Oliver J., and Walter L. Welch. **FROM TIN FOIL TO STEREO: EVOLUTION OF THE PHONOGRAPH.** Indianapolis: Howard W. Sams, 1959, 524 pp.; 1976 [2nd ed.], 550 pp. Standard technical history of recording methods, with material on men, inventions, patents, companies, and products, with many extracts and critical comments. Includes motion picture recording. Corporate genealogy chart. Photos, diagrams, bibliography, notes, index.

10-044 Reiss, Eric L. **THE COMPLEAT TALKING MACHINE: A COLLECTOR'S GUIDE TO ANTIQUE PHONOGRAPHS.** Chandler, AZ: Sonoran, 1996 [2nd ed.], 236 pp.; 1998 [3rd ed.], 249 pp. Illustrated collector's guide. Bibliography.

10-045 Welch, Walter L., and Leah B.S. Burt. **FROM TINFOIL TO STEREO: THE ACOUSTIC YEARS OF THE RECORDING INDUSTRY, 1877–1929.** Gainesville: University Press of Florida, 1994, 200 pp. A partial revision of the initial chapters in Read & Welch, 10-043, but providing less detail. Photos, notes, index.

See Wile, 5-065.

10-046 Wilson, Percy, and G.W. Webb. **MODERN GRAMOPHONES AND ELECTRICAL REPRODUCERS**. London: Cassell, 1929, 271 pp. An introduction and historical survey. Photos, diagrams, bibliography, index.

C. Electronic/Magnetic Audio Recording

10-047 Angus, Robert. "75 Years of Magnetic Recording," *HIGH FIDELITY* (March 1973), pp. 42-50. See next entry.

10-048 _____. "A History of Magnetic Recording," *AUDIO* (August 1984), pp. 27-33. Brief survey of highlights.

10-049 * Begun, Semi J. **MAGNETIC RECORDING**. New York: Murray Hill Books, 1949, 242 pp. A basic book. Chapter One (to p. 15) offers historical survey. Photos, diagrams, bibliography, index.

10-050 Bubbers, John J. "Magnetic Recording," *JOURNAL OF THE AUDIO ENGINEERING SOCIETY*, 46:32–36 (January–February 1998). Brief historical survey.

10-051 Campbell, G.A., and O.J. Zobel. "Electrical Recording," *BELL SYSTEM TECHNICAL JOURNAL*, 2:1–46, 95–113 (January and April 1923). Early Bell Labs work.

10-052 * Chanan, Michael. **REPEATED TAKES: A SHORT HISTORY OF RECORDING AND ITS EFFECT ON MUSIC**. New York: Verso, 1995, 204 pp. Informal narrative of the major technologies and their changing impacts. Notes, index.

10-053 Clark, Mark. "Suppressing Innovation: Bell Laboratories and Magnetic Recording," *TECHNOLOGY AND CULTURE*, 34:516–538 (July 1993). Concerns about connecting devices to Bell system lines helped limit Labs' work.

10-054 * Clark, Mark, and Henry Nielsen. "Crossed Wires and Missing Connections: Valdemar Poulson, the American Telegraphone Company and the Failure to Commercialize Magnetic Recording," *BUSINESS HISTORY REVIEW*, 69:1–41 (Spring 1995). Discusses three key reasons for the failure of the 1898 Danish invention.

10-055 Cousino, Bernard A., and Ralph E. Cousino. "A Continuous Loop Magnetic Tape Cartridge," *JOURNAL OF THE AUDIO ENGINEERING SOCIETY*, 5:49–57 (January 1958). The two brothers patented one version of what became the radio tape cartridge. Diagrams, photos.

10-056 ** Daniel, Eric D., et al., eds. **MAGNETIC RECORDING: THE FIRST 100 YEARS**. New York: IEEE Press, 1999, 360 pp. Traces development of watershed products and technical breakthroughs including discussion of recording of sound, early devices, audio and video recording methods, and magnetic storage. An important historical anthology. Photos, diagrams, references, index.

10-057 ** Dearling, Robert, and Celia Dearling. **THE GUINNESS BOOK OF RECORDED SOUND**. Enfield, England: Guinness Books, 1984, 225 pp. The story of recordings from wax cylinders to laser discs and the people and companies (mainly British) who made it happen. Photos, bibliography, index.

10-058 "1888–1988: A Hundred Years of Magnetic Sound Recording," *JOURNAL OF THE AUDIO ENGINEERING SOCIETY*, 36:170–174+ (March 1988). A brief survey history.

10-059 ** Ford, Peter. "History of Sound Recording," *RECORDED SOUND*, as follows:
    "I: The Age of Empiricism," 7:221–119 (Summer 1962).
    "II: The Evolution of the Microphone and Electrical Disk Recording," 8:266–276 (Autumn 1962).
    "III: The Evolution of Magnetic Recording," 9:115–123 (April–July 1963).
    "IV: Motion Picture and Television Sound Recording," 10–11:146–154 (October 1963).
    "V: The Evolution of Stereophonic Sound Techniques," 13:181–188 (April 1964).
Similar to the 36-part series by the same author in *HI-FI NEWS* from January 1960 to July 1963. Solid technical history. Photos, references.

10-060 Hammar, Peter, and Don Ososke, "The Birth of the German Magnetophon Tape Recorder, 1928–1945," *db*, 15:24–29 (March 1982). The pioneering magnetic recording process adopted in the U.S. in postwar years.

10-061 Jorysz, Alfred. "Bibliography of Magnetic Recording, 1900–1953," *JOURNAL OF THE AUDIO ENGINEERING SOCIETY*, 2:183–199 (July 1954). Includes 353 items.

10-062 Kraft, James P. **STAGE TO STUDIO: MUSICIANS AND THE SOUND REVOLUTION, 1890–1950.** Baltimore: Johns Hopkins University Press "Studies in Industry and Society," 1996, 255 pp. Primarily a study of how recording technology impacted music unions. Tables, notes, essay on sources, index.

10-063 Lafferty, William. "The Use of Steel Tape Magnetic Recording Media in Broadcasting," *SMPTE JOURNAL*, 94:676–682 (June 1985). Illustrations, bibliography.

10-064 Lane, B. "75 Years of Magnetic Recording," *WIRELESS WORLD*, (March–April 1975), pp. 102–105, 161–164. Brief survey covering all formats.

10-065 "Magnetic Tape from the Earliest Days to the Present," *JOURNAL OF THE AUDIO ENGINEERING SOCIETY*, 36:606–616 (July–August 1988). Brief survey.

10-066 McGinn, Robert E. "Stokowski and the Bell Telephone Laboratories: Collaboration in the Development of High-Fidelity Sound Reproduction," *TELECOMMUNICATION AND CULTURE*, 24:38–75 (January 1983). Concerns a series of 1939–40 experiments with high fidelity recording.

10-067 McProud, C.G. "20 Years of Audio," *AUDIO*, 51:25–36 (May 1967). A review by the publisher. Photos.

10-068 Mooney, Mark Jr. "The History of Magnetic Recording," *HI-FI TAPE RECORDING* (February 1958), pp. 21–37. Useful survey of wire and tape recording devices. Photos, chronology.

10-069 Morton, David."The Rusty Ribbon: John Herbert Orr and the Making of the Magnetic Recording Industry, 1945–1960," *BUSINESS HISTORY REVIEW*, 67:589–622 (Winter 1993). Concerns Orradio company, purchased by Ampex in 1959.

10-070 _____. "Armour Research Foundation and the Wire Recorder: How Academic Entrepreneurs Fail," *TECHNOLOGY AND CULTURE*, 39:213–244 (April 1998). Photos, notes.

10-071 Mullin, John T. "Creating the Craft of Tape Recording," *HIGH FIDELITY*, 18:4:62–67 (April 1976).

10-072 O'Connell, Joseph. "The Fine-Tuning of a Golden Ear: High-End Audio and the Evolutionary Model of Technology," *TECHNOLOGY AND CULTURE*, 33:1–37 (January 1992). Discusses developments since 1970.

10-073 Read, Oliver J. **THE RECORDING AND REPRODUCTION OF SOUND**. Indianapolis: Howard W. Sams, 1949, 364 pp. Chapter Two covers history of acoustical recording. Photos, diagrams, bibliography of magnetic recording, index.

10-074 Spencer, Kenneth J. **HIGH FIDELITY: A BIBLIOGRAPHY OF SOUND REPRODUCTION**. London: Iota Services, 1958, 325 pp. Wide coverage with some 2,700 items, many of them annotated.

10-075 Thiele, Heinz H.K. "Sound Recording in Europe Up to 1945," *JOURNAL OF THE AUDIO ENGINEERING SOCIETY*, 36:396–408 (May 1988). Includes disc and early wire and tape methods.

10-076 ** Van Praag, Phil. **EVOLUTION OF THE AUDIO RECORDER: THE "VINTAGE" YEARS: LATE '40s–EARLY '70s**. Genesee Depot, WI: EC Designs, 1997, 518 pp. Designed for collectors of reel-to-reel machines, this is a fine example of hobby publication with value beyond the collector. Chapters discuss the theory of recording, analysis of tape recorder development, repair and restoration, and offer a guide to production models. Text, diagrams, and photos (most in color) depict wire and tape machine output of the major American firms.

10-077 * Wilson, Carmen F. **MAGNETIC RECORDING, 1900–1949**. Chicago: John Crerar Library, 1950, 61 pp. Some 339 annotated bibliographic entries, including patents. See next entry.

10-078 * _____. "Magnetic Recording, 1888–1952," *TRANSACTIONS OF THE IRE*, AU-4:53–81 (May–June 1956). A complete bibliography, including 38 important patents from 1899. In all, some 650 items arranged by year from 1888, most annotated. See previous entry.

D. Motion Picture Sound

(There is a substantial literature on this topic—this is but a sampling which emphasizes technological development of film recording methods.)

10-079 Cameron, Evan William, William F. Wilbert, and Joan Evans-Cameron, eds. **SOUND AND THE CINEMA: THE COMING OF SOUND TO AMERICAN**

FILM. Pleasantville, NY: Redgrave Publishing, 1980, 232 pp. Consists largely of proceedings of a symposium held at the International Museum of Photography, Rochester, N.Y., 1973. First 67 pp. includes three relevant papers on the technical development of sound by Fielding, Gomery, and Stewart. Photos, bibliography, index.

10-080 Cowan, Lester, ed. **RECORDING SOUND FOR MOTION PICTURES**. New York: McGraw-Hill, 1931, 404 pp. One of the first technical handbooks—very useful for details on early methods. Covers equipment, film recording, studio acoustics and techniques, and sound reproduction. Photos, diagrams, glossary, index.

10-081 ** Crafton, Donald. **THE TALKIES: AMERICAN CINEMA'S TRANSITION TO SOUND, 1926–1931**. New York: Scribner's "History of the American Cinema, Vol. 4," 1997 (reprinted by University of California Press, 1999), 639 pp. Definitive historical assessment of the crucial change in 21 chapters which in part one compares and contrasts the competing systems and companies, turning later to the impact of sound on film content and audiences. Photos, notes, appendices, bibliography, index.

10-082 * Eyman, Scott. **THE SPEED OF SOUND: HOLLYWOOD AND THE TALKIE REVOLUTION, 1926–1930**. New York: Simon & Schuster, 1997 (reprinted by Johns Hopkins University Press, 1999), 413 pp. Largely concerned with social impact. Bibliography, photos, index.

10-083 Franklin Harold B. **SOUND MOTION PICTURES: FROM THE LABORATORY TO THEIR PRESENTATION**. Garden City, NY: Doubleday, Doran, 1929, 401 pp. First 34 pages include history while remainder details equipment and techniques and makes reference to television. Photos, diagrams, index.

10-084 ** Frayne, John G., et al. "A Short History of Motion-Picture Sound Recording in the United States," *SMPTE JOURNAL*, 85:515–528 (July 1976). Useful survey from the 1920s on. Diagrams, references. This is followed by two very brief additional papers: Loren L. Ryder, "Magnetic Sound Recording in the Motion-Picture and Television Industries," pp. 528–530; and Hans Wohlrab, "Highlights of the History of Sound Recording on Film in Europe," pp. 531–533. See also Kellogg, 10-088 and Sponable, 10-092.

10-085 * Geduld, Harry M. **THE BIRTH OF THE TALKIES: FROM EDISON TO JOLSON**. Bloomington : Indiana University Press, 1975, 337 pp. The first scholarly study, this begins with the invention of the phonograph and its application to films. Appendices on sound film patents, notes, index.

10-086 Gomery, Douglas. "The Coming of Sound: Technological Change in the American Film Industry," in Tino Balio, ed. **THE AMERICAN FILM INDUSTRY**. Madison: University of Wisconsin Press, 1985 [2nd ed.], pp. 229–251. Useful, brief survey with bibliographic note of the author's related writings.

10-087 Green, Fitzhugh. **THE FILM FINDS ITS TONGUE**. New York: Putnam, 1929, 316 pp. Popular early survey of talking pictures and especially Warner Bros. work. Photos.

10-088 Harvey, F.K. "Mementos of Early Photographic Sound Recording," *SMPTE JOURNAL*, 91:237–244 (March 1982). Includes work of Bell at Volta Laboratory (see Newville, 10-040) and early film recording. Photos, references

10-088a** Kellogg, Edward W. . "History of Sound Motion Pictures," *SMPTE JOURNAL*, 64:291–302; 356–374; and 422–437 (June, July, and August 1955). A basic three-part paper providing a thorough survey, primarily of American developments from the 1920s. Diagrams, references (totaling 406 items from 1878). Reprinted in Fielding, 12-115. See also Frayne, 10-084, and Sponable, 10-092.

10-089 * Limbacher, James L. **FOUR ASPECTS OF THE FILM**. New York: Brussel & Brussel, 1969. Part four (pp. 197–229) focuses on the development of sound. An appendix (pp. 363–372) lists pioneer sound films and their type of sound. Photos, notes, index.

10-090 Miehling, Rudolph. **SOUND PROJECTION**. New York: Mancall, 1929, 528 pp. One of the first detailed works on motion picture sound systems making this a useful contemporary work. Includes chapters on studio and recording techniques, sound pic-up devices, vacuum tubes, amplifiers, sound reproducers, and acoustics. Twelve chapters treat operating and related procedures while five focus on specific equipment and systems.

10-091 Pitkin, Walter. **THE ART OF SOUND PICTURES**. New York: Appleton, 1930. An early study of how the coming of sound would change the Hollywood product, written while the transition was still in its early stages.

10-092 * Sponable, E.I. "Historical Development of Sound Films," *SMPTE JOURNAL*, 48:275–303 (April 1947). Considerable technical detail on the first two decades. Photos, diagrams, references. See also Frayne, 10-084, and Kellogg, 10-088.

10-093 SYNCHRONIZED REPRODUCTION OF SOUND AND SCENE. New York: Bell Telephone Laboratories, 1929, 44 pp. Reprint of seven articles from the November 1928 *BELL LABORATORIES RECORD*, based in turn on Bell Labs presentations at a Society of Motion Picture Engineers conference. Includes discussion of wax recording, sound recording with the light valve, speech control for a sound-picture system, sound projector systems, and Western Electric projection equipment.

10-094 Tibbetts, John C, and Gregory D. Black. **INTRODUCTION TO THE PHOTOPLAY.** Los Angeles: University of Southern California, College of Letters, Arts, and Sciences, 1929 (reprinted with added subtitle, **1929, A CONTEMPORARY ACCOUNT OF THE TRANSITION TO SOUND IN FILM,** by National Film Society, 1977), 383 pp. Based on a series of lectures presented at the University in 1929 by representatives of the Academy of Motion Picture Arts and Sciences. Bibliography, index.

10-095 Walker, Alexander. **THE SHATTERED SILENTS : HOW THE TALKIES CAME TO STAY.** London : Elm Tree Books / New York: William Morrow, 1978, 218 pp. Largely focused on contemporary views of sound's impact. Photos, notes, bibliography, 1925–29 chronology, index.

E. Stereo Sound and Recording
(Includes stereo broadcasting; for other materials on FM, see 9-D.)

10-096 Boehm, George A.W. "Stereo Goes to Market," *FORTUNE*, 43:108–111 (August 1958). Discusses initial marketing of stereo disc recordings.

10-097 Braun, Mark J. **AM STEREO AND THE FCC: CASE STUDY OF A MARKET-PLACE SHIBBOLETH.** Norwood, NJ: Ablex, 1994, 206 pp. Insightful study of the FCC's changing role in adopting industry technical standards. Charts, tables, references, index.

10-098 Crowhurst, Norman H. **STEREOPHONIC SOUND.** New York: John Rider "Publication No. 209," 1957, 118 pp.; 1961 [2nd ed.], 136 pp. Photos, diagrams, bibliography, index. Basic guide designed for hi-fidelity afficianados.

10-099 _____. **FM STEREO MULTIPLEXING.** New York: John Rider "Publication No. 282," 1961, 66 pp. The various industry-tested FM standards and the final FCC-approved FM standard are described as are two-station systems. Photos, diagrams, index.

10-100 de Boer, K. "Stereophonic Sound Production," *PHILIPS TECHNICAL REVIEW*, 5:107–144 (April 1940). See next entry.

10-101 _____. "Experiments with Stereophonic Records," *PHILIPS TECHNICAL REVIEW*, 5:182–186 (June 1940). Reports on early experimental work.

10-102 "FCC Approves FM Stereophonic Broadcasting," *BROADCAST ENGINEERING* (May 1961), pp. 28–40. Reproduces most of the actual FCC decision and notes.

10-103 Fletcher, Harvey. "Stereophonic Sound Film System—General Theory," *JOURNAL OF THE ACOUSTICAL SOCIETY OF AMERICA*, 13:89–99 (October 1941).

See Ford, Part V, 10-059.

10-104 Frayne, John G., and W.W. Templin. "Stereophonic Recording and Reproducing Equipment," *SMPTE JOURNAL*, (September 1953). An early review.

10-105 Garner, Louis E. Jr. "Stereo Then and Now," *RADIO-ELECTRONICS MAGAZINE* (March 1959). Brief survey.

10-106 * Hilliard, John K "The History of Stereophonic Sound Reproduction," *PROCEEDINGS OF THE IRE*, 50:776–780 (May 1962). Primarily a condensed narrative bibliography with 121 references from about 1940.

10-107 Hines, Burt. "FM Multiplexing: A New Approach to Stereophonic Sound," *RADIO AND TELEVISION NEWS* (November 1955), pp. 55–57, 188–189. Includes some historical discussion.

10-108 Snyder, R.H. "History and Development of Stereophonic Sound Recording," *JOURNAL OF THE AUDIO ENGINEERING SOCIETY,* 1:176–179 (April 1953).

10-109 * Sunier, John. "A History of Binaural Sound," *AUDIO* (March 1986), pp. 36–46. See the same author's book, next entry.

10-110 ** Sunier, John. **THE STORY OF STEREO: 1881–**. New York: Gernsback "Gernsback Library Book No. 98," 1960, 160 pp. History of varied methods and approaches. Photos, diagrams, chapter bibliographies, index.

10-111 Tinkham, Russell J. "Anecdotal History of Stereophonic Recording," *AUDIO,* 46:25–31, 92–93 (May 1962). Personal reminiscences with quotes and abstracts. Photos, references.

10-112 Torick, Emil. "Highlights in the History of Multichannel Sound," *JOURNAL OF THE AUDIO ENGINEERING SOCIETY,* 46:27–31 (January–February 1998).

F. Video Recording

10-113 ** Abramson, Albert. "A Short History of Television Recording," *SMPTE JOURNAL,* 64:72–76 (February 1955). Historical development from 1927 to modern video recording. See next entry for continuation. References. Reprinted in Fielding, 12-115, and in Shiers, 12-128.

10-114 *** _____. "A Short History of Television Recording, Part II," *SMPTE JOURNAL,* 82:188–197 (March 1973). Continues previous entry. Concise survey of changing methods. Photos, diagrams, references. Reprinted in Shiers, 12-128.

10-115 Baldwin, J.L.E. "Digital Television Recording, History and Background," *SMPTE JOURNAL,* 95:1206–1214 (December 1986); reprinted in *SMPTE JOURNAL,* 105:632–640 (October 1996). Continuation of entry above. Diagrams, bibliography.

10-116 Not used.

10-117 THE CHANGING PICTURE IN VIDEO TAPE FOR 1959–1960: A REVIEW FOR THE TELEVISION INDUSTRY. Minneapolis: 3M Company, 1959 [2nd ed.], 62 pp. Promotional piece, but one with considerable value today with sections on working with videotape, prospects for videotape, and several case studies of pioneering use in the late 1950s. Photos, notes.

10-118 * Ginsburg, C.P. **THE BIRTH OF VIDEO RECORDING**. Redwood City, CA: Ampex Corporation, 1981 (reprinting a 1957 study). Available on-line at Ampex web site.

10-119 Hammar, P. "The Birth of Helical Scan Videotape Recording," *BROADCAST ENGINEERING,* 27:86–94 (May 1985).

10-120 ** Marlow, Eugene, and Eugene Secunda. **SHIFTING TIME AND SPACE: THE STORY OF VIDEOTAPE**. New York: Praeger, 1991, 174 pp. First book devoted to the subject, including discussion of videotape's impact on advertising, programs, and the home consumer market. Bibliography, index.

10-121 McLean, Donald F. "Computer-Based Analysis and Restoration of Baird 30-Line Television Recordings," *JOURNAL OF THE ROYAL TELEVISION SOCIETY,* 22:87–94 (April 1985). See also next two entries.

10-122 _____. "The Recovery of Phonovision," **THIRD INTERNATIONAL CONFERENCE ON IMAGE PROCESSING AND ITS APPLICATIONS**. London: IEE Conference Publication 307, 1989, pp. 300–304. This and the previous entry describes the finding of old disc recordings and how they were "read" with modern equipment. The results can also be viewed on the Internet (see 15-062). See also next entry.

10-123 * _____. "Restoration of Baird Mechanical Television Recordings." *ELECTRONICS WORLD*. Series of articles as follows:
 "Dawn of Television," 104:745–749 (September 1998).
 "Restoring Baird's Image," 104:823–829 (October 1998).
 "First Frames," 104:943–946 (November 1998).
 "Looking In," 104:1031–1034 (December 1998).
 "Digital Vision," 105:160–165 (February 1999).
Together, these five articles detail the author's work to achieve modern copies (and thus an ability to view) Baird disc recordings made in the late 1920s to the early 1930s using largely mechanical technology. Photos.

10-124 * Nwungwun, Aaron Foisi. **VIDEO RECORDING TECHNOLOGY: ITS IMPACT ON MEDIA AND HOME ENTERTAINMENT**. Stoneham, MA: Focal Press, 1990, 289 pp. Details the multination rise of audio and then video recording system technologies. Photos, diagrams, bibliography, index.

10-124a Reynolds, Keith. "The Evolution of Digital Audio and Video Format Conversions." *SMPTE JOURNAL,* 103:644–147 (October 1994). Converting from analog to digital and vice versa in broadcast and other operations. Photos, diagrams, references.

10-125 * Roizen, Joe. "The History of Videotape Recording," *TELEVISION: JOURNAL OF THE ROYAL TELEVISION SOCIETY,* 16:15–21 (January–February 1976). The Schoenberg Memorial Lecture before the Royal Institution in December 1975. Photos, diagram.

10-126 Schneider, Arthur. **JUMP CUT! MEMOIRS OF A PIONEER TELEVISION EDITOR**. Jefferson, NC: McFarland, 1997, 208 pp. Long-time NBC engineer relates the changing technology and application of videotape from analog black-and-white to increasingly digital color systems. Photos, index.

10-127 * Stanton, Julia A., and Michael J. Stanton. "Video Recording: A History," *SMPTE JOURNAL,* 96:253–263 (March 1987). Makes references to "James" rather than John Logie Baird's early work, but otherwise a useful survey. References.

10-128 Sugaya, Hiroshi. "The Videotape Recorder: Its Evolution and the Present State of the Art of VTR Technology," *SMPTE JOURNAL*, 97:301–309 (March 1988). Diagrams, references.

10-129 _____. "The Past Quarter-Century and the Next Decade of Videotape Recording," *SMPTE JOURNAL*, 101:10–13 (January 1992). Emphasis on increasing recording density. Diagrams, references.

10-130 TWENTY YEARS OF VIDEO TAPE. St. Paul, MN: 3M Company, 1976, 20 pp. Useful brief chronology from 1877 on.

10-130a Umemoto, Masuo, et al. "Digital Video Recording," *PROCEEDINGS OF THE IEEE*, 83:1044–1054 (July 1995). Discusses the mutual technology transfer process between consumer and professional video recorders over two decades. References.

G. Video Disc and Cassette Recorders

10-131 Alvarado, Manuel, ed. **VIDEO WORLDWIDE: AN INTERNATIONAL STUDY.** London: John Libbey, 1988, 328 pp. A UNESCO-sponsored study, this reviews the expansion of videocassette use in countries around the world, both developed and developing. Tables, charts, no index.

10-132 Boyd, Douglas A., Joseph D. Straubhaar, and John A. Lent. **VIDEOCASSETTE RECORDERS IN THE THIRD WORLD.** White Plains, NY: Longman "Communications," 1989, 292 pp. Development, applications, and implications of the technology in Arab World, Asia, Latin America, and the Caribbean. Tables, notes, references, index.

10-133 Cusumano, Michael A., Yiorgos Mylonadis, and Richard S. Rosenbloom. "Strategic Maneuvering and Mass Market Dynamics: The Triumph of VHS over Beta," *BUSINESS HISTORY REVIEW*, 66:51–94 (Spring 1992). A video "battle of the speeds" to most consumers.

10-134 Dobrow, Julia R., ed. **SOCIAL AND CULTURAL ASPECTS OF VCR USE.** Hillsdale, NJ: Lawrence Erlbaum "Communications," 1990, 219 pp. Eleven research papers including: discussion of an historical perspective of VCR use; audience measurement of VCR use; VCR libraries; use of VCR's to time-shift and re-view programs; and camcorders and the personalization of television. Tables, notes, index.

10-135 Ganley, Gladys D. **THE EXPLODING POLITICAL POWER OF PERSONAL MEDIA.** Norwood, NJ: Ablex "Communication and Information Science," 1992, 181 pp. Largely devoted to uses of videotapes and videocassettes, but including some comment on personal computers and facsimile as well. Notes, index.

10-136 Ganley, Gladys D., and Oswald H. Ganley. **GLOBAL POLITICAL FALLOUT: THE FIRST DECADE OF THE VCR, 1976–1985.** Norwood, NJ: Ablex, 1987, 166 pp. The development and expansion of VCR use, government concerns about use of the technology and efforts to control or limit VCR expansion, and implications of VCR use. Notes, appendices, index.

10-137 ** Graham, Margaret B.W. **RCA AND THE VIDEODISC: THE BUSINESS OF RESEARCH**. New York: Cambridge University Press, 1986, 258 pp. Valuable and unique description of the research process to create and the eventual market failure of RCA's "Selectavision" project, based on internal corporate documents. Bibliography, index.

10-138 Lardner, James. **FAST FORWARD: HOLLYWOOD, THE JAPANESE AND THE VCR WARS.** New York: Norton, 1987, 344 pp. Rise of the videocassette recorder from initial work in the U.S. and Japan to the legal battles (especially the Supreme Court "Betamax" decision) over potential copyright infringement that the technology made possible. References, index.

10-139 Levy, Mark R., ed. **THE VCR AGE: HOME VIDEO AND MASS COMMUNICATION**. Newbury Park, CA: Sage "Focus Editions," 1989, 274 pp. Fourteen research papers on the growth and expansion of home video in the U.S. and Britain, using the VCR, the VCR and individual media use patterns, and the VCR in groups and societies. Notes, references, no index.

10-140 McCabe, E. "The Video Cassette: 1928–1971," *TRAINING IN BUSINESS AND INDUSTRY* (October 1971), pp. 43–55. Historical survey.

10-141 Rice, P., and R.F. Dubbe. "The Development of the First Optical Videodisc," *SMPTE JOURNAL*, 91:277–184 (March 1982). The work of J.L. Baird and later innovators.

10-142 Rosenbloom, Richard S., and Michael A. Cusumano. "Technological Pioneering and Competitive Advantage: The Birth of the VCR Industry," *CALIFORNIA MANAGEMENT REVIEW*, 29:51–76 (Summer 1987). With American ideas, Japanese firms created a worldwide demand.

10-143 Shima, Shigeo. "The Evolution of Consumer VTRs—Technological Milestones," in Stephen P. Ronsheimer, ed. **"A History of Consumer Electronics,"** *IEEE TRANSACTIONS ON CONSUMER ELECTRONICS*, CE-30: 66–80 (May 1984). Brief survey. Extensive bibliography.

10-144 Shiraishi, Yuma. "History of Home Videotape Recorder Development," *SMPTE JOURNAL*, 94:1257–1263 (December 1985).

10-145 Sigel, Efrem, et al., eds. **VIDEO DISCS: THE TECHNOLOGY, THE APPLICATIONS AND THE FUTURE**. White Plains, NY: Knowledge Industry Publications, 1980, 182 pp. Nine chapters assess what was then a promising consumer product. First chapter provides extensive historical treatment. Diagrams, photos, index.

10-146 "Videodisc," *RCA REVIEW*, 39:3–228 (March 1978) is a special issue with 10 papers exploring all aspects of the RCA system's development and operation. Photos, diagrams, references.

10-147 "The Videodisc: The Next Step in the Communications Evolution," *SMPTE JOURNAL*, 83:553–587 (July 1974). Panel of seven papers and related discussion. Photos, diagrams, references.

Chapter 11

Electron Tubes and Solid State Devices

While highlights of the history of electron tubes are mentioned in many accounts, especially in radio histories (see chapter 9), a full study up to modern times has yet to appear. The beginnings of serious research on cathode rays is recorded in the papers of Michael Faraday and others in England and in those of several researchers in Germany and France from the late 1850s on (see Shiers, 1-017, the first edition of this book, at pp. 125–130). By 1900 these researchers and others had produced a clearer understanding of electrical discharges in gases and of the particulate nature of cathode emissions, which resulted in the first practical cathode-ray indicator tube introduced by Karl Ferdinand Braun in 1897 (see 5-075). This chapter is *highly* selective as the literature on all of these topics—especially the solid state revolution—is huge and the focus here is on the most useful historical material.

This chapter begins with (A) a general survey of the field, and continues with (B) developments to 1920, (C) since 1920, and (D) cathode-ray tubes. Between 1900 and 1930, the low-voltage cathode-ray tube and associated circuits slowly developed into practical oscilloscopes and became the heart of electronic television receivers. Section (E) reviews the development of microwave tubes (see 8-D for related material). The last section (F) focuses on semiconductor devices to 1948 ("crystal detectors"), and transistors and solid state devices from 1948.

A. General Surveys

11-000 * Babani, Bernard B., ed. **INTERNATIONAL RADIO TUBE ENCYCLOPEDIA**. London: Bernards, 1949, 410 pp. "Operating characteristics and pin connections of some 15,000 radio tubes of all types manufactured throughout the world." Basically a large comparative table, this provides chapters on radio receiving tubes, triode transmitter tubes, transmitting tetrodes and pentodes, rectifiers and hyratrons, regulator and control tubes, tuning indicators, cathode-ray tubes, photo tubes, and rare types and their equivalents. Valuable for its census of postwar tube manufacture.

11-001 * Haantjes, J., and H. Carter. **CLASSIFICATION OF ELECTRON TUBES.** New York: Macmillan, 1960, 1962 [2nd ed.], 100 pp. Very useful and color fully illustrated guide to the many kinds including thermionic tubes, cathode-ray tubes, and many other types. Useful guide just as tubes were beginning to give way to solid state devices. Diagrams, photos, glossary, two foldout charts.

11-002 Hong, Sungook. "From Effect to Artifact (II): The Case of the Thermionic Valve," *PHYSIS: REVISTA INTERNAZIONALE DE STORIA DELLA SCIENZA,* 33:85–124 (1996). Argues that the popularly accepted story of J.A. Fleming's invention of the valve is at least misleading.

11-003 _____. "Syntony and Credibility: John Ambrose Fleming, Guglielmo Marconi, and the Maskelyne Affair," in J.Z. Buchwald, ed. **SCIENTIFIC CREDIBILITY AND TECHNICAL STANDARDS IN 19TH AND EARLY 20TH CENTURY GERMANY AND BRITAIN.** Dordrecht: Kluwer Academic, 1996, pp. 157–176.

11-004 Kogan, Philip, and Joan Pick. **THE CATHODE RAY REVOLUTION: FOUNDATIONS OF ELECTRONICS.** London: Sampson Low, Marston / Boston: Ginn & Co. "Foundations of Science Library," 1966, 128 pp. Though designed for younger readers, this is a remarkably clear discussion of the full development of tubes. Despite the title, it is arranged in three parts: introducing valves (tubes), semiconductors and transistors, and the cathode-ray revolution (the last 20 pages). Diagrams, photos, index.

11-005 Mouromtseff, Ilia E. "Electronics and Development of Electronic Tubes," *JOURNAL OF THE FRANKLIN INSTITUTE,* 240:171–192 (September 1945). Complete survey from the 19th century with full sources. Photos, references. A similar paper appeared in *PROCEEDINGS OF THE IRE,* 33:223–233 (April 1945).

11-005a Sibley, Ludwell. **TUBE LORE: A REFERENCE FOR USERS AND COLLECTORS.** Flemington, NJ: Author, 1996, 186 pp. "Guide to virtually every tube ever produced and sold in North America" including their specifications and brief histories.

11-006 *** "Special Issue: Historical Notes on Important Tubes and Semiconductor Devices," *IEEE TRANSACTIONS ON ELECTRON DEVICES,* ED-23:597–788 (July 1976). Twenty historical papers, a few of which are focused on tubes while most discuss such solid-state devices as the transistor, integrated circuits and the like. Photos, diagrams, references.

11-007 *** Stokes, John W. **70 YEARS OF RADIO TUBES AND VALVES: A GUIDE FOR ELECTRONIC ENGINEERS, HISTORIANS, AND COLLECTORS.** Vestal, NY: Vestal Press, 1982, 247 pp. An overall history of vacuum tubes developed and used in radio in America, Britain, Canada, and Australia into the 1960s. Emphasis on the 1927–1937 period. Diagrams, charts, photos, glossary, chapter references, index.

11-008 ** **THERMIONIC VALVES 1904–1954: THE FIRST FIFTY YEARS.** London: Institution of Electrical Engineers, 1955, 69 pp. Commemorative lectures and exhibition including catalogue of the latter. Photos, diagrams.

11-009 * Thrower, Keith R. **HISTORY OF THE BRITISH RADIO VALVE TO 1940**. Ropley, England: MMA International 1993, 213 pp. Covers British work with some comparative comment on efforts in America and Germany from 1913 to 1940. Photos, diagrams.

11-010 *** Tyne, Gerald F.J. **SAGA OF THE VACUUM TUBE**. Indianapolis, IN: Howard W. Sams, 1977, 494 pp. Probably the definitive treatment of tube developments and types in all countries to about 1930. Details on tube types and variations with specific data and illustrations highlight this important book. Based in part on a 23-part series that appeared in *RADIO NEWS* from 1943 to 1946, this book version fills in details for the 1920s. Photos, chapter references, index.

11-011 White, William C. "Evolution of Electronics," *ELECTRONICS*, 25:98–99 (September 1952) includes very useful family tree diagram showing roots and branches of commercial tubes.

B. Developments to 1920

11-012 * Armstrong, Edwin H. "Operating Features of the Audion," *ELECTRICAL WORLD*, 64:1149–1152 (December 12, 1914). First disclosure to correctly describe how a three-electrode tube operates as a detector and as a high-frequency amplifier. Diagrams, graphs, photos.

11-013 *_____. "Some Recent Developments in the Audion Receiver," *PROCEEDINGS OF THE IRE*, 3:215–247 (September 1915); reprinted in same journal, 51:1083–1097 (August 1963). Based on an earlier paper, this includes discussion by de Forest (see 11-016) and Armstrong that reveals the contentious atmosphere that perpetually surrounded these men. Reprinted in Shiers, 9-080.

11-014 Bucher, Elmer E. **VACUUM TUBES IN WIRELESS COMMUNICATION: A PRACTICAL TEXT BOOK FOR OPERATORS AND EXPERIMENTERS**. New York: Wireless Press, 1918, 202 pp. This pioneer text describes basic tubes with two, three, or four elements and discusses practical applications including amplification, regeneration, detection, and modulation. Full coverage of contemporary practices. Diagrams, graphs.

11-015 * Chipman, Robert A. "De Forest and the Triode Detector," *SCIENTIFIC AMERICAN*, 212:92–100 (March 1965). Background and early radio detectors from coherer to triode. A critical review of de Forest's invention, his patent claims, and his understanding of the tube. Photos, diagrams.

11-016 ** de Forest, Lee. "The Audion: A New Receiver for Wireless Telegraphy," *PROCEEDINGS OF THE AIEE*, 25:719–747 (1906). Pioneering paper on the three-element tube. De Forest makes clear he is aware of Fleming's prior work but declares his device operates in a different way—as a relay. In attempting a scientific explanation he finds the phenomena "are exceedingly puzzling to explain." Reprinted in Shiers, 9-080.

11-017 _____. "The Audion–Detector and Amplifier," *PROCEEDINGS OF THE IRE*, 2:24–30 (March 1914), reprinted in *IEEE SPECTRUM*, 1:22–29

(March 1964). On the first public demonstration of the Audio tube as a telephone amplifier. Photos, diagram. See Armstrong, 11-012.

11-018 ** Fleming, John A. "On the Conversion of Electric Oscillations into Continuous Currents by Means of a Vacuum Valve," **PROCEEDINGS OF THE ROYAL SOCIETY,** 74:476–487 (February 1905). First publication on the "oscillation valve" or thermionic diode. Diagrams, graphs, tables.

11-019 ***_____. **THE THERMIONIC VALVE AND ITS DEVELOPMENTS IN RADIOTELEGRAPHY AND TELEPHONY.** London: Wireless Press, 1919, 279 pp.; 1924 [2nd ed.], 438 pp. This important book by the inventor of the valve (see 5-124) provides a thoroughly documented record of the history of the electron tube applied to radio applications. Diagrams, notes, appendix on litigation, index.

11-020 _____. "How I Put the Electrons to Work in the Radio Bottle," **POPULAR RADIO,** 3:175–182 (March 1923). Popular story of his experiments from 1882 to 1904. Photos, diagrams.

11-021 Scott-Taggart, John. **THERMIONIC TUBES IN RADIO TELEGRAPHY AND TELEPHONY.** London: Wireless Press, 1921, 424 pp. An exhaustive treatment of radio techniques as well as radio tubes. Descriptive and partly tutorial, this book documents the progress of radio tube technology during and just after World War I. Photos, diagrams, notes, index.

11-022 ** Shiers, George. "The First Electron Tube," **SCIENTIFIC AMERICAN,** 220:104–112 (March 1969). Relates the chronology of the invention from Edison in 1879 to de Forest in 1907. Photos, diagram (including double-page chart showing chronology of invention).

11-023 Smith-Rose, Reginald L. "The Evolution of the Thermionic Valve," **JOURNAL OF THE IEE,** 56:253–266 (April 1918). Thorough survey and includes elementary thermionics, Fleming oscillation valve, de Forest Audion detector and amplifier, Lieben-Reisz valve, the pure electron discharge, and other uses. Diagrams, references. Reprinted in Shiers, 9-080.

11-024 Van der Bijl, H.J. **THE THERMIONIC VACUUM TUBE.** New York: McGraw-Hill, 1920, 391 pp. A pioneer book on the physics and characteristics of thermionic tubes and their uses as rectifiers, amplifiers, oscillators, modulators, detectors, and other purposes. Photos, diagrams, notes, index.

C. Developments after 1920

11-025 de Forest, Lee. "Evolution of the Vacuum Tube," **RADIO NEWS,** 11:990–991, 1039 (May 1930). The inventor revisits what was by then a controversial and legally debated history. Photos, diagrams.

11-026 _____. "How the Radio Tube Grew Up," **RADIO NEWS,** 13:23, 74–75 (July 1931). On electron-tube developments during the early 1900s and the manufacture of radio tubes. Photos.

Electron Tubes and Solid State Devices

11-027 ELECTRON TUBES. Princeton, NJ: RCA Review, 1949, 2 vols. RCA science papers. Diagrams, photos, references.

> **VOL. 1: 1935–1941.** (475 pp.) Reprints 19 papers with summaries of 32 more.
>
> **VOL. 2: 1942–1948.** (454 pp.) Reprints 21 papers with summaries of 20 more.

11-028 Fink, Donald G. "Electron Tubes 1930 to 1950," *ELECTRONICS*, 23:66–69 (April 1950). Useful though very brief survey. Chronological list of new tube types. Tables.

11-029 Gorham, John E. "Electron Tubes in World War II," *PROCEEDINGS OF THE IRE*, 35:295–301 (March 1947). Survey of advances, under the pressures of wartime demand, of electron tubes of all types. Photos, tables.

11-030 Gosling, B.S. and M. Thompson. "The Development of Valves for Wireless," *WORLD POWER*, 3:195–203, 333–339; 4:147–154 (April, June, and September, 1925). British study emphasizing developments there. Illustrations, references.

11-031 Ingram, S.B. "A Decade of Progress in the Use of Electronic Tubes, Part 1—In the Field of Communication," *ELECTRICAL ENGINEERING*, 59:643–649 (December 1940). Includes review articles on communication, broadband carrier systems, ultra-shortwave tubes, beam power tubes, wideband amplifier tubes, and others. Bibliography of 88 items.

11-032 MacArthur, Elmer D. "An Indexed Bibliography of Electron Tubes and Their Applications," *GENERAL ELECTRIC REVIEW*, 41:455–460 (October 1938). 138 entries from 1921.

11-033 Mouromtseff, Ilia E. "A Quarter Century of Electronics," *ELECTRICAL ENGINEERING*, 66:171–177 (February 1947). Brief review of major trends emphasizing developments in tube design and use.

11-034 Mullard, S.R. "The Development of the Receiving Valve," *JOURNAL OF THE IEE*, 76:10–16 (1935). General survey from 1922. Diagrams, photos.

11-035 Sharp, Clayton H. "The Edison Effect and its Modern Applications," *JOURNAL OF THE AIEE*, 41:68–78 (January 1922). Historical survey of electron-tube developments: diodes, triodes, power rectifiers, and other types. Photos, graphs, diagrams.

11-036 * TUBES AND TRANSISTORS: A COMPARATIVE STUDY. New York: Electron Tube Information Council, 1960, 62 pp. Interesting industry promotional piece, with some history, attempting to preserve a market for tubes in the face of expanding application of transistors. Photos, charts, diagrams, bibliography.

11-037 Wood, James. "Beating the Transistor: Developments in UHF Tube Technology," *IEE REVIEW*, 38:107–110 (March 1992). Improvements in UHF tubes. Illustrations, bibliography.

D. Cathode-Ray Tubes
(See also chapter 12-C through 12-E.)

11-038 * Ardenne, Manfred von, trans. by G.S. McGregor and R.C. Walker. **CATHODE-RAY TUBES.** London: Pitman, 1939, 530 pp. English translation of a 1936 German volume which adds considerable updating due to rapid developments. Useful indicator of prewar progress and status. Photos, diagrams, tables, index.

11-039 _____. "Evolution of the Cathode-Ray Tube: A Survey of Developments over Three Decades," *WIRELESS WORLD*, 66:28–32 (January 1960). A general survey of cathode-ray tubes in early television and in oscillographs with an emphasis on the author's pioneer contributions from 1928. Photos, diagrams, references.

11-040 Herold, E.W. "History and Development of the Color Picture Tube," *PROCEEDINGS OF THE SOCIETY OF INFORMATION DISPLAY*, 15:4:141–149 (Fall 1974). Survey of developments in different firms and nations.

11-041 _____. "A History of Color Television Displays," *PROCEEDINGS OF THE IEEE*, 64:1331–1338 (September 1976). Much the same as entry immediately above.

11-042 Kaplan, S.H. "The History of Color Picture Tubes and Some Future Projections," *SMPTE JOURNAL*, 99:396–400 (May 1990). Research trends and likely developments are assessed. Photos, references.

See Klemperer, 12-056.

11-043 Law, Harold B. "The Shadow Mask Color Picture Tube: How It Began—An Eyewitness Account of Its Early History," *IEEE TRANSACTIONS*, 23:752–759. (July 1976). A member of the developmental team relates how the design of a crucial part of the NTSC system was derived. Photos. See also 12-140.

11-044 McGee, James D., and Hans G. Lubszynski. "EMI Cathode-Ray Television Transmission Tubes," *JOURNAL OF THE IEE*, 84:468–475 (April 1939). A short history of the development of the Emitron and Super-Emitron camera tubes, with technical details and brief historical survey from 1908. Photos, diagrams, references.

11-045 McGee, James D., "Distant Electric Vision," *PROCEEDINGS OF THE IRE*, 39:596–608 (June 1950). Brief history followed by descriptions of the design and development of EMI camera tubes: Emitron, Super-Emitron, and the CPS Emitron. Reprinted in Shiers, 12-128.

11-046 _____. "A Review of Some Television Pick-Up Tubes," *PROCEEDINGS OF THE IEE*, 97 (Pt. III):377–392 (November 1950). Appears to be a close parallel to item immediately above. Diagrams, references.

11-047 _____. "The History of Electronic Imaging," in T.P. McLean and P. Schagen, eds. **ELECTRONIC IMAGING.** New York: Academic Press, 1979, pp. 11–54. Based on papers delivered before the Royal Society in London in 1978.

11-048 McGregor-Morris, John T. "The History and Development of the Cathode Ray Tube," *JOURNAL OF THE TELEVISION SOCIETY*, 2 (Series 2):257–262 (June 1937). Brief survey of highlights. Photos.

11-049 Neuhauser, Robert G. "Color Television Tubes: A History but Not Yet an Obituary," *SMPTE JOURNAL*, 99:708–722 (1990). Suggesting the type has been developed as far as possible with analog technology.

11-050 Parr, Geoffrey, "The History of the Cathode-Ray Tube," *TELEVISION AND SHORT-WAVE WORLD*, 10:85–87 (February 1937). An early account of how the British CRT was developed. Photos, diagrams.

11-051 _____. THE LOW VOLTAGE CATHODE-RAY TUBE AND ITS APPLICATIONS. London: Chapman & Hall, 1937, 177 pp.; with O.H. Davie, 1952 [2nd ed.], 1959 [3rd ed.], 433 pp. Emphasizes applications and measurements, useful today for its early contemporary viewpoint. Comprehensive bibliography. Photos, diagrams, index.

11-052 Townsend, Frederick H. "Historical Review of the Cathode Ray Tube As a Television Display Device," *BROADCAST JOURNAL* (March–April 1969), pp. 22-26. Short survey. Photos, diagrams.

11-053 Weimer, Paul K. "A Historical Review of the Development of Television Pickup Devices (1930–1976)," *IEEE TRANSACTIONS*, 23:739–752 (July 1976). One of the better and more complete surveys.

11-054 * Zworykin, Vladimir K. "The Iconoscope—A Modern Version of the Electric Eye," *PROCEEDINGS OF THE IRE*, 22:16–32 (January 1934). Basic paper by the inventor on the theory, construction, characteristics, and operation of the new camera tube. Photos.

E. Microwave Tubes
(For microwaves, see 8-D.)

11-055 Boot, Henry A.H., and John T. Randall. "The Cavity Magnetron," *JOURNAL OF THE IEE*, 93 (Pt. 3A):928–938 (March 1946). An account by the inventors. See also Randall, 11-062. Reprinted in 11-006.

11-056 Brittain, James. "The Magnetron and the Beginnings of the Microwave Age," *PHYSICS TODAY*, 38:60–67 (June 1985). Covers events from 1916 through 1939 as a case study of the interconnections between science, technology, and institutional environments. Photos, diagrams, references.

11-057 Harvey, Arthur F. "Microwave Tubes: An Introductory Review with Bibliography," *PROCEEDINGS OF THE IEE*, 107 (Pt. C):29–59 (March 1960). Three parts: grid-controlled space-charge wave and traveling-wave tubes; crossed-field tubes; and noise in electron tubes. Bibliography of nearly 700 items.

11-058 Kompfer, Rudolf. "Traveling-Wave Tubes," *REPORTS ON PROGRESS IN PHYSICS*, 15:275–327 (1952). Survey and development, mathematical treatment with much historical background. Diagrams, charts, references.

11-059 _____. THE INVENTION OF THE TRAVELING WAVE TUBE. San Francisco: San Francisco Press, 1964, 30 pp. A first-hand story giving step-by-step details of the process of invention as well as of the invention itself to the mid-1940s. Diagrams, references.

11-060 Lyon, Ed. "The Real Story of the Magnetron," *THE A.W.A. REVIEW*, 9:181–204 (1995). Surveys complete development. Photos, diagrams, notes.

11-061 Megaw, Eric C.S. "The High-Power Pulsed Magnetron: A Review of Early Developments," *JOURNAL OF THE IEE*, 93: (Pt. 3A):977–984 (March 1946).

11-062 Randall, John T. "The Cavity Magnetron," *PROCEEDINGS, PHYSICAL SOCIETY*, 58:247–252 (May 1946). On early work in England from late 1939 to 1942. Illustrations. See also Boot, 11-055.

11-063 Rodrique, G.P. "A Generation of Microwave Ferrite Devices," *PROCEEDINGS OF THE IEEE*, 76:121–137 (February 1988). Developments since the late 1970s. Diagrams, bibliography.

11-064 Varian, Russell H., and Sigurd F. Varian. "A High Frequency Oscillator and Amplifier," *JOURNAL OF APPLIED PHYSICS*, 10:321–327 (May 1939). First article by the Varian brothers (see 4-166) on their klystron tube. Photos, references.

11-065 Varian, Russell H., "The Invention and Development of the Klystron," *MILITARY AUTOMATION* (September–October 1957), pp. 256–259. Personal account of the circumstances that led to the first operable tube. See also 4-165.

11-066 Wathen, Robert L. "Genesis of a Generator: The Early History of the Magnetron," *JOURNAL OF THE FRANKLIN INSTITUTE*, 255:271–287 (April 1953). A survey from 1921 through the English work during the early years of World War II. Photos, diagrams, references.

11-067 _____. "The Traveling Wave Tube—A Record of Its Early History," *JOURNAL OF THE FRANKLIN INSTITUTE*, 256:429–442 (December 1954). On the development of theory and experimental results in England, France, and the United States from 1935 to 1947. Illustrations, references.

11-068 White, William C. "Some Events in the Early History of the Oscillating Magnetron," *JOURNAL OF THE FRANKLIN INSTITUTE*, 254:197–204 (September 1952). Primarily about the early work of Japanese researchers up to 1930. Photos, references.

F. Solid-State Devices

(There is a huge and growing literature on this topic. What follows is restricted to a selection of the best material providing historical insight.)

11-069 Bello, Francis. "The Year of the Transistor," *FORTUNE*, 47:128–133, 162–168 (March 1953). Survey of the technology and its development and applications. Photos, diagrams.

Electron Tubes and Solid State Devices

11-070 Bown, Ralph. "The Transistor As an Industrial Research Episode," *SCIENCE MONTHLY,* 80:40–46 (January 1955). A personal story of transistor research and development within Bell Labs. Photos, diagrams.

11-071 ** Braun, Ernest, and Stuart Macdonald. **REVOLUTION IN MINIATURE: THE HISTORY AND IMPACT OF SEMICONDUCTOR ELECTRONICS.** Cambridge, England: Cambridge University Press, 1978, 1982 [2nd ed.], 247 pp. Perhaps the best scholarly history of the invention and the industry. Tables, charts, references, index.

11-072 Brinkman, William F., et al. "A History of the Invention of the Transistor and Where It Will Lead Us," *IEEE JOURNAL OF SOLID STATE CIRCUITS,* 32:1858–1865 (December 1997). A brief modern survey.

11-073 Hazewindus, Nico, with John Tooker. **THE U.S. MICROELECTRONICS INDUSTRY: TECHNICAL CHANGE, INDUSTRY GROWTH AND SOCIAL IMPACT.** New York: Pergamon Press, 1982, 199 pp. Useful treatment of the interaction between technology and policy. Charts, tables, references, index.

11-074 Herold, Edward W. "Semiconductors and the Transistor," *JOURNAL OF THE FRANKLIN INSTITUTE,* 259:87–106 (February 1955). A survey of the history, semiconductors, modern techniques, commercial devices, and laboratory developments. Photos, tables, references.

11-075 * Krull, Alan R., comp. **TRANSISTORS AND THEIR APPLICATIONS: A BIBLIOGRAPHY, 1948–1953.** Evanston: Northwestern University, Technological Institute Library, 1954, 77 pp. Over 900 references to work literature, with some titles from 1938. Reprinted in *TRANSACTIONS OF THE IRE* ED-1:40–70 (August 1954).

11-076 ** Malone, Michael S. **THE MICROPROCESSOR: A BIOGRAPHY.** New York: Springer-Verlag/TELOS, 1995, 333 pp. Nontechnical history of the quarter-century revolution which both describes how the microprocessor works and its historical development and improvement. Photos, glossary, notes, index.

11-077 * Morris, P.R. **A HISTORY OF THE WORLD SEMICONDUCTOR INDUSTRY.** London: Peter Peregrinus and IEE "IEE History of Technology Series 12," 1990, 171 pp. From the development of the valve (tube) to early semiconductor research, development of the transistor, and factors in growth of the U.S., Asian, and European semiconductor industries. Tables, charts, appendices, bibliography, index.

11-078 * Pearson, Gerald L., and Walter H. Brattain. "History of Semiconductor Research," *PROCEEDINGS OF THE IRE,* 43:1794–1806 (December 1955). Comprehensive story over 120 years. Tables, references, bibliography.

11-079 * Pickard, Greeleaf W., "The Discovery of the Oscillating Crystal," *RADIO NEWS,* 6:1166, 1270 (January 1925). Historical account of the early work on crystal detectors by Eccles and the author in 1910.

11-080 Queisser, Hans. **THE CONQUEST OF THE MICROCHIP.** Cambridge, MA: Harvard University Press, 1988, 200 pp. An early participant relates the rise and role of Silicon Valley. Photos, diagrams, bibliographic note, index.

11-081 * Reid, T.R. **THE CHIP: HOW TWO AMERICANS INVENTED THE MICROCHIP AND LAUNCHED A REVOLUTION.** New York: Simon & Schuster, 1984, 243 pp. The work of Kilby and Noyce and what they helped to spawn, told in nontechnical journalistic fashion. Photos, notes, index.

11-082 Ridenour, Louis N. "A Revolution in Electronics,: *SCIENTIFIC AMERICAN,* 185:13–17 (August 1951). On electron tubes, electronic equipment, and the possibilities for transistors.

11-083 ** Riordan, Michael, and Lillian Hoddeson. **CRYSTAL FIRE: THE BIRTH OF THE INFORMATION AGE.** New York: W.W. Norton, 1997, 352 pp. Part of the Sloan Foundation-sponsored series of technology histories, here a physicist and historian of science combine forces to retell the story of the Bell Labs team that developed the transistor, based on considerable archival research. Photos, notes, index.

11-084 Rockett, Frank H. "The Transistor," *SCIENTIFIC AMERICAN,* 179: 52–55 (September 1948). One of the first general reports on the invention. Photos.

11-085 Rogers, Everett M., and Judith K. Larsen. **SILICON VALLEY FEVER: GROWTH OF HIGH-TECHNOLOGY CULTURE.** New York: Basic Books, 1984, 302 pp. Sociological study of the community and its networks. Photos, notes, index.

11-086 * Seitz, Frederick, and Norman G. Einspruch. **ELECTRONIC GENIE: THE TANGLED HISTORY OF SILICON.** Champaign: University of Illinois Press, 1997, 281 pp. Unique history of the element and its many applications, largely devoted to telecommunication. Seitz was involved in much of the more recent history discussed. Photos, diagrams, index.

11-087 *** "Special Issue: 50th Anniversary of the Transistor!" *PROCEEDINGS OF THE IEEE,* 86:1–308 (January 1998). Important collection of 20 papers, some reprints of classics, and other invited historical overviews. All aspects are covered: technical, business, implications, and applications. Photos, diagrams, references.

11-088 "The Transistor 1948–1958: Ten Years of Progress at Bell Telephone Laboratories," *BELL LABORATORIES RECORD,* 36:190–236 (June 1958). Designs and applications. Photos, diagrams.

11-089 * "The Transistor: Two Decades of Progress," *ELECTRONICS,* 41:77–130 (February 19, 1968). Staff report surveying the industry. Photos, diagrams, charts, tables.

11-090 * "Transistor Issue," *PROCEEDINGS OF THE IRE,* 40:1283–1602 (November 1952); 46:949–1360 (June 1958). Two special issues, both of value for their relatively early analyses of transistor development and applications. The first contains 51 papers—a comprehensive collection on physics, theory, characteristics, and applications. Photos, diagrams, references. The second issue contains 43 papers covering broad range of topics. Photos, diagrams, references.

Chapter 12

Television

Television, like radio, has been well treated in general literature since the mid-1920s. This chapter is highly selective—largely because the massive Shiers compendium (1-018) provides a definitive annotated bibliography to 1940 on this topic, including citations to material in other languages.

Drawing substantially from Shiers work is Section A's chronological listing of the first 15 years of television books—taking the story up to regular operation by RCA in the United States, and the wartime closure of the BBC's pioneering system in 1939. Other historical articles and books on television systems appear in Section B, sub-divided into sections on mechanical and electronic television systems. Section C provides overall historical surveys of all of television's technical development. Section D entries concern color television development, and Section E surveys the limited literature available on collecting vintage television equipment. For video recording, see 10-F and 10-G; for cathode-ray tube development, see 11-D; facsimile is covered in 13-A; and high-definition or digital TV is found in 13-G. For related Internet resources, see 15-J.

A. Contemporary Books on Television (1924–1939)

(As with parallel sections in previous chapters, this listing includes all known English-language books and pamphlets [but not articles] devoted to television for the first 15 years of such publication—through 1939. As with those earlier sections, these are listed in chronological order. Contemporary articles on mechanical and electronic television as well as later books appear in Section B. For closely related works on facsimile, see 13-A.)

12-001 Bohringer, Arthur J. **THE A.B.C. OF WIRELESS TELEVISION**. London: Taunton Bros, 1924, 33 pp. First booklet publication in English; includes facsimile. Illustrations.

12-002 ** Dinsdale, Alfred. **TELEVISION: SEEING BY WIRE OR WIRELESS**. London: Pitman, 1926, 62 pp.; and as TELEVISION, Television Press, 1928 [2nd ed.], 180 pp. Appears to be the first English-language book devoted to television (two German volumes appeared the same year—see Shiers, 1-018, p. 103). Focused almost entirely on Baird's system; second edition adds work of AT&T. Photos, diagrams. See also the same author's 12-019.

12-003 Secor, H. Winfield, and J.H. Kraus. **ALL ABOUT TELEVISION**. New York: Experimenter Publishing, 1927, 112 pp. Largely compiled from articles in *RADIO NEWS* and *SCIENCE AND INVENTION*, this is a survey of both phototelegraphy (facsimile) and television, much of it focused on Baird's work. Extensive photos and diagrams.

12-004 Tiltman, Ronald F. **TELEVISION FOR THE HOME: THE WONDERS OF "SEEING BY WIRELESS."** London: Hutchinson, nd but 1927, 106 pp. Primarily about Baird's work with numerous extracts from magazines and newspapers. Photos, no index.

12-005 Lane, Henry M. **THE BOSTON POST BOOK ON TELEVISION**. Boston: Post Publishing, 1928, 35 pp.

12-006 Larner, Edgar T. **PRACTICAL TELEVISION**. London: Ernest Benn / New York: van Nostrand, 1928, 175 pp.; 1929 [2nd ed.], 223 pp. Historical material includes facsimile and mechanical television systems, including European and American work. Emphasis on Baird's work. Photos, diagrams, index.

12-007 * Jenkins, C. Francis. **RADIOMOVIES, RADIOVISION, TELEVISION**. Washington: Jenkins Laboratories, 1929, 143 pp. Details of the author's own equipment, experiments, and demonstrations. Photos, diagrams. See Jenkins, 13-024 for his earlier work on facsimile.

12-008 * Sheldon, H. Horton, and Edgar Norman Grisewood. **TELEVISION: PRESENT METHODS OF PICTURE TRANSMISSION**. New York: van Nostrand, 1929, 194 pp. Historical background includes facsimile. Chapters review key elements of all mechanical systems. One chapter each devoted to the systems of Baird, Bell Labs, Jenkins, and Alexanderson (GE). Photos, diagrams, index.

12-009 Yates, Raymond Francis. **ABC OF TELEVISION, OR SEEING BY RADIO: A COMPLETE AND COMPREHENSIVE TREATISE DEALING WITH THE THEORY, CONSTRUCTION AND OPERATION OF TELEPHOTOGRAPHIC AND TELEVISION TRANSMITTERS AND RECEIVERS**. New York: Norman Henley / London: Chapman & Hall, 1929, 210 pp. A popular exposition on both facsimile and mechanical television, with light technical treatment "written especially for home experimenters, radio fans, and students." Theory, construction, and operations. Photos, diagrams, index.

12-010 Benson, Thomas W. **FUNDAMENTALS OF TELEVISION**. New York: Mancall Publishing, 1930, 145 pp. Includes discussion of Baird, AT&T, and Jenkins systems. Photos, diagrams, index.

12-011 Hutchinson, Robert W. **EASY LESSONS IN TELEVISION.** London: University Tutorial Press, 1930, 175 pp. How to build a mechanical receiver. Diagrams.

12-012 ** Moseley, Sydney A., and H.J. Barton Chapple. **TELEVISION TO-DAY AND TO-MORROW.** London: Pitman, 1930, 130 pp.; 1931 [2nd ed.], 163 pp.; 1933 [3rd ed.], 198 pp.; 1934 [4th ed.], 208 pp.; 1940 [5th ed.], 179 pp. Important book, intended for general reader and amateur constructor, which through its editions helps to trace the development of British television. Almost wholly about Baird and his system in early editions. Photos, diagrams, index. Final edition reflects electronic television developments, and includes foldout map of television service area of Alexandra Palace station.

12-013 Philip, Charles G. **TELEVISION FOR ALL.** London: Percival Marshall, 1930, 82 pp. Focused on mechanical systems, chiefly the work of Baird.

12-014 ROMANCE AND REALITY OF TELEVISION. Boston: Short wave & Television Corp., 1930, 31 pp.; [2nd ed.] 63 pp. A manual for the construction and operation of short wave and mechanical television receiving sets. Diagrams.

12-015 Stranger, Ralph. **SEEING BY WIRELESS.** London: Newnes, 1930, 63 pp. Mechanical television.

12-016 TWO-WAY TELEVISION AND A PICTORIAL ACCOUNT OF ITS BACKGROUND. New York: Bell Telephone Laboratories, 1930, 16 pp. Details mechanical system experiments of 1927–30 in photos and captions. See Ives, 12-067

12-017 Felix, Edgar H. **TELEVISION: ITS METHODS AND USES.** New York: McGraw-Hill, 1931, 272 pp. A general account of contemporary mechanical television systems with some history and a view of future problems and prospects. Photos, diagrams, index.

See Jenkins, 5-145.

12-018 Collins, A. Frederick. **EXPERIMENTAL TELEVISION: A SERIES OF SIMPLE EXPERIMENTS WITH TELEVISION APPARATUS ALSO HOW TO MAKE A COMPLETE HOME TELEVISION TRANSMITTER AND TELEVISION RECEIVER.** Boston: Lothrop, Lee & Shepard, 1932 (reprinted by Lindsay Publications, 1991), 313 pp. One of the earliest American practical guides for home constructors of mechanical receivers. Chapters devoted to experiments with different aspects of television. Diagrams, index.

12-019 *** Dinsdale, A.A. **FIRST PRINCIPLES OF TELEVISION.** New York: John Wiley, 1932 (reprinted by Arno Press "History of Broadcasting," 1971), 241 pp. Invaluable contemporary technical survey of developments which is largely devoted to mechanical systems including those of Baird, AT&T, Jenkins, and others. Perhaps the most inclusive and thus important early book on television. Photos, diagrams, references, index.

12-020 Dunlap, Orrin E. **THE OUTLOOK FOR TELEVISION.** New York: Harper, 1932, 297 pp. Layman's explanation of television taken from the author's columns in the *New York Times,* largely from the 1928-31 period—and thus useful for contemporary viewpoint on television's (then) mechanical future. Photos, chronology, index.

12-021 Hathaway, Kenneth A. **TELEVISION: A PRACTICAL TREATISE ON THE PRINCIPLES UPON WHICH THE DEVELOPMENT OF TELEVISION IS BASED.** Chicago: American Technical Society, 1933, 174 pp.; 1935 [2nd ed., without subtitle], 170 pp. Elementary treatment chiefly on mechanical systems. Photos, diagrams.

See Tiltman, 5-058.

12-022 Camm, Frederick J. **NEWNES TELEVISION AND SHORT-WAVE HANDBOOK.** London: George Newnes Ltd, 1934, 1935 [2nd ed.], 256 pp.; 1937 [3rd ed.]; 1939 [4th ed.]; 1942 [5th ed., as NEWNES TELEVISION HANDBOOK]; 1945 [6th ed.], 224 pp. First 115 pages of early editions devoted largely to mechanical TV systems with useful diagrams and photos. Later editions focused entirely on electronic systems. Glossary, index.

12-023 Chapple, H.J. Barton. **TELEVISION FOR THE AMATEUR CONSTRUCTOR.** London: Pitman, 1933, 233 pp.; 1934 [2nd ed.], 266 pp. Focuses on building a mechanical receiver.

12-024 Reyner, John H. **TELEVISION, THEORY AND PRACTICE.** London: Chapman & Hall, 1934, 196 pp.; 1937 [2nd ed.], 224 pp. An early text focusing on mechanical systems, with a practical approach. Includes chapter on European developments and another on American progress, plus one on cathode-ray tube systems. Photos, diagrams, index.

12-025 Baker W.R.G., and L. Malter. **A SURVEY OF TELEVISION PROGRESS IN AMERICA.** London: George Newnes, 1935. A booklet discussing both mechanical and electronic system developments.

12-026 Chapple, H.J. Barton. **POPULAR TELEVISION: UP-TO-DATE PRINCIPLES AND PRACTICE EXPLAINED IN SIMPLE LANGUAGE.** London: Pitman, 1935, 112 pp. Brief survey published on the eve of regular BBC television transmissions. Two chapters on scanning, others on transmission and reception, receivers, synchronization, and sidelines (applications). Photos, diagrams, index.

12-027 * Dowding, G.V., ed. **BOOK OF PRACTICAL TELEVISION.** London: Amalgamated Press, 1935, 320 pp. 26 chapters by 12 contributors cover all aspects of how "low definition" (mechanical) systems work compared to "high definition" (electronic systems using cathode-ray tubes) and how to make receivers. Photos, diagrams.

12-028 Hutchinson, Robert W. **TELEVISION UP-TO-DATE.** London: University Tutorial Press, 1935, 184 pp.; 1937 [2nd ed.], 211 pp. Includes both mechanical and electronic systems with considerable reference to "high definition" (100 or more scanning lines) television. Second edition adds chapter on BBC television station. Photos, diagrams, index.

12-029 ** Molloy, Edward, general ed. **TELEVISION TODAY: PRACTICE AND PRINCIPLES CLEARLY EXPLAINED.** London: George Newnes, nd but

1935, 2 vols., 776 pp. A comprehensive and authoritative collection of 121 chapters by 39 contributors, most associated with companies in the field. Designed for home constructors and those in the industry, topics range from elementary to more advanced tutorial levels. Covers both mechanical and electronic systems, thus illustrating the transition between them. Originally issued in weekly parts sold on newstands. Photos, graphs, tables, diagrams, index.

12-030 Osborn, E.A. **TELEVISION FOR YOU: BEING A PRACTICAL GUIDE FOR THE MAN-IN-THE-STREET—TO THE MYSTERIES OF TELEVISION UP-TO-DATE.** London: Practical Press, 1935, 78 pp. Includes a rather full summary of the British "Selsdon Report" that led to regular BBC television operations.

12-031 Risdon, P.J. **TELEVISION REALLY EXPLAINED.** London: Foulsham's Wireless Guides, 1935, 94 pp. Both mechanical and electronic means are described.

12-032 Robinson, Ernest H. **TELEVIEWING.** London: Selwyn & Blount "Topical Books," 1935, 288 pp., 1936 [2nd ed.], 288 pp. Compares and contrasts both mechanical and electronic systems. Diagrams.

12-033 Scroggie, M[arcus] G. **TELEVISION.** London: Blackie & Son "Technique Series," 1935, 68 pp.; 1948 [2nd ed.], 77 pp. First edition covers both mechanical and electronic approaches, and concludes with two chapters on the program service and outlook. Second edition deals with postwar BBC operations. Photos, diagrams, index.

12-034 Zworykin, V.K., and G.A. Morton. **POSSIBILITIES OF THE ICONOSCOPE IN TELEVISION.** London: Newnes, 1935. According to Shiers (1-018, p. 435), a brief booklet which assesses the then-new Iconoscope invention.

12-035 * Ardenne, Manfred von., trans. by O.S. Puckle. **TELEVISION RECEPTION: CONSTRUCTION AND OPERATION OF A CATHODE RAY TUBE RECEIVER FOR THE RECEPTION OF ULTRA-SHORT WAVE TELEVISION BROADCASTING.** London: Chapman & Hall, 1936, 121 pp. Based on the author's experience with the initial all-electronic Berlin transmitter and developing British practice. Photos, diagrams, bibliography, index.

12-036 Cannell, John C. **ROMANCE OF TELEVISION.** London: Harrap, 1936.

12-037 * Eckhardt, George H. **ELECTRONIC TELEVISION.** Chicago: The Goodheart-Willcox Co, 1936 (reprinted by Arno Press "Telecommunications," 1974), 162 pp. Important contemporary view of early Farnsworth and Zworykin work. Twenty-one chapters appear in three parts: on pickup and transmission of electronic pictures, reception, and by-products of television research. Photos, diagrams, index.

12-038 Halloran, Arthur Hobart. **TELEVISION WITH CATHODE-RAYS.** San Francisco: Pacific Coast Radio Publishing, 1936, 286 pp. Loose-leaf lecture notes for university extension course. Photos, diagrams (2 folding), notes.

See Lewis, 1-031.

12-039 Moseley, Sydney A., and Herbert McKay. **TELEVISION: A GUIDE FOR THE AMATEUR.** London: Oxford University Press, 1936, 144 pp. Covers both mechanical (Baird, Scophony, Milhaly-Traub) and electronic (EMI, Farnsworth) systems with basic descriptions. Designed for those seeking to build their own receivers. Photos, diagrams, glossary.

12-040 Myers, Leonard M. **ELECTRON OPTICS: AN INTRODUCTION.** London: Pitman, 1936, 338 pp.; 1938 [2nd ed. with revised subtitle: THEORETICAL AND PRACTICAL, Chapman & Hall], 618 pp. Discusses theory of image projection, photometry, the Kerr effect, mechanical optical scanning systems, and electron optical scanning systems. Photos, diagrams, references, index.

12-041 ** TELEVISION... New York (first two vols.) then Princeton, NJ: Radio Corporation of America, 1936–51, 6 vols. Each volume (only first two had subtitles) contains many technical papers by RCA engineers providing a picture of television research status at that time. Photos, diagrams, references.

> **VOL. 1: COLLECTED ADDRESSES AND PAPERS ON THE FUTURE OF THE NEW ART AND ITS RECENT TECHNICAL DEVELOPMENTS.** (July 1936), 452 pp. 25 papers from 1933 to 1936, including five general and nontechnical analyses.
>
> **VOL. 2: THE FUTURE OF THE NEW ART AND ITS RECENT TECHNICAL DEVELOPMENTS.** (October 1937), 435 pp. 29 papers.
>
> **VOL. 3: 1938–41.** (December 1946), 486 pp. 28 papers in sections on pickup devices (7), transmission (10), reception (8), and general topics (3, including surveys of development). An appendix provides abstracts of the papers in the first two volumes, by then out of print.
>
> **VOL. 4: 1942–46.** (January 1947), 510 pp. Papers in sections on pickup devices (7), transmission (3), reception (10), color television (2), military television (7), and general (3 surveys on the status of the medium in 1946). Nine other papers are summarized and a bibliography of 275 RCA papers from 1929–46 is appended.
>
> **VOL. 5: 1947–48.** (August 1950), 461 pp. Papers and brief summaries on pickup (2), transmission (5), reception (14), UHF (3), color television (2), and general topics (10).
>
> **VOL. 6: 1949–50.** (August 1950), 422 pp. Papers and brief summaries on similar topics as previous volume: pickup (3), transmission (3), reception (6), UHF (5), color television (7), and general (3). Concludes with bibliography of more than 500 RCA television research papers from the 1929–50 period, listed chronologically.

12-042 *** Birkenshaw, Douglas C. **TECHNICAL DESCRIPTION OF THE MARCONI–E.M.I. SYSTEM OF TELEVISION AT THE LONDON TELEVISION STATION.** London: BBC Engineering Division, 1937–46, 306 pp. Divided into 12 sections and intended for internal use (though many copies survive), this notebook dates primarily to 1936–39 but some material was reissued and updated in 1946 when the BBC resumed television transmissions. Known informally as "the black book" this described virtually every circuit in the Alexandra Palace (London) facilities.

12-043 * Garratt, G.R.M. **TELEVISION: AN ACCOUNT OF THE DEVELOPMENT AND GENERAL PRINCIPLES OF TELEVISION...** London: HMSO, 1937, 64 pp. Guide to a Science Museum exhibition, this provides a short survey from the 1880s up to the opening of regular BBC telecasts in late 1936. Diagrams, photos, references, index. Reprinted in Shiers, 12-128.

See Parr, 11-051.

12-044 Tyers, Paul D. **TELEVISION RECEPTION TECHNIQUE.** London: Pitman "Telecommunications Series," 1937, 144 pp. Engineering text on then-current British practice with electronic television. Photos, diagrams (one foldout), index.

12-045 ** Wilson, John C. **TELEVISION ENGINEERING.** London: Pitman, 1937, 492 pp. A comprehensive text devoted almost wholly to mechanical systems (odd given publication date), with full analyses of principles and techniques covering a wide range of proposals, inventions, and methods. Perhaps the best technical assessment of mechanical systems at the zenith of their development. Photos, diagrams, tables (some foldout), references, appendices, chapter references, index.

See Witts, 1-036

12-046 AND NOW—THE BBC PRESENTS TELEVISION TO THE WORLD. London: British Broadcasting Corporation, 1938, 32 pp. Promotional booklet featuring Alexandra Palace facilities and personalities. Photos.

12-047 * Ashbridge, Sir Noel. **TELEVISION.** London: The Institution of Mechanical Engineers "The Twenty-Fifth Thomas Hawksley Lecture," 1938, 36 pp. A key BBC engineering official describes the pioneering BBC efforts. Photos, diagrams.

12-048 Maloff, I.G., and D.W. Epstein. **ELECTRON OPTICS IN TELEVISION, WITH THEORY AND APPLICATION OF TELEVISION CATHODE-RAY TUBES.** New York: McGraw-Hill, 1938, 299 pp. Two RCA engineers provide an advanced technical analysis making considerable use of mathematics. Photos, diagrams, index.

12-049 Moseley, Sidney A., and H.J. Barton Chapple. **A SIMPLE GUIDE TO TELEVISION.** London: Pitman, 1938, 43 pp. Electronic television. Photos, diagrams.

12-050 Reyner, J.H. **TESTING TELEVISION SETS.** London: Chapman & Hall, 1938, 128 pp. Designed for amateurs and repair personnel. Photos, diagrams.

12-051 Waldrop, Frank C., and Joseph Borkin. **TELEVISION: A STRUGGLE FOR POWER.** New York: William Morrow, 1938, 299 pp. An economic and social study of the American industry, especially business versus public interest. Focus on interaction of government regulation, business planning, and technical developments. Notes, bibliography, index.

See Ardenne, 11-038.

12-052 Beitman, Morris. N. **CYCLOPEDIA OF TELEVISION FACTS**. Chicago: Supreme Publications, 1939, 58 pp. The work of RCA and other laboratories. Diagrams.

12-053 Corbishley, Harold, ed. **TELEVISION AND SHORT-WAVE WORLD PRACTICAL HANDBOOK**. London: Television and Short-Wave World, 1939. Drawn from articles in the British periodical, this emphasizes home constructor projects. See also West and Osman, 12-060.

12-054 Harding, Charles Francis, et al. **THE PURDUE UNIVERSITY EXPERIMENTAL TELEVISION SYSTEM**. Lafayette, IN: Purdue University Engineering Experiment Station, Research Series No. 65, 1939, 53 pp. General description of the system, the transmitting station (which included scanning disc technology), the receiver, and picture and signal reception experience. Photos, diagrams.

12-055 Kerby, Philip. **THE VICTORY OF TELEVISION**. New York: Harper, 1939, 120 pp. Includes chapters on early inventors, studios, lighting, television compared to cinema and radio, music, sponsor problems, public service programs, and an appendix of FCC regulations and terms. Photos, index.

12-056 Klemperer, O. **ELECTRON OPTICS**. London: Cambridge University Press, 1939, 107 pp.; 1953 [2nd ed.], 471 pp. By the research staff of Electric and Musical Industries (EMI), London, and focusing on their work with CRTs.

12-057 * **THE LONDON TELEVISION STATION**. London: BBC, 1939, 39 pp. Description of the facilities at Alexandra Palace which began regular operation in November 1936. Photos. See also Birkenshaw, 12-042.

12-058 Siegel, Norman. **TELEVISION: A SERIES OF ARTICLES**. New York: RCA, 1939, 26 pp. Reprinted from the reporter's stories in the *New York World-Telegram* in June.

12-059 Sleeper, William B. **THE TELEVISION HANDBOOK: LOOK AND LISTEN**. New York: Norman Henley, 1939, 96 pp. Describes 1939 television practice for repairmen, set-builders, and students as well as the status of television as of April 30, 1939. Photos, diagrams.

12-060 West, S., and D.E. Osman. **BUILDING TELEVISION RECEIVERS AT HOME: A PRACTICAL HOME-CONSTRUCTOR HANDBOOK FOR EVERYBODY**. London: Television and Short-Wave World, 1939, 112 pp. Articles reprinted, with revisions, from Corbishley, 12-053.

B. Other Historical Books and Articles

(Includes articles from before 1939 as well as both books and articles published since. Arranged by author.)

Mechanical Systems

(Until about 1935, various mechanical means of showing pictures and motions, most built around spinning discs, dominated television development. Key centers of activity included Baird in London, Jenkins in Washington, and AT&T researchers in the New

York region. Note: nearly all of the titles listed in Section A also focus on mechanical television. Most of what follows is retrospective history.)

12-061 Bray, W.J. "Post Office Contributions to the Early History of the Development of Television in the United Kingdom," *BT TECHNICAL JOURNAL,* 5:64–77 (January 1987).

12-062 * Bridgewater, T.H. **JUST A FEW LINES...** London: British Vintage Wireless Society, 1992, 20 pp. Issued as a supplement to the society's *BULLETIN,* this focuses on the 1932–33 work of Baird engineers (the author was one—and the last survivor) with the BBC. Photos, diagrams, notes.

12-063 Burns, R.W. "The Contributions of the Bell Telephone Laboratories to the Early Development of Television," *HISTORY OF TECHNOLOGY,* 13:181–213 (1991).

12-064 Clarke, Basil. "Pioneering Days: The *First* Live TransAtlantic Link," *INTERNATIONAL TV TECHNICAL REVIEW,* 3:498–501 (October 1962). Baird work in 1931 as seen by a participant. Photos. For part two see Clarke, 12-082.

12-065 Dinsdale, A.A. "Television in America To-Day," *JOURNAL OF THE TELEVISION SOCIETY,* 1:137–149 (1932). Covers both mechanical and early electronic systems, and largely duplicates two chapters in 12-019. Photos, diagrams. Reprinted in Shiers, 12-128.

12-066 ** Hogan, John V.L. "The Early Days of Television," *JOURNAL OF THE SMPTE,* 63:169–173 (November 1954). On the mosaic and scanning disc systems with a brief survey of methods tried out during the 1920s. Reprinted in Fielding, 12-115, and Shiers, 12-128.

12-067 * Ives, Herbert E., et al. "Television," *BELL SYSTEM TECHNICAL JOURNAL,* 6:551–652 (October 1927). General survey of the problem and four papers outlining Bell System research and experiments, including television signals, synchronization, and both wire and wireless transmission. Photos, graphs, diagrams, map. See also TWO-WAY TELEVISION, 12-016.

12-068 Not used.

12-069 ** Knapp, J. George, and Julian E. Tebo. "The History of Television, Part I: Mechanical and Semi-Mechanical Systems," *IEEE COMMUNICATIONS,* 16:8–22 (May 1978). Emphasis on American work patented by Bell Labs, but should be used carefully as has numerous errors. Photos, diagrams, references.

12-070 * Kurtz, E.B. **PIONEERING IN EDUCATIONAL TELEVISION, 1932–1939 (A DOCUMENTARY PRESENTATION).** Iowa City: University of Iowa, 1959, 166 pp. Unique published record of mechanical television experimentation on W9XK, especially useful as it reprints contemporary reports and correspondence and a wealth of photographs.

12-071 * Shiers, George. "Early Schemes for Television," *IEEE SPECTRUM,* 7:24–34 (May 1970). Surveys the mechanical precursors (1877–84) of modern television. Photos, references. Reprinted in Shiers, 12-128.

12-072 ** _____. "Historical Notes on Television before 1900," *JOURNAL OF THE SMPTE*, 86:129–137 (March 1977). Details the work of researchers in 11 countries. Chronological table, photos, references. List of errata published in July issue, p. 500.

12-073 _____. "Television in Prospect, 1873–1927," *JOURNAL OF THE ROYAL TELEVISION SOCIETY*, 16:8–12 (September–October 1977). Discusses prophecies and opinions about the possibilities for seeing by electricity and its applications. Photos, references.

12-074 ** _____. "The Rise of Mechanical Television, 1901–1930," *JOURNAL OF THE SMPTE*, 90:508–521 (June 1981). Summary survey of people, inventions, and events. Tables, chart, bibliography.

12-075 Van-den Ende, Jan, et al. "Shaping the Early Development of Television," *IEEE TECHNOLOGY AND SOCIETY MAGAZINE*, 16:13–26 (Winter 1997/1998). Illustrations, diagrams, bibliography.

12-076 * Yanczer, Peter F. **THE MECHANICS OF TELEVISION: THE STORY OF MECHANICAL TELEVISION**. St. Louis: Author, 1987, 170 pp. Detailed assessment of the many mechanical systems, clearly explained and well illustrated. Diagrams.

Electronic Television (including Broadcasting)

(Though long theorized as possible, electronic television system development centered on cathode-ray tubes began in earnest only in the 1920s and edged mechanical systems aside by the mid-1930s. Many of the later works in Section A at least introduce electronic methods. What follows combines contemporary reports, generally published from the 1930s onward, with more recent studies—all arranged by author.)

12-077 Ashbridge, Noel. "Television in Great Britain," *PROCEEDINGS OF THE IRE*, 25:697–707 (June 1937). Survey from the first Baird demonstrations with the BBC in 1929. Photos. Reprinted in Shiers, 12-128.

See BBC, 4-188, 4-190.

12-078 Bitting, Robert C. Jr. "Creating an Industry," *JOURNAL OF THE SMPTE*, 74:1015–1023 (November 1965). The rise of electronic television research at RCA based on the author's MIT master's thesis. References. Reprinted in Shiers, 12-128.

12-079 Blumlein, Alan D, et al. "The Marconi-EMI Television System," *JOURNAL OF THE IEE*, 83:758–792, discussion 793–801 (December 1938). Transmitted wave forms, vision input equipment, and the transmitter. Photos, diagrams.

12-080 Burns, R.W. "The History of Television for Public Showing in Cinemas in the United Kingdom," *PROCEEDINGS OF THE IEE*, Part A, 132:653–663 (December 1985). Photos, diagrams, bibliography. See author's related book, 12-113.

Television

12-081 _____. "Seeing by Electricity," *PROCEEDINGS OF THE IEE* Part A, 133:27–37 (January 1986). Photos, diagrams, bibliography. See author's related book, 12-113.

12-082 Clarke, Basil. "Pioneering Days: The Quest for High-Definition," *INTERNATIONAL TV TECHNICAL REVIEW,* 3:546–549 (November 1962). Continues 12-064 and relates mid-1930s cathode-ray work with the BBC, in which the author participated. Photos.

12-083 *_____. "Perspectives on Television: The Role Played by the Two NTSC's in Preparing Television Service for the American Public," *PROCEEDINGS OF THE IEEE,* 64:1322–1331 (September 1976). Covers both National Television System Committees—in 1939–40 to develop black-and-white, and the early 1950s to develop color standards.

12-084 _____. "The Forces at Work Behind the NTSC Standards," *JOURNAL OF THE SMPTE,* 90:498–502 (June 1981). Brief historical survey of both NTSC groups, that for black-and-white, and the later committee for color television.

12-085 De Forest, Lee. **TELEVISION TODAY AND TOMORROW.** New York: Dial Press, 1942, 361 pp. Designed for the lay reader, this survey of electronic television is by the inventor of the Audion tube. Diagrams, notes, index.

12-086 DuMont Laboratories. **PIONEERING THE CATHODE-RAY AND TELEVISION ARTS.** Passaic, NJ: Allen B. DuMont Laboratories, 1941, 40 pp. Illustrated historical survey of a decade's work. Photos, list of company patents.

12-087 Dunlap, Orrin E. **THE FUTURE OF TELEVISION.** New York: Harper, 1942, 194 pp.; 1947 [2nd ed.], 194 pp. Largely nontechnical, this is a useful wartime survey of American TV practice and prospects. Photos, chronology, list of operating stations, index.

12-088 Farnsworth, Philo T. "Television by Electron Image Scanning," *JOURNAL OF THE FRANKLIN INSTITUTE,* 218:411–444 (October 1934). Conversion of optical image to an electron signal, image dissector tube, theory, construction, and operation. Diagrams, graphs, photos. Reprinted in Shiers, 12-128.

12-089 ** Fink, Donald G. **PRINCIPLES OF TELEVISION ENGINEERING.** New York: McGraw-Hill, 1940, 541 pp.; 1952 [2nd ed.], 721 pp. Technical and practical treatment of methods and equipment, image analysis, camera tubes, scanning beams, the video signal, amplification, carrier transmission, broadcast practice, receivers. Photos, diagrams, chapter references, index.

12-090 ***_____. **TELEVISION STANDARDS AND PRACTICE: SELECTED PAPERS FROM THE PROCEEDINGS OF THE NATIONAL TELEVISION SYSTEM COMMITTEE AND ITS PANELS.** New York: McGraw-Hill, 1943, 405 pp. Full description of the system approved by the FCC in 1941. Diagrams, notes, appendices, list of NTSC technical papers, index. First three chapters and list of papers reprinted in Shiers, 12-128.

12-091 *_____. "Television Broadcasting Practice in America: 1927–1944," *JOURNAL OF THE IEE,* 92 (Pt. III):145–164 (September 1945). An overview including standards, program service, specifications, channel frequencies, theater television, color television. Diagrams, photos, tables, references. Reprinted in Shiers, 12-128. See also the same author's "Television Broadcasting in the United States, 1927–1950," *PROCEEDINGS OF THE IRE,* 39:116–123 (February 1951).

12-092 Gomery, Douglas. "Theater Television: A History," *SMPTE JOURNAL,* 98:120-123 (February 1989). Brief survey of use of television exhibition in theater settings, primarily in the 1930s and 1940s, but with some "pay TV" settings much later. Bibliography.

12-093 Hylander, C.J., and Robert Harding Jr. **AN INTRODUCTION TO TELEVISION.** New York: Macmillan, 1941, 207 pp. A survey of the technology for laypersons just as American television stations were beginning operation. Photos, diagrams, index.

12-094 Jones, Glyn. "From Attics to Ignominy," *NEW SCIENTIST,* 120:39–43 (December 10, 1988). Illustrations.

See Kempner, 1-030.

12-095 * Lohr, Lenox R. **TELEVISION BROADCASTING: PRODUCTION, ECONOMICS, TECHNIQUE.** New York: McGraw-Hill, 1940, 274 pp. Valuable for details on prewar operations and program production facilities and equipment developed by NBC and RCA, as described by the president of the network. Photos, diagrams, index.

12-096 Loughlin, B.D. "The Revolution and Evolution from Dot Sequential to NTSC," *IEEE TRANSACTIONS ON CONSUMER ELECTRONICS,* CE-30:2:18–23 (May 1984). A Hazeltine engineer who participated describes the trend from RCA's 1949 dot-sequential color system to the NTSC-recommended standard of 1951. References.

12-097 ** MacNamara, T.C., and D.C. Birkinshaw, "The London Television Service," *THE JOURNAL OF THE IEE,* 83:504:729–801 (December 1938). One of the fuller technical descriptions. Photos, diagrams, references.

12-098 PIONEERING IN TELEVISION: PROPHECY AND FULFILLMENT. New York: RCA, 1946, 98 pp.; 1956 [4th ed.], 177 pp. Nearly 100 excerpts from speeches and statements by David Sarnoff concerning the company's television work. List of RCA firsts in television.

12-099 Preston, S.J. "The Birth of a High Definition Television System," *JOURNAL OF THE TELEVISION SOCIETY,* 7:115–126 (1953). On British development of the EMI electronic system, early tubes and equipment. Decisions on an electronic system, technical details, choice of 405-line standard, postwar advances to 1949. Photos. Reprinted in Shiers, 12-128.

12-100 "RCA's Television," *FORTUNE,* 38:80–85, 194–204 (September 1948). History, production, business, and market aspects. Photos.

12-101 Ritchie, Michael. **PLEASE STAND BY: A PREHISTORY OF TELEVISION.** Woodstock, NY: Overlook Press, 1994, 247 pp. Popular narrative with much emphasis on what was aired prior to 1948 inception of network service in the U.S. Focus is on British and American pioneering services. Appendices, photos, notes, bibliography, index.

12-102 * Singleton, Thomas. **THE STORY OF SCOPHONY**. London: Royal Television Society, 1988, 152 pp. Details an early British all-electronic system and the key people behind it. Photos, charts, index.

12-103 Swift, John. **ADVENTURE IN VISION: THE FIRST TWENTY-FIVE YEARS OF TELEVISION**. London: John Lehmann, 1950, 223 pp. Overview for laymen focusing on British progress: Baird, the BBC, people and programs, with mention of activities in France, Germany, and the U.S. Photos, folding studio plan, diagrams, index.

12-104 ** Swinton, A.A. Campbell. "Presidential Address," *JOURNAL OF THE RÖNTGEN SOCIETY*, 8:1–15 (January 1912). This historic document has a schematic diagram of an electric television system as previously suggested by the author. Reprinted in Shiers, 12-128.

12-105 _____. "The Possibilities of Television," *WIRELESS WORLD*, 14:51–56, 82–84. (April 9, 16, 1924). Updates author's scheme (previous entry) with three diagrams for using either wire or radio transmission.

12-106 Teer, K. "Looking Back at Distant Vision: Television Technology from 1936 to 1986," *PHILIPS TECHNICAL REVIEW*, 10:12:297–311 (September 1986). Useful survey focusing on receiver tubes. Photos, notes.

12-107 ** UNESCO. **TELEVISION: A WORLD SURVEY and SUPPLEMENT**. Paris: UNESCO, 1953, 184 pp.; 1955, 51 pp. (reprinted in one volume by Arno Press "International Propaganda and Communications," 1972). The two reports together provide a unique record of the rise of television in developed (and a few developing) nations. In each case information is provided on history, stations and transmitters, relay facilities, production of equipment, and plans for growth. The original book covers 44 countries while the Supplement adds 14 and updates the rest. Tables, maps. For related UNESCO studies, see 3-072 and 3-073.

12-108 * Zworykin, Vladimir K. "Television," *JOURNAL OF THE FRANKLIN INSTITUTE*, 217:1–37 (January 1934). Outline of the all-electronic system developed by RCA. Includes photos of received images and experimental receiver.

12-109 *** Zworykin, Vladimir K., and George A. Morton. **TELEVISION: THE ELECTRONICS OF IMAGE TRANSMISSION**. New York: John Wiley, 1940, 646 pp.; 1954 [2nd ed.], 1,037 pp. A basic and important text with emphasis upon RCA developments. Second edition adds material on color television. Photos, diagrams, tables, references, index.

C. Historical Surveys

(Supplementing the entries above, these works generally cover both mechanical and electronic systems. Arranged by author.)

12-110 * Abramson, Albert. **ELECTRONIC MOTION PICTURES: A HISTORY OF THE TELEVISION CAMERA**. Berkeley: University of California Press, 1955, 212 pp. Thorough survey of mechanical and electronic eras in many countries, somewhat superceded by next entry. Photos, diagrams, chapter notes, index.

12-111 ** _____. **THE HISTORY OF TELEVISION, 1880 TO 1941**. Jefferson, NC: McFarland, 1987, 354 pp. Valuable overall source for prewar development of American and world television. Useful for specifics on inventors and patents as well as general trends in the technical development of both mechanical and electronic systems. Photos, notes, bibliography, glossary, index.

12-112 Arnold, Erik. **COMPETITION AND TECHNICAL CHANGE IN THE TELEVISION INDUSTRY: AN EMPIRICAL EVALUATION OF THEORIES OF THE FIRM**. London: Macmillan, 1985, 234 pp. Detailed study of the British industry in the 1970s and into the 1980s. Tables, charts, bibliography, index.

12-113 *** Burns, R.W. **BRITISH TELEVISION: THE FORMATIVE YEARS**. London: Peter Peregrinus with Science Museum "IEE History of Technology Series 7," 1986, 488 pp. Definitive narrative of the 1923–39 period, tracing the critical changeover from mechanical systems (chiefly Baird) to electronic means (chiefly EMI), ending with the initiation of the world's first regularly scheduled television service (November 1936). Tables, charts, maps, photos, chapter references, appendices, index. See also Norman, 12-126, and next citation.

12-114 *** _____. **TELEVISION: AN INTERNATIONAL HISTORY**. London: IEE "IEE History of Technology Series 22," 1997, 661 pp. Discusses developments to 1940 in the U.S., Britain (see related book immediately above), Germany, and France. This is the definitive treatment to date, providing comparisons of developments in different countries in text and tables. Photos, diagrams, references, bibliography, index.

12-115 Fielding, Raymond, ed. "Historical Papers—Television," in **A TECHNOLOGICAL HISTORY OF MOTION PICTURES AND TELEVISION**. Berkeley: University of California Press, 1967, pp. 227–254. Four SMPTE papers on television are included, all listed elsewhere in the present bibliography. Photos, diagrams, references.

12-116 * Fisher, David E., and Marshall Jon Fisher. **TUBE: THE INVENTION OF TELEVISON**. Washington: Counterpoint "Sloan Technology Series," 1996, 427 pp. One of a series of popular histories funded by the Sloan Foundation, this is a useful modern survey of the medium—but must be used carefully on some matters of detail. Other than a final section on recent developments (such as HDTV), this is largely a pre-1940 story of the trend from mechanical to electronic systems. Photos, diagrams, chronology, notes, bibliography, index.

12-117 Fisk, Jim, and Dave Ingram. "50 Years of Television," *HAM RADIO* (February 1976), pp. 36–46. Useful survey. Photos, diagrams, references, bibliography.

12-118 Gaggioni, H.P. "The Evolution of Video Technologies," *IEEE COMMUNICATIONS MAGAZINE*, 25:20–36 (November 1987).

12-119 *** Garratt, G.R.M., and A.H. Mumford. "The History of Television," *PROCEEDINGS OF THE IEE*, 99 (Pt.IIIA):25–42 (May 1952). A basic paper providing a broad review of early facsimile, discovery of the value of selenium, early mechanical systems, electronic systems, and public television service in Britain. Diagrams, photos, references. Reprinted in Shiers, 12-128. (Note: this entire issue of 866 pp. is devoted to the "British Contribution to Television" and includes 85 papers.)

12-120 Gorokhov, P.K. "The Origins of Modern Television," *RADIO ENGINEERING*, 16:104–116 (June 1961) as reprinted from *RADIOTEKNIKA*, 16:71–80 (1961). Early history with emphasis on Rosing's work, development of mechanical systems, and Russian contributions up to 1935. References. Reprinted in Shiers, 12-128.

12-121 Hills, Raymond C. "Fifty Years of High Definition Television Transmission," *JOURNAL OF THE IEEE*, 56:1–15 (January 1986). Illustrates how the definition of "high" has changed over the years.

12-122 * THE HISTORY OF TELEVISION FROM EARLY DAYS TO THE PRESENT. London: IEE "Conference Publication Number 271," 1986, 185 pp. Contents includes: survey papers (3), video recording methods (4), transmitters and receivers (3), planning and technical standards (4), program production (3), and current trends and the future (3). Diagrams, charts, tables, photos, references.

12-123 Hubbell, Richard W. **4000 YEARS OF TELEVISION: THE STORY OF SEEING AT A DISTANCE.** New York: Putnam, 1942, 256 pp. Popularly written nontechnical account of the history of television, largely focused on American development but with chapters on foreign systems and television in wartime. Diagrams, index.

**12-124 ** Jensen, Axel G. "The Evolution of Modern Television," *JOURNAL OF THE SMPTE*, 63:174–188 (November 1954); reprinted 100:357–371 (May 1991). A basic paper providing full survey of electronic systems with emphasis on American work. Brief mention of early mechanical proposals. References. Reprinted in Fielding, 12-115, and Shiers, 12-128.

12-125 Layer, Harold A. "From Radiovision to Video... and Television Between," *AUDIOVISUAL INSTRUCTION*, 20:5:6–11 (May 1975). Brief review of the medium's development, with short bibliography of the first 12 years of television." Photos, diagrams.

12-126 * Norman, Bruce. **HERE'S LOOKING AT YOU: THE STORY OF BRITISH TELEVISION, 1908–1939.** London: BBC and Royal Television Society, 1984. 224 pp. Popular history especially useful for its illustrations. Diagrams, photos, bibliography, index. For definitive treatment of this same period, compare with Burns, 12-113.

**12-127 ** O'Brien, Richard S., et al. "101 Years of Television Technology," *JOURNAL OF THE SMPTE*, 85:457–480 (July 1976); reprinted 100:606–629 (August 1991). Very good yet concise basic paper covering all highlights. Photos, diagrams, two-page chronological chart, bibliography.

12-128 *** Shiers, George, ed. **TECHNICAL DEVELOPMENT OF TELEVISION**. New York: Arno Press "Documents in the History of Telecommunications," 1977, about 500 pp. Some 30 historical papers and reports from 1878 to 1973 (many are listed elsewhere in this chapter) in sections on general history, mechanical systems, electronic systems, and color. Photos, diagrams, tables, references.

12-129 TWENTY-FIVE YEARS OF WORLD TELEVISION: FIRST FROM THE START. Hayes, England: EMI Electronics "An International TV Technical Review Publication," 1961, 58 pp. Focuses on inception and early years of British television beginning with regular BBC service in 1936. Photos.

12-130 * Udelson, Joseph H. **THE GREAT TELEVISION RACE: A HISTORY OF THE AMERICAN TELEVISION INDUSTRY, 1925–1941**. University: University of Alabama Press, 1981, 197 pp. Includes both technology and business aspects of the prewar story.

D. Color Television
(See also 11-D for further citations on cathode-ray tubes.)

12-131 Bello, Francis. "Color TV: Who'll Buy a Triumph?" *FORTUNE*, 52:136–139, 201–206 (November 1955). Brief history, description of several systems, company efforts, industry problems and prospects. Photos, diagrams.

12-132 * Bourton, K. **A BIBLIOGRAPHY OF COLOUR TELEVISION**. London: Television Society, 1954–56, 34 pp. Some 650 unannotated items arranged chronologically (and including related topics) beginning in 1860.

12-133 COLOUR 69: AN ADVERTISER'S GUIDE TO COLOUR TELEVISION. London: British Bureau of Television Advertising Ltd., 1969, 112 pp. Despite title, this is a useful description of the PAL color system as it began implementation in Britain, including chapters on major equipment and operations. Photos, diagrams, appendices, map.

12-134 ** Color Television Issues of the *PROCEEDINGS OF THE IRE*. Details development of the NTSC color television system. Photos, diagrams, references. As follows:

> "Color Television Issue," 39:1123–1360 (October 1951), 20 papers describing experimental results by RCA, CBS, and others.

> "Second Color Television Issue," 43:1–348 (January 1954), 49 papers, 22 about or by the NTSC and 28 papers by other individuals.

12-135 * Crane, Rhonda J. **THE POLITICS OF INTERNATIONAL STANDARDS: FRANCE AND THE COLOR TV WAR**. Norwood, NJ: Ablex, 1979, 123 pp. Only book-length treatment of why there are three major world color television standards (NTSC, PAL, and SECAM). Map, notes, bibliography, index.

12-136 *** Fink, Donald G., ed. **COLOR TELEVISION STANDARDS: SELECTED PAPERS AND RECORDS OF THE NATIONAL TELEVISION**

Television

SYSTEM COMMITTEE. New York: McGraw-Hill, 1955, 520 pp. Key documents from the second NTSC that developed color television standards. Photos, diagrams, tables, appendices, index.

12-137 Gabor, Dennis. "Colour TV," *ENDEAVOUR*, 21:25–34 (January 1962). On the development of various color systems and tubes with some historical background. Diagrams, references.

12-138 * Goldmark, Peter C., et al. "Color Television," *PROCEEDINGS OF THE IRE*, 30:162–182 (April 1942) and *JOURNAL OF THE SMPTE*, 38:311–353 (April 1942). Details the CBS partially mechanical system. Part 1 includes photos, diagrams, graphs, references. Part 2, *PROCEEDINGS OF THE IRE*, 31:465–478 (September 1943). Photos, references.

12-139 Gouriet, G.G. AN INTRODUCTION TO COLOUR TELEVISION. London: Published for the Television Society by Norman Price, Publishers, 1955, 72 pp. (also issued in paperback format "for the use of BBC staff only"). Substance of 1954 lectures to the society by various authorities describing the American system. Appears to be first British book on the subject. Diagrams.

12-140 Law, Harold B. "The Shadow Mask Color Picture Tube: How It Began—An Eyewitness Account of Its Early History," *JOURNAL OF THE SMPTE*, 86:214–221 (April 1977). By a member of the RCA technical staff. Photos, diagrams, references. See also 11-043.

12-141 Patchett, G.N. "Colour Television," *JOURNAL OF THE BRITISH IRE*, 16:591–620 (November 1956). Description of various systems, transmitting and receiving equipment, the NTSC and British standards, with some historical material. References.

12-142 Radio Corporation of America. Selected color television publications. New York: 1951–54. Many were issued in this period of high-profile promotio—these are examples.

COLOR TELEVISION DEVELOPMENTS AS VIEWED BY THE PRESS and **RCA COLOR TELEVISION PRESS REPORTS**. 1951, 30 pp. each. Interesting collection of press reports (and RCA press releases) when RCA was contesting FCC decision approving CBS incompatible color system.

RCA AIDS INDUSTRY IN ADVANCING COMPATIBLE COLOR TELEVISION. November 1953, 28 pp. Brief narrative, with chronology, of RCA work in this field.

RCA COLOR TELEVISION. December 1953, 34 pp. Company promotional piece for general audiences issued right after FCC approved RCA-developed standards. Photos.

"Color Television Issue," *BROADCAST NEWS* 77:1–80 (January–February 1954). First detailed description for engineers of the newly-approved RCA-developed color standards and equipment. Photos, diagrams, tables.

Chapter 12

12-143 * Tyler, Kingdon S. **TELECASTING AND COLOR**. New York: Harcourt, Brace, 1946, 213 pp. First book devoted to the topic, this focuses almost entirely on the partially mechanical CBS system (author was a CBS engineer). Photos, diagrams, bibliography, index.

12-144 * U.S. Senate Committee on Interstate and Foreign Commerce. **THE PRESENT STATUS OF COLOR TELEVISION: REPORT OF THE ADVISORY COMMITTEE ON COLOR TELEVISION...** 81st Cong., 2nd Sess., Senate Document No. 197, July 1950, 63 pp. Important survey of technical developments just as FCC selected the incompatible CBS-developed standard. Tables. Reprinted in Shiers, 12-128.

E. Vintage Televisions
(Collecting old television sets has not caught on like the old radio receiver hobby, but appears to be gathering steam.)

12-145 Bennett-Levy, Michael. **HISTORIC TELEVISIONS AND VIDEO RECORDERS**. Musselburgh, Scotland: MLB Publications, 1993, 58 pp. Color photos by Ivor Tetteh-Larty.

12-146 _____. **TV IS KING: EXHIBITION AT SOTHEBY'S... AUGUST 1994**. Musselburgh, Scotland: MLB Publications, 1994, 64 pp. Catalogue for an exhibition.

12-147 Collins, Philip. **THE GOLDEN AGE OF TELEVISIONS**. Santa Monica, CA: General Publishing Group, 1997, 132 pp. Lavish color album of classic receivers with occasional captions and backgrounds related to the period of the set shown. Chronology, index.

12-148 Emmerson, Andrew. **OLD TELEVISION**. Aylesbury, England: Shire Publications, 1995.

See Geddes and Bussey, 4-059.

12-149 Heightman, D.W. "Television Reception 1925–1975," *RADIO & ELECTRONIC ENGINEERING,* 45:559–569 (October 1975). A history of receiver technology.

12-150 Mothersole, Peter L. **THE EVOLUTION OF THE DOMESTIC TELEVISION RECEIVER**. London: Royal Television Society, 1988, 38 pp. Brief monograph tracing development of British receivers. Photos, diagrams. See also Geddes and Bussey, 4-059.

12-151 Ward, Scott. **CLASSIC TVs: PRE-WAR THRU 1950s**. Gas City, IN: L-W Books Sales, 1996, 92 pp. First such guide for collectors—it includes old advertisements and catalogue pages plus photos of various models and current prices.

12-152 * **WATCHING TV: HISTORIC TELEVISIONS AND MEMORABILIA FROM THE MZTV MUSEUM**. Toronto: Royal Ottawa Museum, 1995, 48 pp. Museum exhibit catalogue useful primarily for good photos of historical receivers. Photos, bibliography.

Chapter 13

Newer Media Technologies

For this and the next chapter, we adopt a much looser definition of "history." While typically defined as referring to events at least a generation (25 years) old, many of the technologies discussed here and in chapter 14 have only moved out of the laboratory in the last couple of decades or even more recently. Thus, history purists may well argue that none of these services have been around long enough to have developed a true "history." But their development is already of growing importance—and cannot be ignored in a work surveying the literature on telecommunication technology development.

With one exception, the media technologies detailed in this chapter have developed since the 1950s. The single exception—which appears first—is facsimile, which in its primitive original form dates back to the 19th century. The early history of phototelegraphy includes the inventions of Alexander Bain (1843), Frederick Bakewell (1948), the first commercial system of Giovanni Caselli (1861), and other copying or writing telegraphs for reproducing script. The material in section A describes the dramatic changes in fax to the widely utilized service of today.

The chapter continues with (B) general "new media" surveys, (C) cable television, (D) television-based teletext/videotex (for Internet, see 14-E), (E) remote control devices and interactive services, (F) digital audio/radio, and (G) high-definition/digital television which began a limited regular service in the United States only late in 1998. None of these yet have formal "histories" and thus much of the material in this chapter provides what will become historically useful snapshots of their changing technical status.

A. Facsimile

(The best historical treatment of early phototelegraphy is a German text by Alexander Korn and G. Glatzel, HANDBUCH DER PHOTOTELEGRAPHIE UND TELEAUTOGRAPHIE [Leipzig, 1911.] There still being no adequate overall history of the technology, most of what follows reflects contemporary thinking.)

Chapter 13

13-001 Armagnat, Henri. "Phototelegraphy," **SMITHSONIAN INSTITUTION ANNUAL REPORT FOR 1908**. Washington: Government Printing Office, 1909, pp. 197–207. Translated from the French, this is a survey of then-current systems by Korn, Berjonneau, Carbonelle, and Senlecq-Tival.

13-002 * Baker, T. Thorne. **THE TELEGRAPHIC TRANSMISSION OF PHOTOGRAPHS**. London: Constable / New York: van Nostrand, 1910, 141 pp. A survey of phototelegraphic systems, particularly that of Korn and the author. One chapter covers wireless telegraphy systems with details of Knudson's work and the author's experiments. Photos, diagrams, no index.

13-003 _____. "The Telegraphy of Photographs, Wireless and by Wire," **SMITHSONIAN INSTITUTION ANNUAL REPORT FOR 1910**. Washington: Government Printing Office, 1911, pp. 257–274. Includes details of several English experiments. Photos, diagrams.

13-004 * _____. **WIRELESS PICTURES AND TELEVISION: A PRACTICAL DESCRIPTION OF THE TELEGRAPHY OF PICTURES, PHOTOGRAPHS, AND VISUAL IMAGES**. London: Constable, 1926 / New York: van Nostrand, 1927, 188 pp. Almost entirely devoted to facsimile, despite title. Considerable historical material by a pioneer in phototelegraphy. Photos, diagrams, index.

13-005 _____. "Phototelegraphy," *EXPERIMENTAL WIRELESS*, 4:229–238 (1927). On the work of Arthur Korn and his current system.

13-006 Bidwell, Shelford. "Selenium and Its Applications to the Photophone and Telephotography," *PROCEEDINGS OF THE ROYAL INSTITUTION*, 9:524–535 (1881). On selenium, its characteristics, construction of cells, and demonstration of experimental apparatus. Diagrams.

13-007 Bond, Donald S., and Vernon J. Duke, "Ultrafax," *RCA REVIEW*, 10:99–115 (March 1949); reprinted in *JOURNAL OF THE BRITISH IRE*, 9:146–156 (April 1949). Technical description of the high-speed facsimile system employing electronic scanning and other television techniques. Photos, diagrams, table. References. See also ULTRAFAX, 13-035

13-008 Callahan, John L., J.N. Whitaker, and Henry Shore. "Photoradio Apparatus and Operating Technique Improvements," *PROCEEDINGS OF THE IRE*, 23:1441–1482 (December 1935). Detailed description of the RCA system, equipment, machines, circuits, and operation. Photos, diagrams, graphs, tables, references.

13-009 Cole, A.W., and J.A. Smale. "The Transmission of Pictures by Radio, *PROCEEDINGS OF THE IEE*, 99 (Pt.III):325–335, discussion pp. 359–363 (November 1952). Review since 1842, particularly of radio methods after 1924, with details of British apparatus. Photos, references.

13-010 * Coopersmith, Jonathan. "Facsimile's False Start," *IEEE SPECTRUM*, 30:46–49 (February 1993). Reviews the build-up and decline of facsimile fascination in the 1940s.

Newer Media Technologies

13-011 _____. "Losing the Race: The British Post Office and Picture Telegraphy," *ESSAYS IN ECONOMICS AND BUSINESS HISTORY,* 13:71–82 (1995).

13-012 Costigan, Daniel M. **FAX: THE PRINCIPLES AND PRACTICE OF FACSIMILE COMMUNICATION.** Philadelphia: Chilton, 1971, 270 pp. First chapter provides survey history. Photos, diagrams, notes, bibliography, index.

13-013 *_____. **ELECTRONIC DELIVERY OF DOCUMENTS AND GRAPHICS.** New York: van Nostrand Reinhold, 1978, 344 pp. Updated and expanded version of title above, retaining historical material. Photos, diagrams, notes, index.

13-014 Davis, Watson. "Seeing by Radio," *POPULAR RADIO,* 3:266–275 (April 1923). Description of the Jenkins apparatus employing prismatic rings for transmitting pictures by radio. See also Jenkins, 13-024. Photos.

13-015 "Facsimile Telegraphy and Phototelegraphy," *SCIENTIFIC AMERICAN,* 112:571–573 (June 5, 1915). A summary of progress from 1843 with about 70 names, dates, and brief descriptions.

13-016 * Goldsmith, Alfred N., et al., eds. **RADIO FACSIMILE.** New York: RCA Institutes Technical Press, 1938, 353 pp. Twenty papers by RCA Laboratories engineers. Six historical papers through p. 111, with remainder devoted to status of technology in 1938. Many papers reprinted from *PROCEEDINGS OF THE IRE* or *RCA REVIEW.* Photos, diagrams, references, bibliography of 261 items.

13-017 * Hills, Lee, and Timothy J. Sullivan. **FACSIMILE.** New York: McGraw-Hill, 1949, 319 pp. Two newspaper facsimile editors detail developments to that point and offer many case studies of what then appeared to be a technology on the verge of widespread adoption. Photos, glossary, index.

13-018 Hogan, John V.L. "Facsimile and Its Future Uses," *ANNALS OF THE AMERICAN ACADEMY OF POLITICAL AND SOCIAL SCIENCE,* 213:162–169 (January 1941). A general nontechnical survey of prewar thinking.

13-019 _____. "Facsimile: What It Can Do Now," *FREQUENCY MODULATION,* 1:2:5–7, 30–31 (March 1946). Brief survey, focusing on postwar newspaper applications. Photos.

13-020 Horton, J. W. "The Electrical Transmission of Pictures and Images," *PROCEEDINGS OF THE IRE,* 17:1540–1563 (1929). Summary of period practice. Photos.

13-021 * Hunkin, Tim. "Just Give Me the Fax," *NEW SCIENTIST,* 137:33–37 (February 13, 1993). 150 years are reviewed.

13-022 Isakson, D.W. "Developments in Telephotography," *TRANSACTIONS OF THE AIEE,* 41:794–801 (August 1922). A brief historical survey is followed by description of the Leisman coded facsimile system.

13-023 Ives, Herbert E., et al. "The Transmission of Pictures over Telephone Lines," *BELL SYSTEM TECHNICAL JOURNAL,* 4:187–214 (April 1925). The work of

AT&T researchers, including examples of reproductions (the frontispiece is an example of a three-color reproduction). Photos, diagrams, references.

13-024 * Jenkins, C. Francis. **VISION BY RADIO, RADIO PHOTOGRAPHS.** Washington: Jenkins Laboratories, 1925, 140 pp. Privately printed illustrated survey of the state of the art, including the author's work. Photos (of apparatus and sample facsimile messages), diagrams, list of patents, no index. See also Jenkins, 12-007.

13-025 ** Jones, Charles R. **FACSIMILE.** New York: Murray Hill Books, 1949, 422 pp. A complete review of then-current systems with technical details and descriptions of commercial services. First chapter provides brief history. Photos, diagrams, references, glossary, index.

13-026 Korn, Arthur. "The Transmission of Photographs and Drawings by Wireless Telegraphy," *WIRELESS WORLD,* 1:353–357 (September 1913). Photos, diagram.

13-027 Lister, W.C. "The Development of Photo-Telegraphy," *ELECTRONIC ENGINEERING,* 19:37–43 (February 1947). Brief review of some early historical apparatus and 20th-century systems. Photos, diagrams, no references.

13-028 MacCafferty, Maxine, comp. **FAX AND TELETEXT.** London: Aslib "Aslib Bibliography No. 5," 1977, 77 pp. A bibliography.

13-029 * Martin, Marcus J. **WIRELESS TRANSMISSION OF PHOTOGRAPHS.** London: Wireless Press, 1916, 117 pp.; 1919 [2nd ed.], 145 pp. General discussion with mention of other systems but mainly on the writer's work and machines. Reprinted from articles in *WIRELESS WORLD.* Photos, diagrams, appendices, index.

13-029a McConnell, Kenneth R., et al. **FAX: DIGITAL FACSIMILE TECHNOLOGY AND APPLICATIONS.** Norwood, MA: Artech House "Telecommunications Library," 1989, 222 pp.; 1992 [2nd ed.]; 1999 [3rd ed.], 460 pp. A standard handbook of the service, this includes useful historical material, expanded in the latest edition. Photos, charts, references, index.

13-030 Peterson, M.J. "The Emergence of a Mass Market for Fax Machines," *TECHNOLOGY IN SOCIETY,* 17:469–482 (1995). One of the handful of historical studies concerning the "modern" fax era, this focuses on the various barriers to the introduction of fax services. References.

13-031 Ranger, Richard H. "Transmission and Reception of Photodiagrams," *PROCEEDINGS OF THE IRE,* 14:161–180 (April 1926). Brief history and details of RCA equipment. Photos, diagrams.

13-032 _____. "Photoradio Developments," *PROCEEDINGS OF THE IRE,* 17:966–984 (June 1929). Review of progress during 1926–28 with mention of various systems—AT&T, Barlane, Karolus-Telefunken, Marconi, Jenkins, Dieckmann, Cooley, RCA.

13-033 _____. "Mechanical Developments of Facsimile Equipment," *PROCEEDINGS OF THE IRE,* 17:1564–1575 (1929). Further RCA developments.

13-034 Reynolds, F.W. "A New Telephotograph System," *ELECTRICAL ENGINEERING*, 55: 996–1007 (September 1936). Technical details of the Bell system. Map, photos, diagrams, references.

13-034a ** Schmidt, Susanne K., and Raymund Werle, "Facsimile," Chapter 8 in **COORDINATING TECHNOLOGY: STUDIES IN THE INTERNATIONAL STANDARDIZATION OF TELECOMMUNICATIONS**. Cambridge, MA: MIT Press, 1998, pp. 185–228. Insightful assessment of facsimile standards-setting development from Group 1 through Group 4 technologies. Tables, photos, references.

13-035 ULTRAFAX. New York: RCA, 1948, 32 pp. Company brochure with brief description of the system. Photos, diagrams. See also Bond, 13-007.

13-036 Velie, Lester. "A Million Words a Minute," *COLLIER'S*, 122:18:13–14, 98–101. The RCA "Ultrafax" system. Photos.

13-036a * Wenger, Pierre-Andre. "The Future Also Has a Past: The Telefax, A Young 150-Year-Old Service," **TELECOMMUNICATIONS JOURNAL**, 56:777–782 (December 1989). Useful survey of the entire story of its development. Photos, references.

13-037 Zworykin, Vladimir. "Facsimile Picture Transmission," *PROCEEDINGS OF THE IRE*, 17:536–550 (1929). The television pioneer's early work at Westinghouse is described.

B. General New Media Surveys

(This material dates from the past three decades and virtually none was designed to be historical. Indeed, many were designed to project likely futures. Still, these citations are—or will be—of growing historical value and offer useful insights into contemporary thinking about expanding delivery options.)

13-038 Arlen, Gary H. **TOMORROW'S TVs: A REVIEW OF NEW TV SET TECHNOLOGY, RELATED VIDEO EQUIPMENT AND POTENTIAL MARKET IMPACTS, 1987-1995**. Washington: National Association of Broadcasters "CommTech Series," 1987, 120 pp. Now useful as a snapshot of how television's future was seen in the mid-1980s as digital options were just coming on line. Photos, diagrams, tables, appendices.

13-039 * Bagdikian, Ben H. **THE INFORMATION MACHINES: THEIR IMPACT ON MEN AND THE MEDIA**. New York: Harper & Row, 1971, 359 pp. One of the more thoughtful journalists of his time assesses the projected impact of changing technology on traditional print and broadcast news media. While some of the analysis is economic and organizational, the root of this volume concerns expanding technological options including an appendix projecting what would be available. Tables, charts, index.

13-040 * Bretz, Rudy. **COMMUNICATIONS MEDIA: PROPERTIES AND USES**. Santa Monica, CA: RAND Corp. Memorandum RM-6070-NLM/PR, 1969, 114 pp. Later reprinted as A TAXONOMY OF COMMUNICATION MEDIA

(Englewood Cliffs, NJ: Educational Technology Publications, 1970). Attempt to develop seven classes of telecommunications and media services into a taxonomy to allow easier comparisons. Chart, references.

13-041 *CHANNELS* **FIELD GUIDE.** New York: Media Commentary Council, later Act III Publishing, 1983–91, annual. Nine issues (published by the monthly *CHANNELS* magazine) are very useful today to trace a period rife with changing industries, technological innovations, and major corporate players in consumer electronics and services. Photos, charts, tables.

13-042 Dholakia, Ruby Roy, et al., eds. **NEW INFOTAINMENT TECHNOLOGIES IN THE HOME: DEMAND-SIDE PERSPECTIVES.** Mahwah, NJ: Lawrence Erlbaum "Communications," 1996, 291 pp. Fourteen scholarly papers discuss forecasting demand for newer services, developing strategies for innovation and marketing, entertainment as the driver of most new consumer services, changing definitions of literacy, case studies in introducing new electronic information services. Diagrams, tables, references, index.

13-043 * Gerbner, George, et al., eds. **COMMUNICATION TECHNOLOGY AND SOCIAL POLICY: UNDERSTANDING THE NEW "CULTURAL REVOLUTION."** New York: John Wiley, 1973, 573 pp. Edited volume of conference proceedings, this includes considerable discussion of "new" technologies including trends in switched services, broadcasting, mobile systems, communication satellites, interconnection, and computer-telecommunications. Diagrams, tables, references (but sadly, no index).

13-044 * Grant, August, et al., eds. **COMMUNICATION TECHNOLOGY UPDATE.** Newton, MA: Focal Press, 1992–96, annual. Useful and concise survey that was regularly revised by a host of authorities, with more than 30 chapters including a dozen on various electronic media, and as many on telecommunication services. These annual "snapshots" will be of increasing historical interest. Tables, charts, bibliography, glossary.

13-045 THE HOME VIDEO & CABLE YEARBOOK. White Plains, NY: Knowledge Industry Publications, 1981, 248 pp.; 1982, 262 pp. Only two volumes were issued and today they provide a valuable record of what was then breaking technology. Among other topics, includes cable, satellites, over-the-air subscription services, MDS, LPTV, VCRs and other recording modes, and videotex. Focuses on U.S. but surveys developments in Europe and Japan. Tables, index.

13-046 Kamen, Ira. **QUESTIONS AND ANSWERS ABOUT PAY TV.** Indianapolis: Howard W. Sams, 1973, 158 pp. An excellent survey of the different technologies proposed for over-the-air subscription television (STV). Maps, photos, diagrams.

13-047 Mahoney, Sheila, et al. **KEEPING PACE WITH THE NEW TELEVISION.** New York: VNU Books, 1980, 281 pp. Includes useful surveys on the status of domestic satellites, cable, over-the-air pay TV, home video, and videotex when network television still ruled the roost. Glossary, index.

13-048 McGee, William L., et al. **CHANGES, CHALLENGES AND OPPORTUNITIES IN THE NEW ELECTRONIC MEDIA.** San Francisco: BMC Publications,

1982, ca 600 pp. A large red binder seemingly intended for regular updating, this includes considerable detail on the development, then-current status, and prospects of competing means of transmitting pay television, low-power television, direct broadcast satellite, videotex and teletext, and home video systems. Diagrams, charts, tables.

13-048a NEW TECHNOLOGIES AFFECTING RADIO AND TELEVISION BROADCASTING. Washington: National Association of Broadcasters, 1981, 75 pp. Interesting for its "protectionist" view of burgeoning technological options often seen as threats to existing broadcast service. Tables, charts, bibliography.

13-049 Newman, Joseph, ed. **WIRING THE WORLD: THE EXPLOSION IN COMMUNICATIONS.** Washington: U.S. News & World Report, 1971, 207 pp. Popular treatment of cable television, videocassettes (long before they were commonly available), picture telephones, satellites, and pay TV systems. Photos, tables, charts, appendix on history of electrical communication, glossary, index.

13-050 Rice, Ronald E., ed. **THE NEW MEDIA: COMMUNICATION, RESEARCH, AND TECHNOLOGY.** Beverly Hills, CA: Sage, 1984, 352 pp. Twelve papers focus largely on new methods of research into diversifying technologies. Tables, charts, notes, references, index.

13-051 Robertson, Angus, ed. **FROM TELEVISION TO HOME COMPUTER: THE FUTURE OF CONSUMER ELECTRONICS.** Poole, England: Blandford Press, 1979, 323 pp. Seventeen chapters assess the then-likely future of the field as seen from a British perspective on the eve of introduction of the personal computer. Useful details on what was then available and presumed to be coming in the 1980s. Photos, diagrams, glossary, index.

13-052 * Rogers, Everett M. **COMMUNICATION TECHNOLOGY: THE NEW MEDIA IN SOCIETY.** New York: Free Press, 1986, 273 pp. Theoretically-based analysis of changing (and increasing) technological choices in communication, their adoption and implementation, social impacts, and new means of researching them. References, index.

13-053 * Salvaggio, Jerry L., and Jennings Bryant. **MEDIA USE IN THE INFORMATION AGE: EMERGING PATTERNS OF ADOPTION AND CONSUMER USE.** Hillsdale, NJ: Lawrence Erlbaum, 1989, 315 pp. Need for new means of researching changing technologies, adopting and using newer communication media, and emerging models of media use in the information age. Includes specific media elsewhere: teletext in Britain, home computers in Canada, and videocassettes in the Third World. Tables, charts, references, index.

13-054 Silverstone, Roger, and Eric Hirsch, eds. **CONSUMING TECHNOLOGIES: MEDIA AND INFORMATION IN DOMESTIC SPACES.** London: Routledge, 1992, 241 pp. Thirteen papers explore newer home services and patterns of and factors in their innovation. References, index.

13-055 THE VIDEO AGE: TELEVISION TECHNOLOGY AND APPLICATIONS IN THE 1980s. White Plains, NY: Knowledge Industry Publications, 1982, 264 pp. Includes chapters on the home video industry, cable, video discs, videotex, and various nonbroadcast applications, plus a final paper on the future of television technology. Photos, tables (no index).

13-056 Williams, Frederick. **THE COMMUNICATIONS REVOLUTION.** Beverly Hills, CA: Sage, 1982, 289 pp. Broad survey of "what was coming," the impact of newer technologies on older ones, and likely applications of newer systems. References, index.

C. Cable Television

(There is still no good overall history of the medium's half century of operation, let alone a broad assessment of its technical development. The following entries shed some light on that development.)

13-057 Baldwin, Thomas E., and D. Stevens McVoy. **CABLE COMMUNICATION.** Englewood Cliffs, NJ: Prentice-Hall, 1983, 416 pp.; 1987 [2nd ed.]. This survey text ranges over all aspects of cable, focusing on cable technology in the first several chapters. Photos, diagrams, references, appendices, index.

13-058 * Bartlett, Eugene R. **CABLE TELEVISION TECHNOLOGY AND OPERATIONS: HDTV AND NTSC SYSTEMS.** New York: McGraw-Hill, 1990, 421 pp. Describes the building and maintenance of cable television systems in the context of analog NTSC and what was then projected as analog HDTV (see 13-G). Chapters were drawn from a series of training lectures. Photos, diagrams, index.

13-059 Black, Sharon K. **CABLE TELEVISION FOR EUROPE.** Washington: Government Printing Office, Department of Commerce, Office of Telecommunications OT Report 74-28, 1974, 101 pp. Very useful early view of how cable might develop in Western Europe including evolving technology and equipment.

13-060 CABLE OPERATOR'S HANDBOOK. Blue Ridge Summit, PA: TAB Books, 1967; 1973 [2nd ed.], 352 pp. Designed for system engineers and system operators, this broadly surveys all the technical elements of pre-satellite cable systems. Photos, diagrams.

13-061 "Cable: The First Forty Years," *BROADCASTING* (November 21, 1988), pp. 35–49. Journalistic survey with a useful chronology. Photos.

13-061a Chin, Felix. **CABLE TELEVISION: A COMPREHENSIVE BIBLIOGRAPHY.** New York: IFI/Plenum, 1978, 286 pp. Annotated selection, arranged topically, from the estimated 7,000 cable articles issued between 1950 and 1977. Appendices, index.

13-062 Coll, D.C., and K.E. Hancock. ""A Review of Cable Television: The Urban Distribution of Broad-Band Visual Signals," *PROCEEDINGS OF THE IEEE*, 73:773–788 (April 1985). Using examples from the U.S., Britain, and especially Canada, this paper explores the technical structure of cable systems, including early use of fiber optic cables.

13-063 Easton, K[enneth] J. **THIRTY YEARS IN CABLE TV: REMINISCENCES OF A PIONEER.** Mississauga, ON: Pioneer Publications, 1980, 207 pp. Chapters discuss the early development of the cable industry and its technology in Britain, the U.S., and Canada, long-distance reception, pay television, and related topics.

13-064 "A History of Community Antenna Television," *COMMUNITY ANTENNA TELEVISION JOURNAL* (March 1975) pp.10–56; (April 1975)

Newer Media Technologies

pp.11–59. A cable industry assessment from the 1940s on, with some discussion of the technological basics. Maps, tables.

13-065 Hollins, Timothy. **BEYOND BROADCASTING: INTO THE CABLE AGE**. London: British Film Institute, 1984, 385 pp. Compares American and British progress and policies on expanding cable television availability including discussion of technology basics and impact on other industries. Tables, appendices, index.

13-066 * Knox, William T. "Cable Television," *SCIENTIFIC AMERICAN*, 235:4:22–29 (October 1971). Useful early survey prior to the rise of satellites and fiber optic applications. Diagrams.

13-067 Herbert W. Land Assoc. **TELEVISION AND THE WIRED CITY: A STUDY OF THE IMPLICATIONS OF A CHANGE IN THE MODE OF TRANSMISSION**. Washington: National Association of Broadcasters, 1968, 256 pp. Broadcast industry's jaundiced view of cable expansion. Includes some discussion of technological capabilities of early cable systems. Tables, chapter bibliographies.

13-068 Phillips, Mary Alice Mayer. **CATV: A HISTORY OF COMMUNITY ANTENNA TELEVISION**. Evanston, IL: Northwestern University Press, 1972, 209 pp. The only formal history of the industry, this focuses primarily on policy and regulation. Notes, bibliography, index.

13-069 Pilnick, Carl, and Walter S. Baer. **CABLE TELEVISION: A GUIDE TO THE TECHNOLOGY**. Santa Monica, CA: Rand Corp. Report R-1141-NSF, 1973, 65 pp. Useful handbook of pre-satellite cable system structures and enabling technologies with sections on headends, coaxial cable, microwave signal distribution, etc. Diagrams.

13-070 ** Smith, E. Stratford. "The Emergence of CATV: A Look at the Evolution of a Revolution," *PROCEEDINGS OF THE IEEE*, 58:967–982 (July 1970). Covers both the technology and the policy elements of cable expansion. Tables, notes.

13-071 Smith, Ralph Lee. "The Wired Nation: A Special Issue," *THE NATION* (May 18, 1970), 210:582–606. A classic early analysis of cable's expansion, prospects, and problems.

13-072 _____. **THE WIRED NATION: CABLE TV, THE ELECTRONIC COMMUNICATIONS HIGHWAY**. New York: Harper & Row, 1972, 128 pp. Expansion of work above. Bibliography, index.

13-073 Southwick, Thomas P. **DISTANT SIGNALS: HOW CABLE TV CHANGED THE WORLD OF TELECOMMUNICATIONS**. Overland Park, KS: PRIMEDIA Intertec, 1999, 358 pp.

13-074 ** Ward, John E. **PRESENT AND PROBABLE CATV/BROADBAND-COMMUNICATIONS TECHNOLOGY**. New York: Alfred P. Sloan Foundation, 1971, ca 70 pp. Useful survey of pre-satellite cable technology. Diagrams, tables. A shorter version appears as Appendix A in ON THE CABLE: THE TELEVISION OF ABUNDANCE—REPORT OF THE SLOAN COMMISSION ON CABLE COMMUNICATIONS. New York: McGraw-Hill, 1971, pp. 179–212.

D. Teletext and Videotex

(The use of television receivers for analog transmission of text and graphic information had a brief period in the sun in the late 1970s and early 1980s. Although there is a large periodical literature, it is not covered here, given the rapid demise of the process in the face of the digital computer-based Internet—for which see 14-E.)

13-075 * Bloom, L.R. **VIDEOTEX SYSTEMS AND SERVICES.** Washington: National Telecommunications and Information Administration Report 80-50, 1980, 159 pp. Comparison of the basic technologies and standards of different national systems. Diagrams, references, bibliography.

13-076 Bouwman, H., and M. Christofferson. "The Internet and the Meltdown of Audiotext and Videotext in Europe," *RÉSEAUX: THE FRENCH JOURNAL OF COMMUNICATION,* 3:63–74 (1995). How "new" media technologies are themselves soon replaced.

13-077 Fedida, Sam, and Rex Malik. **VIEWDATA REVOLUTION.** New York: John Wiley/Halstead Press, 1979, 186 pp. Fedida was a key figure in the British Post Office "Prestel" system. Primarily concerns applications, present and projected. Index.

13-078 * Martin, James. **VIEWDATA AND THE INFORMATION SOCIETY.** Englewood Cliffs, NJ: Prentice-Hall, 1982, 293 pp. Perhaps the best study of the system's technology. Photos, charts, tables, glossary, index.

13-079 Mosco, Vincent. **PUSHBUTTON FANTASIES: CRITICAL PERSPECTIVES ON VIDEOTEX AND INFORMATION TECHNOLOGY.** Norwood, MA: Ablex, 1982, 195 pp. Interesting political economy approach to videotex that seeks to place the technology within the context of older electronic services. Tables, diagrams, notes, bibliography, index.

13-080 PRESTEL 1980. London: British Post Office, 1980, 112 pp. Album based on exhibition that year on all aspects of the British videotex system. Photos, tables.

13-080a * Schmidt, Susanne, K., and Raymund Werle. "Interactive Videotex," Chapter 7 in **COORDINATING TECHNOLOLOGY: STUDIES IN THE INTERNATIONAL STANDARDIZATION OF TELECOMMUNICATIONS.** Cambridge, MA: MIT Press, 1998, pp. 146–184. Interesting assessment of the attempt to set international standards and promote the technology swept away by the Internet in the late 1990s. Photos diagrams, references.

13-081 *** Tydeman, John, et al. **TELETEXT AND VIDEOTEX IN THE UNITED STATES: MARKET POTENTIAL, TECHNOLOGY, PUBLIC POLICY ISSUES.** New York: McGraw-Hill, 1982, 314 pp. Best single survey of the technology when its outlook appeared bright. Includes several chapters (4 of 15) specifically devoted to technology. Photos, diagrams, bibliography, index.

E. Remote Control and Interactive Media

(Excluded here is the extensive literature on teleconferencing, virtually all of which focuses on specific applications rather than technology development.)

13-082 Bellamy, Robert V. Jr., and James R. Walker, eds. **THE REMOTE CONTROL IN THE NEW AGE OF TELEVISION**. Westport, CT: Praeger, 1993, 271 pp. Sixteen research papers discuss the development of the device, individual use, group viewing, and impact of the device on the industry. Charts, tables, notes, references, index.

13-083 _____. **TELEVISION AND THE REMOTE CONTROL: GRAZING ON A VAST WASTELAND**. New York: Guilford, 1996, 192 pp. Unlike the previous entry, a more integrated assessment of the development of the remote control and its many implications. Tables, chapter notes, references, index.

13-084 * Bretz, Rudy, with Michael Schmidbauer. **MEDIA FOR INTERACTIVE COMMUNICATION**. Beverly Hills, CA: Sage, 1983, 264 pp. Includes applications of audio and video systems, alphameric systems, data response, teletext, videotex, and developments to come. Photos, diagrams, notes, glossary, references, index.

13-085 Connell, Stephen, and Ian A. Galbraith. **ELECTRONIC MAIL: A REVOLUTION IN BUSINESS COMMUNICATIONS**. White Plains, NY: Knowledge Industry Publications "Office Productivity Series,", 1980, 139 pp. Refers to *pre*-Internet-based systems, including combinations of telecommunication and mail service. Diagrams, appendices, index.

13-086 "Interactive Media—Special Report," *IEEE SPECTRUM* (April 1996), 23:4:20–33. Several articles assess developments thus far, trials, the likely future, and the role of the Internet.

13-087 Wilson, Kevin G. **TECHNOLOGIES OF CONTROL: THE NEW INTERACTIVE MEDIA FOR THE HOME**. Madison: University of Wisconsin Press "Studies in Communication and Society," 1988, 180 pp. Early analysis of Internet networking (termed videotex here) and its likely home impact. References, index.

F. Digital Audio/Radio

(While service began in Europe around 1995, U.S. technical standards were still under consideration as this book went to press and regular digital broadcasting had yet to begin.)

13-088 Anglin, R.J. Jr. "Digital Audio Broadcasting: United States Technologies and Systems, Terrestrial and Satellite," *ELECTRONICS INFORMATION AND PLANNING*, 23:3:129–140 (December 1995). Useful survey of U.S, in-band, on-channel (IBOC) developments in this Indian technical journal. Diagrams.

13-089 DeSonne, Marcia. **DIGITAL AUDIO BROADCASTING: STATUS REPORT AND OUTLOOK**. Washington: National Association of Broadcasters, 1990, 69 pp. Compares and contrasts the competing systems and reviews their status. Diagrams, notes.

13-090 DIGITAL AUDIO BROADCASTING. London: IBC Technical Services, 1995, 376 pp. Conference proceedings volume includes 16 papers assessing developments in Europe, North America, and elsewhere. Diagrams, tables.

13-091 * Dolby, Raymond. "Sound Recording—Will There Be Progress Forever?" *PROCEEDINGS OF THE IEEE*, 86:2469–2472 (December 1998). Argues future improvements can only be at the margins given that it is hard to discern output from input signals. "We are essentially at the end of the line in recording and reproduction perfection," and at the beginning of a new multichannel format era.

13-092 Gleave, Michael. "Digital Radio Takes Off," *IEE REVIEW*, 43:239–242 (November 20, 1997). Reviews progress of initial DAB service in Britain and elsewhere, and the role of the Eureka 147 technical standard as a growing world standard.

13-093 Henderson, Ken. "A Rebirth in Digital Recording," *IEE REVIEW*, 45:13–16 (January 21, 1999). Suggests the Philips MiniDisc format is reviving in Britain after its early 1990s' failure.

13-094 Jurgen, Ronald K. "Broadcasting with Digital Audio," *IEEE SPECTRUM*, 33:52–59 (March 1996). Worldwide system testing, primarily with Eureka 147 standard.

13-095 Kumin, Daniel. "Home Recording for the Digital Millennium," *AUDIO* (February 1999), pp. 32–39. Evaluates the various competing technical standards.

13-096 Lavin, P.A. "DAB–Is It Already Out of Date?" *EBU TECHNICAL REVIEW*, 278:10–14 (Winter 1998). Argues the European Eureka 147 system may be behind the times.

13-097 Not used.

13-098 Monforte, John "The Digital Reproduction of Sound," *SCIENTIFIC AMERICAN*, 251:6:78–84 (December 1984). Careful comparison of analog and digital recording methods in text, photos, and charts.

13-099 Muller-Romer, F. "DAB Progress Report—1997," *EBU TECHNICAL REVIEW*, 274:12–22 (Winter 1997). Assessment of European developments with Eureka 147 standard.

13-100 O'Leary, T. "Terrestrial Digital Audio Broadcasting in Europe," *EBU TECHNICAL REVIEW*, 255:19–26 (Spring 1993). Summarizes launch plans then envisioned by several different countries.

13-101 * Springer, Kenneth. **UNDERSTANDING DAB: A GUIDE FOR BROADCAST MANAGERS AND ENGINEERS**. Washington: National Association of Broadcasters, 1992, 120 pp.; 1994 [2nd ed.], 167 pp. Basics of proposed systems. Tables, charts, diagrams, references, index.

13-102 Tuttlebee, W.H.W., and D.A. Hawkins. "Consumer Digital Radio: From Concept to Reality," *ELECTRONICS & COMMUNICATIONS ENGINEERING JOURNAL*, 10:6:263–276 (December 1998). Discusses the BBC experience with the Eureka 147 system.

13-103 Viterbi, Andrew J. "The Evolution of Digital Wireless Technology from Space Exploration to Personal Communication Services," *IEEE TRANSACTIONS*

ON VEHICULAR TECHNOLOGY, 43:638–643 (August 1994). Includes DAB services.

G. High-Definition/Digital Television

(HDTV history breaks into analog and digital periods around 1990. Service began in late 1998 in the U.S.—and an analog system was in service in Japan in the 1980s.)

13-104 ** Advisory Committee on Advanced Television Service. **FINAL REPORT AND RECOMMENDATIONS**. Washington: Federal Communications Commission, November 28, 1995, ca 250 pp. Conclusion of an eight-year process, this is the final digital technical standard recommendation of the industry committee to the FCC. Among the several appendices is the standard itself. Charts, tables.

13-105 Beltz, Cynthia A. **HIGH-TECH MANEUVERS: INDUSTRIAL POLICY LESSONS OF HDTV**. Washington: AEI Press, 1991, 142 pp. Survey from a conservative think tank of how government actions or inactions affected the development of HDTV and what government's role might be in getting such technology introduced. Useful discussion of technical standards. Charts, tables, notes, no index.

13-106 * Benson, K. Blair, and Donald G. Fink. **HDTV: ADVANCED TELEVISION FOR THE 1990s**. New York: McGraw-Hill, 1991, ca 300 pp. Perhaps the best technical survey just as digital thinking was taking over. Diagrams, index.

13-107 ** **THE BIG PICTURE: HDTV AND HIGH-RESOLUTION SYSTEMS: BACKGROUND PAPER**. Washington: Government Printing Office, Office of Technology Assessment, 1990, 108 pp. Policy issues, standard-setting process, historical development of HDTV, high-definition and communications distribution technologies, and linkages between HDTV and other industries. Valuable pre-digital survey. Charts, tables, glossary, appendices.

13-108 ** Brinkley, Joel. **DEFINING VISION: THE BATTLE FOR THE FUTURE OF TELEVISION**. New York: Harcourt Brace, 1997, 402 pp. A *New York Times* reporter reviews the on-again, off-again, history of the eventual American digital standard. Very useful record of people and organizations which shows considerable insight on changing strategies of the key players during the shift from analog to digital systems. Photos, index.

13-109 * CasaBianca, Lou, ed. **THE NEW TV: A COMPREHENSIVE SURVEY OF HIGH DEFINITION TELEVISION**. Westport, CT: Meckler, 1992, 166 pp. Reprints of articles from *HDTV World Review*, including several taking an historical perspective. Bibliography, index.

13-110 Cohen, Robert B., and Kenneth Donow. **TELECOMMUNICATIONS POLICY, HIGH DEFINITION TELEVISION, AND U.S. COMPETITIVENESS**. Washington: Economic Policy Institute, 1989, 50 pp. Focuses on comparing a lack of concerted U.S. policy over adopting an HDTV standard with the industrial policies of Japan and Europe. Bibliography.

13-111 * de Bruin, Ronald, and Jan Smits. **DIGITAL VIDEO BROADCASTING: TECHNOLOGY, STANDARDS, AND REGULATIONS**. Norwood, MA: Artech

House "Digital Audio and Video Series," 1999, 315 pp. With a strong European orientation, this provides an interesting combination of history, comparative policy, and comparative technology approaches to HDTV. The technical chapters deal largely with European practice, with some comparative material concerning U.S. practices. Photos, diagrams, references, glossary, index. See also Marsdan and Verhulst, 13-120.

13-112 *** Dupagne, Michel, and Peter B. Seel. **HIGH-DEFINITION TELEVISION: A GLOBAL PERSPECTIVE**. Ames: Iowa State University Press, 1998, 392 pp. Perhaps the first extensive scholarly assessment of HDTV development and prospects, this includes both analog and digital developments. Chapters review the technical and then policy developments in the U.S., Japan, and Europe. Tables, diagrams, chronology, notes, references, index.

13-113 * Freeman, John P. "The Evolution of High-Definition Television," *SMPTE JOURNAL,* 93:492–501 (May 1984). Assessment of some 50 competing analog approaches. Tables, references.

13-114 Habermann, W., and D. Wood. "Images of the Future: The EBU's Part to Date in HDTV System Standardization," *EBU TECHNICAL REVIEW,* 219:267–280 (1986). Useful for tracing the early analog HDTV standards process in Europe.

13-115 HDTV PROCEEDINGS. Washington: National Association of Broadcasters, 1991–1993, annual, 3 vols. published, about 250 pp. each. Proceedings of the HDTV sessions at the annual NAB convention, including discussions of both technological development and potential applications. Photos, tables, charts, diagrams, references.

13-116 Jones, Graham. "Design and Implementation of the ATSC Demonstration of HDTV at NAB '97," *SMPTE JOURNAL* 108:202–208 (April 1999). The first industry demonstration of the ATSC standard. Diagrams, references.

13-117 Krivocheev, Mark I. "Current CCIR Activities in HDTV," *TELECOMMUNICATION JOURNAL* 58:699–709 (December 1991). The growing role of the ITU body in seeking a standard in the analog-to-digital period.

13-118 * _____, edited by S.N. Baron. "The First Twenty Years of HDTV: 1972–1992," *SMPTE JOURNAL,* 102:913–930 (October 1993). Focuses on the work of the CCIR study group the author chaired dealing with, among other projects, the work of NHK and others in Japan. Diagrams, tables, notes, bibliography of CCIR documents.

13-119 Lim, Jae S. "Digital Television Here at Last," *SCIENTIFIC AMERICAN,* 278:5:78–83 (May 1998). Good survey of the impending startup of American HDTV that fall—and decisions that led to that innovation.

13-120 Marsden, Christopher T., and Stefaan G. Verhulst. **CONVERGENCE IN EUROPEAN DIGITAL TV REGULATION**. London: Blackstone Press "Law in its Social Setting," 1999, 247 pp. While primarily a policy analysis, this sheds considerable light on how Europe has approached the setting of technical standards for HDTV systems. Notes, references, index. See also de Bruin and Smits, 13-111.

Newer Media Technologies

13-121 * NAB GUIDE TO ADVANCED TELEVISION SYSTEMS. Washington: National Association of Broadcasters, 1989, 150 pp.; 1991 [2nd ed.], 160 pp. Very useful and detailed directories of then-current thinking. The first edition describes the pre-digital systems, and the second reviews the scene after the conversion to digital systems. Covers individual systems, industry testing, chronology, FCC advisory groups, glossary, etc. Photos, tables, diagrams, charts.

13-122 NHK Science and Technical Research Laboratories, trans. by Jams G. Parker. **HIGH DEFINITION TELEVISION: HI-VISION TECHNOLOGY.** New York: van Nostrand Reinhold, 1993, 288 pp. The best English-language technical description of the Japanese MUSE analog standard. Diagrams, bibliography, index.

13-123 Niblock, Michael. **THE FUTURE FOR HDTV IN EUROPE: THE ROLE OF BROADCASTERS IN THE COMMERCIAL DEVELOPMENT OF A EUROPEAN STANDARD FOR HIGH-DEFINITION TELEVISION.** Manchester, England: University of Manchester, European Institute for the Media, 1991, 157 pp. One of the few in-depth studies of the European search for their "own" system standard to protect their consumer electronics industry from predation by Japan or the U.S. Discusses varied systems under development and their likely costs. Charts, tables, appendices.

13-124 Nickelson, R.L. "HDTV Standards—Understanding the Issues," *TELECOMMUNICATION JOURNAL* 57:301–312 (June 1990). As seen when HDTV consideration was still focused on analog means. Diagrams, references.

13-125 Prentiss, Stan. **HDTV: HIGH-DEFINITION TELEVISION.** Blue Ridge Summit, PA: TAB Books 1990, 232 pp.; 1993 [2nd ed.], 322 pp. Sums up the scramble to develop technical standards on the eve of digital means overtaking analog approaches. Useful review of initial system proposals in the United States. Second edition focuses on digital systems. Diagrams, tables, glossary, index.

13-126 Rice, John F., ed. **HDTV: THE POLITICS, POLICIES, AND ECONOMICS OF TOMORROW'S TELEVISION.** New York: Union Square Press, 1990, 332 pp. Despite the title, there is considerable discussion of the viability of technologies as then understood in the pre-digital era. Useful for the large number of contributors who were playing active roles in HDTV decision-making. Charts, no index.

13-127 Saffady, William. **HIGH DEFINITION TELEVISION: A BIBLIOGRAPHY.** Westport, CT: Meckler, 1990, 121 pp. Unannotated guide divided into sections on books/reports/journals; papers and conference proceedings; and newspaper and news magazine articles. See also Sudalnik and Kuhl, 13-129.

13-128 Sobel, Alan. "Television's Bright New Technology," *SCIENTIFIC AMERICAN*, 278:5:70–77 (May 1998). Reviews flat panel display technologies for digital television. Photos, diagrams.

13-129 ** Sudalnik, James E., and Victoria A. Kuhl. **HIGH-DEFINITION TELEVISION: AN ANNOTATED MULTIDISCIPLINARY BIBLIOGRAPHY, 1981–1992.** Westport, CT: Greenwood "Bibliographies and Indexes in Science and Technology No. 8," 1994, 347 pp. Extensive record of the early literature arranged by subject and then chronologically—the analog period and the early digital research era. Index.

13-130 Tirman, W. Robert. **THE ELEPHANT AND THE BLIND MEN: THE PHENOMENON OF HDTV AND ITS WOULD-BE STAKEHOLDERS.** Cambridge, MA: Harvard University Program on Information Resources Policy, 1991, 150 pp. Focuses on the equipment providers, the users, and the broader applications of an eventual HDTV system. Useful snapshot of thinking about a year after the digital watershed. Charts, diagrams, tables, glossary, notes.

13-131 * U.S. Congress, Office of Technology Assessment. **THE BIG PICTURE: HDTV AND HIGH-RESOLUTION SYSTEMS—BACKGROUND PAPER.** Washington: Government Printing Office, 1990, 108 pp. Useful survey of analog HDTV developments and potential and impact on related fields. Charts, tables.

13-132 Wassiczek, N., et al. "European Perspectives in the Development of HDTV Standards," *TELECOMMUNICATION JOURNAL*, 57:313–321 (June 1990). The interface of technical, political, and economic issues in the analog HDTV era.

13-133 Weiss, S. Merrill, and Rupert L. Stow. **NAB GUIDE TO HDTV IMPLEMENTATION COSTS.** Washington: National Association of Broadcasters, 1993, ca 140 pp. The economic side of a major shift in technology is assessed outlining broadcast station choices in services to be provided, simulcasting, equipment required, etc. Charts, tables.

13-134 Wilmotte, Raymond M. **TECHNOLOGICAL BOUNDARIES OF TELEVISION.** Washington: Federal Communications Commission, Office of the Chief Engineer, 1974, 3 vols. Interesting assessment undertaken just as the Japanese were beginning research into HDTV.

> **VOL. 1: FINDINGS AND RECOMMENDATIONS.** (70 pp.) Overall survey of what might be done to improve NTSC television systems.
>
> **VOL. 2: SIMULATED TV DISPLAYS OF DIFFERENT SIZE AND SHARPNESS.** (25 pp.) Of special interest for its photographs demonstrating what would later become known as "high-definition" television.
>
> **VOL. 3: APPENDICES TO VOLUME 1.** (114 pp.) Includes technical appendices and references.

Chapter 14

Transmission: Mobile, Satellite, Fiber Optic, and the Internet

While the transmission modes reviewed in this chapter are recent innovations—having appeared in the marketplace since the first edition of this book—they have begun to attract studies of their development and background. As with the "contemporary" book sections of chapters 6, 7, 9, and 12, most of the books cited here were published as current surveys and were not *intended* as histories. Still, they provide material of growing historical interest. (See the introduction to chapter 13 for a comment on the application of the term "history.")

Section A includes surveys covering more than one type of transmission. This is followed by specific modes: (B) mobile systems including analog and more recent digital personal communication service operations, (C) communication satellites, including both domsats and direct broadcast satellite (DBS) studies, (D) fiber-optic modes, and (E) the Internet, which includes a few company-specific histories.

A. General Surveys

14-001 Belitsos, Byron, and Jay Misra. **BUSINESS TELEMATICS: CORPORATE NETWORKS FOR THE INFORMATION AGE**. Homewood, IL: Dow Jones Irwin, 1986, 460 pp. Broad-ranging survey is especially interesting for its several case studies of application of different modes of telecommunication. Photos, charts, notes, index.

14-001a Brown, Ronald. **TELECOMMUNICATIONS: THE BOOMING TECHNOLOGY**. Garden City, NY: Doubleday Science Series, 1970, 191 pp. Useful snapshot of the status of the field and many of its specific delivery modes as of the time of publication. Photos, diagrams, index.

14-002 * Crombie, Douglass D., ed. **LOWERING BARRIERS TO TELECOMMUNICATIONS GROWTH.** Washington: Government Printing Office, "Office of Telecommunications Special Publication 76-9," 1976, 260 pp. Reviews direct satellite communications, land mobile radio, broad-band networks, and fiber optic communications—all then in their initial stages of implementation. Numerous appendices. Diagrams, maps.

14-003 Dewalt, Bryan. **BUILDING A DIGITAL NETWORK: DATA COMMUNICATIONS AND DIGITAL TELEPHONY, 1950–1990.** Ottawa: National Museum of Science and Technology, 1992, 70 pp. Based on a museum exhibition. Photos.

14-004 FUTURE OF COMMUNICATIONS TECHNOLOGY. Ottawa: Information Canada "Telecommission Study 4(a)," 1971, 233 pp. One of a series of studies done for Canadian policymakers, this one assesses then-current technology (including common carrier, broadcast, and mobile) and its existing penetration and implications. Tables, charts, notes.

14-005 Huurdeman, Anton A. **GUIDE TO TELECOMMUNICATIONS TRANSMISSION SYSTEMS.** Norwood, MA: Artech "Telecommunications Library," 1997, 395 pp. While largely devoted to current applications, this includes historical material at the beginning of each chapter. Photos, diagrams, tables, index.

14-006 Maddox, Brenda. **BEYOND BABEL: NEW DIRECTIONS IN COMMUNICATIONS.** New York: Simon & Schuster, 1972, 288 pp. Surveys in a nontechnical fashion the rise and application of satellites, cable television, and telephones. Bibliography, index.

14-007 ** Martin, James. **TELECOMMUNICATIONS AND THE COMPUTER.** Englewood Cliffs, NJ: Prentice-Hall "Series in Automatic Computation," 1969, 470 pp.; 1976 [2nd ed.], 670 pp.; 1990 [3rd ed.], 720 pp. Multiple chapters on transmission, switching, and "imperfections." One of the standard descriptive surveys for two decades. Diagrams, tables, photos, references, index.

14-008 *** _____. **FUTURE DEVELOPMENTS IN TELECOMMUNICATIONS.** Englewood Cliffs, NJ: Prentice-Hall "A James Martin Book," 1971, 413 pp.; 1977 [2nd ed.], 668 pp. Influential work which surveyed what existed and what would likely develop, and why. Its 33 chapters appear in sections on major means of transmission, synthesis issues, and basic technologies utilized. Diagrams, tables, photos, references, index.

14-009 * _____. **THE WIRED SOCIETY: A CHALLENGE FOR TOMORROW.** Englewood Cliffs, NJ: Prentice-Hall "A James Martin Book," 1978, 300 pp.; revised as **TELEMATIC SOCIETY: A CHALLENGE FOR TOMORROW,** 1981, 244 pp. A popularization of his many other more technical works, this focused on likely developments. References, index.

14-010 Pelton, Joseph N. **WIRELESS AND SATELLITE TELECOMMUNICATIONS: THE TECHNOLOGY, THE MARKET & THE REGULA-**

Transmission: Mobile, Satellite, Fiber Optic, and the Internet 269

TIONS. Upper Saddle River, NJ: Prentice Hall PTR "Digital and Wireless Communications," 1995, 277 pp. Includes useful discussion of the development and expansion of such services and related technologies. Tables, charts, bibliography, glossary, index.

B. Mobile Communications
(Includes both analog cellular and digital personal communication services.)

14-011 ** Bowers, Raymond, et al., eds. **COMMUNICATIONS FOR A MOBILE SOCIETY: AN ASSESSMENT OF NEW TECHNOLOGY.** Beverly Hills, CA: Sage, 1978, 432 pp. One historical paper; the remaining 15 papers now form a useful historical record of early mobile communication development around the world and in many different applications. Diagrams, tables, notes, index.

14-012 * **BRINGING INFORMATION TO PEOPLE: CELEBRATING THE WIRELESS DECADE.** Washington: Cellular Telecommunications Industry Association, 1993, 72 pp. Useful historical brochure issued by a trade association to review the first decade of U.S. cellular service. Photos, chronology.

14-013 Brodsky, Ira. **WIRELESS: THE REVOLUTION IN PERSONAL TELE-COMMUNICATIONS.** Norwood, MA: Artech "Mobile Communication Series," 1995, 276 pp. Offers historical material throughout including discussion of mobile data, PCS, spectrum concerns, and satellite transmission. Photos, charts, glossary, index.

14-014 * Calhoun, George. **DIGITAL CELLULAR RADIO.** Norwood, NJ: Artech "Telecommunications Library," 1988, 448 pp. Includes useful historical survey material in chapters 1–3 (covering mobile radio before cellular and the development of cellular), and chapter 5 (on the analog to digital transition). Tables, diagrams, chapter references, index.

14-015 _____. **WIRELESS ACCESS AND THE LOCAL TELEPHONE NETWORK.** Norwood, MA: Artech "Telecommunications Library," 1992, 595 pp. Primarily a descriptive snapshot of the industry's technological status in the early 1990s, with some historical material. Diagrams, tables, notes, index.

14-016 Day, Frederick J., and Huong N. Tran. **REGULATION OF WIRELESS COMMUNICATIONS SYSTEMS.** Rockville, MD: Government Institutes, 1997, 353 pp. Includes historical background throughout including one specific historical survey chapter. Notes, appendices, index.

14-017 * Garard, Garry A. **CELLULAR COMMUNICATIONS: WORLD-WIDE MARKET DEVELOPMENT.** Norwood, MA: Artech "Mobile Communications Series," 1998, 514 pp. Considerable historic background is provided here on radio before cellular, the development of cellular and the trial systems, initial development of cellular in Europe, the development of the Global System for Mobile Communications digital standard, digital developments (including PCS) in the U.S. Charts, tables, index.

14-018 Huang, Derrick C. **UP IN THE AIR—NEW WIRELESS COMMUNI-CATIONS.** Cambridge, MA: Harvard University Program on Information Resources

Policy," 1992, 133 pp. Changing technologies and the industries developing them including the technical factors involved in market development. Tables, notes.

14-019 Jagoda, A., and M. de Villepin. **MOBILE COMMUNICATIONS**. New York: Wiley "Series in Communication and Distributed Systems," 1993, 180 pp. Translation of a 1991 French publication, this is a useful historical survey of (primarily European) applications. Tables, glossary, index.

14-020 ** Meurling, John, and Richard Jeans. **THE MOBILE PHONE BOOK: THE INVENTION OF THE MOBILE TELEPHONE INDUSTRY**. London: Communications Week International, 1994, 233 pp. Informal history (no references) of the cellular business in Europe and North America. See also the same authors' earlier history of the Ericsson digital switch, 4-203. Glossary, diagrams, no index.

14-021 * Noble, Daniel E. "The History of Land-Mobile Radio Communications," *PROCEEDINGS OF THE IRE,* 50:1405–1414 (May 1962). Good technical overview of early analog wireless developments. Photos.

14-022 Paetsch, Michael. **MOBILE COMMUNICATIONS IN THE U.S. AND EUROPE: REGULATION, TECHNOLOGY, AND MARKETS**. Norwood, MA: Artech "Mobile Communications Series," 1993, 417 pp. Primarily descriptive rather than historical, this does provide useful comparisons between first and second generation cellular, and the rise of digital PCS. Charts, tables, chapter notes, bibliography, index.

14-023 Schneiderman, Ron. **WIRELESS PERSONAL COMMUNICATIONS: THE FUTURE OF TALK**. New York: IEEE Press, 1994, 195 pp. Useful for 1980s and early 1990s developments across a variety of services. Photos, charts, glossary, directory, index.

14-024 ** U.S. Congress, Office of Technology Assessment. **WIRELESS TECHNOLOGIES AND THE NATIONAL INFORMATION INFRASTRUCTURE**. Washington: Government Printing Office, 1995, 290 pp. One of the best surveys of the development and status of the various technologies and their application. Charts, diagrams, tables, appendices, index.

C. Satellite Communications

(The role of government research, development, and procurement dominated satellite communication—not surprising given both the huge expenditures involved and Cold War–driven priorities until the early 1990s. Largely excluded here are satellite public policy studies.)

General

14-025 * Brown, Martin P. Jr. **COMPENDIUM OF COMMUNICATION AND BROAD-CAST SATELLITES 1958 TO 1980**. New York: IEEE Press/ John Wiley, 1981, 375 pp. Diagrams and photos detail the first two generations of such satellites launched by any country. Tables, diagrams, references. See also Caprara, 14-027, and Martin, 14-044.

14-026 *** Butrica, Andrew J., ed. **BEYOND THE IONOSPHERE: FIFTY YEARS OF SATELLITE COMMUNICATION**. Washington: Government Printing Office (NASA History Office, "The NASA History Series," NASA SP-4217), 1997, 321 pp. Twenty-two detailed chapters by as many contributors divided into sections on passive military satellite origins, military and commercial developments in the U.S. and Canada, European satellites, developing the world satellite system, and applications. An important collection. Photos, notes, appendices (chronology, bibliography), glossary, index.

14-027 * Caprara, Giovanni. **THE COMPLETE ENCYCLOPEDIA OF SPACE SATELLITES: EVERY CIVIL AND MILITARY SATELLITE OF THE WORLD SINCE 1957**. New York: Portland House, 1986, 219 pp. Translated from the original Italian edition, this devotes more than 40 pages to telecommunications satellites. Each is described in brief narrative supplemented with a clear line drawing. Photos, diagrams, bibliography, index. See also Brown, 14-025, and Martin, 14-044.

14-028 Carter, Leonard J., advisory ed. **SYMPOSIUM ON COMMUNICATIONS SATELLITES**. New York: Academic Press, 1962. 202 pp. Proceedings of the 1961 London meetings hosted by the British Interplanetary Society. Photos, diagrams, maps, bibliographies.

14-029 *** Clarke, Arthur C. "Extra-Terrestrial Relays: Can Rocket Stations Give World-Wide Radio Coverage?" **WIRELESS WORLD**, 51:305–308 (October 1945). The classic pioneer paper proposing use of the geostationary orbit (GSO). Diagrams, references. Widely reprinted.

14-030 Collette, René. "Space Communications in Europe: How Did We Make It Happen?" *HISTORY AND TECHNOLOGY,* 9:83–93 (1992). Assesses the cooperative nature of the European Space Agency.

14-031 Committee on Satellite Communications, Space Applications Board. **FEDERAL RESEARCH AND DEVELOPMENT FOR SATELLITE COMMUNICATION**. Washington: National Academy of Sciences, 1977.

See Dunlap, 3-032.

14-032 Elder, Donald C. **OUT FROM BEHIND THE EIGHT-BALL: A HISTORY OF PROJECT "ECHO."** San Diego: American Astronautical Society "AAS History Series, Vol. 16," 1995, 162 pp. *Echo* was a large passive satellite ca. 1960 that relied on reflecting signals back from a huge balloon in low Earth orbit. Photos, diagrams, bibliography, index.

14-033 Galloway, Jonathan F. **THE POLITICS AND TECHNOLOGY OF SATELLITE COMMUNICATIONS**. Lexington, MA: Lexington Books, 1972, 247 pp. Bibliography, index.

14-034 Gatland, Kenneth W., ed. **TELECOMMUNICATIONS SATELLITES**. Englewood Cliffs, NJ: Prentice-Hall, 1964. Chapters explore various American and European projects.

14-035 * Gavaghan, Helen. **SOMETHING NEW UNDER THE SUN: SATELLITES AND THE BEGINNING OF THE SPACE AGE.** New York: Copernicus, 1998, 300 pp. One of the Sloan Foundation-sponsored historical studies, this explores the early history of satellite applications in navigation, meteorology, and (more briefly) communications. Photos, chronology, notes and bibliography, index.

14-036 Gould, R.G., and Y.F. Lum. **COMMUNICATION SATELLITE SYSTEMS: AN OVERVIEW OF THE TECHNOLOGY.** New York: IEEE Press, 1976, 164 pp. Basic technical handbook. Diagrams, charts, index. For the bibliography originally intended for this book, see Unger, 14-055.

14-037 *** Hudson, Heather E. **COMMUNICATION SATELLITES: THEIR DEVELOPMENT AND IMPACT.** New York: Free Press, 1990, 338 pp. Useful combination of history, technological description and review of the impact of both domestic and international satellites. Notes, glossary, bibliography, index.

14-038 * International Telecommunication Union. **REPORT BY THE INTERNATIONAL TELECOMMUNICATION UNION ON TELECOMMUNICATION AND THE PEACEFUL USES OF OUTER SPACE.** Geneva: ITU, 1961–date, annual. Best regular source for satellite activities of the world's smaller countries. Tables, notes.

14-039 Jaffe, Leonard. "Communications Satellites," *MECHANICAL ENGINEERING*, 84:12:34–41 (December 1962).

14-040 _____. **COMMUNICATIONS IN SPACE.** New York: Holt, Rinehart & Winston "Holt Library of Science," 1966. An early look at its potential. Diagrams, bibliography, index.

14-041 Lee, Maj. Robert E. **HISTORY OF THE DEFENSE SATELLITE COMMUNICATIONS SYSTEM (1964–1986).** Maxwell AFB, AL: Air University Press "ACSC Report No. 87-1545," 1987. Useful, given the central role of military research in initiating communication satellite technology.

14-042 Lipman, Andrew D., et al., eds. **TELEPORTS AND THE INTELLIGENT CITY.** Homewood, IL: Dow Jones-Irwin, 1986, 410 pp. Case studies of various urban fiber and satellite systems, including the pioneer New York City operation, when attention focused on their satellite facilities. Photos, diagrams, tables, index.

14-043 * Marsten, Richard B. "A Dream Come True: Satellite Broadcasting," *IEEE TRANSACTIONS ON AEROSPACE AND ELECTRONIC SYSTEMS*, 33:360–381 (January 1997). Broad historical survey by a National Academy of Science official. Photos, diagrams, maps, bibliography.

14-044 *** Martin, Donald H. **COMMUNICATION SATELLITES 1958–1995.** El Segundo, CA: The Aerospace Corporation, 1977, 1984, 1996 [3rd ed.], 483 pp. The most comprehensive historical resource, this includes chapters on experimental satellites, international satellites, and those for mobile and military services, those from different parts of the world, and amateur and scientific satellites. Footprint maps, diagrams, tables, references, 65-page topically divided bibliography, index. See also Brown, 14-025 and Caprara, 14-027.

Transmission: Mobile, Satellite, Fiber Optic, and the Internet 273

14-045 * Pierce, John R. "Orbital Radio Relays," *JET PROPULSION,* 25:153–157 (April 1955). Very early paper by a Bell Labs researcher, just a decade following Clarke's initial paper (see 14-029).

14-046 Pierce, John R., and R. Kompfner. "Transoceanic Communication by Means of Satellite," *PROCEEDINGS OF THE IRE,* 47:372–380 (March 1959); reprinted in *PROCEEDINGS OF THE IEEE,* 85:1011–1019 (June 1997). Early projections of what became the AT&T *Telstar* project. Diagrams, bibliography.

14-047 _____. "Communication Satellites," *SCIENTIFIC AMERICAN,* (October 1961), 221:10:90–102. An early Bell Labs authority describes pioneering projects and the outlook. Photos, maps, diagrams.

14-048 *_____. **THE BEGINNINGS OF SATELLITE COMMUNICATION.** San Francisco: San Francisco Press "History of Technology Monographs," 1968, 61 pp. With an introduction by Arthur C. Clarke, two pioneers relate the developments. Bibliography.

14-049 * Pritchard, Wilbur L. "The History and Future of Commercial Satellite Communications," *IEEE COMMUNICATIONS SOCIETY MAGAZINE,* 22:22–37 (May 1984). Overall survey by a long-time authority. Photos, tables, bibliography.

14-050 * Rees, David W.E. **SATELLITE COMMUNICATIONS: THE FIRST QUARTER CENTURY OF SERVICE.** New York: John Wiley "Telecommunications," 1990, 329 pp. Reviews the rise and operation of satellites, and devotes most chapters to applications over time (international, business, broadcast, mobile, and national networks). Diagrams, tables, appendices, glossary, bibliography, index.

14-051 Rosen, Harold A. "Syncom and Its Successors," *PROCEEDINGS OF THE IEEE,* 72:1429–1434 (November 1984). Review of the initial GSO satellite system.

14-052 * Smith, Delbert D. **COMMUNICATION VIA SATELLITE: A VISION IN RETROSPECT.** Leyden, Netherlands: Sijthoff, 1976, 335 pp. Relates development of technology and policy for both domestic and international satellites with both military and civil applications. Notes, bibliography, index.

14-053 "Special Satellite Number," *EBU REVIEW,* 118B:12–81 (November 1969). Thirteen papers assess both technology and policy early in the international satellite era. Maps, diagrams, photos, tables.

14-054 * "The *Telstar* Experiment," *BELL SYSTEM TECHNICAL JOURNAL,* 42: 739–1903 (3 vols., June 1963), reprinted by the National Aeronautics and Space Administration as NASA SP-32 (Washington: Government Printing Office, 1963, with an additional 4th vol. in December 1965). Definitive collection of Bell Labs-authored papers on all aspects of *Telstar* as the pioneering active satellite project. Photos, charts, maps, tables, references.

14-055 * Unger, Jurgen, W.H., comp. **LITERATURE SURVEY OF COMMUNICATION SATELLITE SYSTEMS AND TECHNOLOGY.** New York: IEEE Press, 1976, 409 pp. Probably the definitive survey to this point. Companion volume to Gould and Lum, 14-036. Index.

Chapter 14

14-056 U.S. Congress, House Committee on Government Operations. **SATELLITE COMMUNICATIONS: MILITARY-CIVIL ROLES AND RELATIONSHIPS: SECOND REPORT BY THE COMMITTEE ON GOVERNMENT OPERATIONS: HOUSE REPORT NO. 178.** 89th Cong., 1st Sess. Washington: Government Printing Office, 1965, 160pp. Includes commercial carrier developments, military requirements, government management and international concerns. Tables, notes, appendices (including one of 30 pages on the history of Project *Advent*).

14-057 _____. **GOVERNMENT OPERATIONS IN SPACE (ANALYSIS OF CIVIL-MILITARY ROLES AND RELATIONSHIPS): HOUSE REPORT NO. 445.** 89th Cong., 1st Sess. Washington: Government Printing Office, 1965, 136 pp. Demonstrates the substantial military role and the early years of NASA operations.

14-058 _____. **GOVERNMENT USE OF SATELLITE COMMUNICATIONS: HOUSE REPORT NO. 2318.** 89th Cong, 2nd Sess. Washington: Government Printing Office, 1966, 105 pp. Discusses satellite communication for defense, procurement of satellite services, and civil applications of satellite communications, among other topics. Tables, notes.

14-059 _____. House Committee on Science and Astronautics. **SATELLITES FOR WORLD COMMUNICATIONS: HOUSE REPORT NO. 343.** 86th Cong., 1st Sess. Washington: Government Printing Office, 1959, 9 pp.

14-060 _____, Senate Committee on Aeronautical and Space Sciences. **COMMUNICATION SATELLITES: TECHNICAL, ECONOMIC, AND INTERNATIONAL DEVELOPMENTS: STAFF REPORT.** 87th Cong., 2nd Sess., Committee Print. Washington: Government Printing Office, 1962, 287 pp. Especially useful for its early contemporary point of view. Maps, bibliography.

14-061 U.S. National Aeronautics and Space Administration. **FINAL REPORT ON THE RELAY 1 PROGRAM.** Washington: Government Printing Office "NASA SP-76," 1965, 767 pp.

14-062 _____. **COMMUNICATIONS SATELLITES.** Washington: Government Printing Office, 1966, 16 pp. While very brief, this includes a useful chronological survey of major projects from *Echo* to the Advanced Technology Satellite project. Photos, diagrams.

14-063 Van Horn, Larry. **COMMUNICATIONS SATELLITES: A MONITOR'S GUIDE.** Brasstown, NC: Grove Enterprises, 1987 [3rd ed.], 255 pp. More than the subtitle suggests, this includes both graphic and text background on virtually all domestic and international satellites to that time, including amateur, weather, DBS, military, and the manned programs of both the U.S. and Soviet Union. Photos, maps, diagrams, glossary.

14-064 Williamson, Mark. **THE COMMUNICATIONS SATELLITE.** Bristol, England: Adam Hilger, 1990, 420 pp. Photos, diagrams, bibliography, index.

See Winston, 3-058.

14-065 Wood, James. **SATELLITE COMMUNICATIONS & DBS SYSTEMS.** Oxford, England: Focal Press, 1993, 279 pp. Well-illustrated narrative description of

Transmission: Mobile, Satellite, Fiber Optic, and the Internet   275

the technologies and industries, with some historical material. Photos, diagrams, glossary, references, bibliography, index.

14-066 ** "Worldwide Satellite Communications," *ASTRONAUTICS AND AEROSPACE ENGINEERING,* 1:8:23–87 (September 1963). Eleven articles survey then-current knowledge and projects, both military and civil, including papers on the *Telstar, Relay,* and *Syncom* satellites. Important for its early date. Photos, diagrams, tables, references.

Broadcast/Cable Satellites

(Most of the substantial literature on this topic, and especially on direct broadcast satellites, concerns policy rather than technology. The entries that follow focus on or provide substantial background about the basic technology employed.)

14-067 "After 10 Years of Satellites, the Sky's No Limit," *BROADCASTING* (April 9, 1984), pp. 43–68. Useful journalistic survey of the first decade of cable and broadcast use of satellites. Photos, tables.

14-068 Bahtt, S.C. **SATELLITE INVASION OF INDIA.** New Delhi, India: Gyan House "South Asia Books," 1994, 286 pp. The development and role of satellite broadcasting in India. Bibliography, index.

14-069 Ball, John E.D. "The Planning and Implementation of the Public Television Satellite Interconnection System," *JOURNAL OF THE SMPTE,* 87:825–831 (December 1978). Reviews the five-year process of development to move terrestrial networks to satellite. Photos, map, diagrams, references.

14-070 **BROADCASTING FROM SPACE.** Paris: UNESCO "Reports and Papers on Mass Communication, No. 60," 1971, 65 pp. Proceedings of a 1969 meeting held in Paris which looked toward a service not then available. See also COMMUNICATION IN THE SPACE AGE, 14-072.

14-071 Chayes, Abram. **SATELLITE BROADCASTING.** London: Oxford University Press, 1973, 159 pp. A survey of technology and policy to that point. Charts, tables, bibliography, index.

14-072 **COMMUNICATION IN THE SPACE AGE: THE USE OF SATELLITES BY THE MASS MEDIA.** Paris: UNESCO, 1969, 200 pp. Papers and proceedings of a 1965 conference. See also BROADCASTING FROM SPACE, 14-070.

14-073 Dodd, Mark Reynolds. **SATELLITES AND RADIO BROADCASTING: HISTORICAL REVIEW AND MARKET DEVELOPMENTS.** Washington: National Association of Broadcasters, 1991, 70 pp. The development of analog and digital satellite delivery of audio signals. Tables, bibliography.

14-074 Lessing, Lawrence. "Cinderella in the Sky," *FORTUNE* (October 1967), 76:4:130–133ff. Early assessment of the outlook for broadcast-related applications of satellites. Photos, diagrams.

D. Fiber Optic Systems

14-075 * Chaffee, C. David. **THE REWIRING OF AMERICA: THE FIBER OPTICS REVOLUTION**. Orlando, FL: Academic Press, 1988, 241 pp. Popular narrative describing how the technology developed and was expanded. Photos, maps, diagrams, index.

14-076 ** Faltas, Sami. "The Invention of Fiber-Optic Communications," *HISTORY AND TECHNOLOGY*, 5:31–49 (1988). Important paper outlining the key figures, firms, and developments. References.

14-077 *** Hecht, Jeff. **CITY OF LIGHT: THE STORY OF FIBER OPTICS**. New York: Oxford University Press, 1999, 336 pp. Another in the Sloan Foundation-sponsored historical series, this traces the full history of the technology including key innovators and technical developments as well as applications. Photos, notes, bibliography, index.

14-078 Holonyak, N. Jr. "The Semiconductor Laser: A Thirty-Five Year Perspective," *PROCEEDINGS OF THE IEEE*, 85:1678–1693 (November 1987).

14-079 * Mims, F.M. III. "The First Century of Lightwave Communications," *INTERNATIONAL FIBER OPTICS AND COMMUNICATIONS*, 3:10–26 (February 1982). From Bell's work through the two world wars to recent developments.

14-080 Newman, D.H. "Sources and Detectors for Optical Fibre Communications Applications: The First 20 Years," *PROCEEDINGS OF THE IEEE*, 133J:213–229 (June 1986). Notes how designs changed depending on changing fiber specifications. Photos, references. See next entry.

14-081 * "Special Issue: The First Twenty Years of Optical Communications," *PROCEEDINGS OF THE IEEE*, 133J:189–244 (June 1986). Special theme issue with eight highly technical papers including several on British developments and one on U.S. Navy beginnings in fiber optics. Diagrams, references. See previous entry.

14-082 **WE'VE GOT LIGHT! BUILDING THE NATION'S FIRST COAST-TO-COAST FIBER-OPTIC NETWORK**. New York: Sprint, 1988, 120 pp. Commemorative book describing the process and many of the people involved. Photos.

E. The Internet

(Historical writing about the Internet and World Wide Web was just getting started as this book moved toward the press. As might be expected, some of the best information on this topic is found on the Internet itself—see 15-L.)

14-083 ** Abbate, Janet. **INVENTING THE INTERNET**. Cambridge, MA: MIT Press "Inside Technology," 1999, 264 pp. Solid technical history from packet switching to Advanced Research Projects Agency (ARPANET) to the setting of Internet standards. Bibliography, maps, notes, index.

14-084 Adam, J. "Architects of the Net of Nets," *IEEE SPECTRUM* (September 1996), 33:10:56-63. Focuses on the roles of Bob Kahn and Vinton Cerf.

14-085 * Berners-Lee, Tim. "WWW: Past, Present, and Future," *COMPUTER* (October 1996), pp. 69–77. Details the author's experience at CERN in the early 1980s where the Web protocol was first initiated.

14-085a *_____, with Mark Fischetti. **WEAVING THE WEB: THE ORIGINAL DESIGN AND ULTIMATE DESTINY OF THE WORLD WIDE WEB BY ITS INVENTOR.** San Francisco: HarperSanFrancisco, 1999, 226 pp. How the the Web idea was developed, first at CERN in Europe in the 1980s, and later elsewhere. Index.

14-086 Grey, Victor. **WEB WITHOUT A WEAVER: HOW THE INTERNET IS SHAPING OUR FUTURE.** Concord, CA: Open Heart Press, 1997, 244 pp. Bibliography, index.

14-087 * Hafner, K., and M. Lyon. **WHERE WIZARDS STAY UP LATE: THE ORIGINS OF THE INTERNET.** New York: Simon and Schuster, 1996, 304 pp. Popular narrative about the key figures behind development of the Internet and World Wide Web. Photos, bibliography, index.

14-088 Hauben, Michael, and Ronda Hauben. **NETIZENS: ON THE HISTORY AND IMPACT OF USENET AND THE INTERNET.** Los Angeles: IEEE Computer Press, 1997, 344 pp. Bibliography, index. For an on-line version, see *http://www.columbia.edu/~hauben/netbook*.

14-089 He, J. "Introduction of the Internet and World Wide Web," *EXPERIMENTAL TECHNIQUES* (September–October 1997), pp. 29–33. Includes chronological tables.

14-090 Kirstein, P.T. "Early Experiences with the ARPANET and Internet in the United Kingdom," *IEEE ANNALS OF THE HISTORY OF COMPUTING*, 21:38–44 (March 1999). Reviews both administrative and technical aspects of Web usage.

14-091 Leiner, Barry M., et al. "The Past and Future History of the Internet," *COMMUNICATIONS OF THE ACM* (February 1997), pp. 102–108.

14-092 Reid, Robert A. **ARCHITECTS OF THE WEB.** New York: Wiley, 1997, 288 pp. Eight chapters review specific companies and their founders in the 1993–96 period.

14-093 * Segaller, Stephen. **NERDS 2.0.1: A BRIEF HISTORY OF THE INTERNET.** New York: TV Books, 1998, 352 pp. Companion to a three-hour public television documentary tracing the key people in developing the Web.

Chapter 15

Telecommunications History on the Internet

Since the mid-1990s, a wholly new form of research resource has appeared and blossomed. One good annotated guide to a variety of Internet resources is Jeffrey K. MacKie-Mason and Christopher Lee, **TELECOMMUNICATIONS GUIDE TO THE INTERNET** (Rockville, MD: Government Institutes "Internet Series," 1999, 241 pp.) which details some 250 different web sites and 65 key Usenet groups. See also the same author's website, 15-003. It covers organizations, technology sites, policy and regulation, other resources, and newsgroups.

Thus what follows is only a *partial* listing of useful historical sites, many of them maintained by dedicated individual webmasters. The majority of companies and organizations, it seems, offer little or no historical material on their websites—making those individually maintained sites even more important. These Internet citations (which focus on telecommunications technology history) are arranged by chapters in this book (excepting only chapter 2 which is not included here), and alphabetically within each topical section. Most sites offer links to many others. A word of warning—while the Internet offers a moveable feast of material, it can all too often be highly frustrating as today's site is all gone tomorrow. The following web-site URLs were checked and valid as of August 1999.

A. Reference

15-001 Kidon Media Link by Kees van d Griendt
http://www.dds.nl/~kidon/media-link
Offers many links to all media around the world useful here for its coverage of radio and television (as well as print) media.

15-002 * The Media History Project
http://www.mediahistory.com
Very broad coverage, much of it nontechnical, of all media (including computers but not other telecommunications) with a chronology and a host of links to other sites.

15-003 * Telecommunications Information on the Internet** by Jeffrey Mackie-Mason
http://china.si.umich.edu/telecom/telecom-info.html
Provides links (by a University of Michigan faculty member who also authored the book noted in the headnote to this chapter) to more than 7,500 different sites, and allows ready searching under such terms as "history," among a host of others.

B. General Surveys

15-004 Centre for the History of Defence Electronics by Bournemouth University
http://chide.bournemouth.ac.uk/
Useful British "window" into material on the development of all types of electronics used in air, land, and sea warfare. Offers good link page.

15-005 ** Global Networking: A Timeline by Dr. T. Matthew Ciolek
http://www.ciolek.com/PAPERS/GLOBAL/1800.html
Very detailed chronology of 19th-century developments in pre-telecommunication, telegraph, telephone services and related technologies, including information storage and retrieval.

15-006 History of Communications by BD Interactive
http://www.bdinteractive.com/history/ahistory.html
A web design firm offers brief historical vignettes on print, radio, television, telephone, and satellite history.

15-007 History of International Telecommunications in Japan by KDD
http://www.kdd.co.jp/nenpyo-e/
Extensive illustrated chronology from mid-19th century to 1997.

15-008 ** A History of Modern Communications, Computing and Media
http://www.acclarke.co.uk/shc.html
The Arthur C. Clarke Foundation, with sponsorship by Cable & Wireless, provide an extensive chronology and links.

15-009 * Telecom Timeline
http://www.aronsson.se/hist.html
Reverse chronology (it begins with the most recent events) covering a variety of firms, and including business and policy events as well as technology.

15-010 * Telecommunications History by Webb & Associates
http://www.webbconsult.com/history.html
Handy chronology and especially useful links to many other resources, including several on Bell and other history organizations.

C. Institutional and Company History

Government Bodies—International

15-011 INTELSAT
http://www.intelsat.int/
Can be searched under "history" for information on past Intelsat satellites and services.

15-012 * International Telecommunication Union (ITU)
http://www.itu.int/itudoc/about/itu/history.html
Opening page which provides access to a history of the organization and more detailed information.

Government Bodies—U.S.

15-013 * U.S. Federal Communications Commission (FCC)
http://www.fcc.gov
Allows ready access to the commission and a variety of useful resources including statistics.

15-014 U.S. Postal Service
http://www.ups.com/about/story.html
Relates history of USPS in fairly detailed narrative.

Organizations

15-015 American Radio Relay League (ARRL)
http://www.arrl.org/tis/bibs/
Offers a listing of historical citations from the organization's magazine, *QST*.

15-016 Institution of Electrical Engineers (IEE)
http://www.iee.org.uk/Archives/archives.htm
The British engineering organization holds extensive archives and many exhibitions which are detailed here, plus links to related sites.

15-017 ** IEEE Center for the History of Electrical Engineering
http://www.ieee.org/history_center
Details on the IEEE historical archives and oral histories, research guides, historical articles. Includes ECHOES, a Web-based bulletin board, and very good links.

15-018 Society of Motion Picture and Television Engineers (SMPTE)
http://www.smpte.org/
Not much history, but useful entry to publications and conferences.

Companies—US

15-019 Ampex
http://www.ampex.com/corporatebg/
A brief chronology since 1944 of the firm known especially for audio and video recording tape and devices.

15-020 * AT&T Labs History
http://www.research.att.com/history/index.html
A quite complete site emphasizing work of AT&T Labs over time, including a chronology, highlights of 30 key innovations, and more.

15-021 ** Bell Laboratories
http://www.bell-labs.com/
Now maintained by Lucent Technologies, this is the main page for Bell Labs. Typing "history" on the search line provided leads to dozens of specific entries.

15-022 Collins Radio
http://www.pixi.com/~jenkins/collins/
An unofficial site of information on the company's past products, with links to still more.

15-023 * General Electric
http://www.ge.com/ibhis5.htm
Chronology broken into periods and tied to company-supported Hall of Electricity.

15-024 * Motorola
http://www.mot.com/General/Timeline/timeln24.html
Offers a time line with many clickable items providing more information on specific events and developments.

15-025 Radio Corporation of America (RCA)
http://www.rca-electronics.com/story/story.asp
The highlights of the company from its formation in 1919 to its sale to General Electric in 1985.

15-026 Raytheon
http://www.raytheon.com/histback.html
Company background press release offers a fair bit of detail on the firm's development.

15-027 Westinghouse
http://www.westinghouse.com/corp/history.html
The major American electronics firm is traced with a chronology and other information.

Companies—Foreign

15-028 * Cable & Wireless Ltd.**
http://www.cwhistory.com/
Extensive page with audio and video offers company and general telecommunications history information similar to the firm's CD-ROM history.

15-029 General Electric Company (of Great Britain)
http://www.gec.com/ah1.htm
Multifaceted site with reference to other information from the company's formation in 1886, including material on Marconi.

15-030 Nippon Electric Co. (NEC)
http://www.nec.com/about/index.html
Provides a 100-year chronology.

15-031 Overseas Telecommunications Corp/Telstra (Australia)
http://www.telstra.com.au/archives/
Interesting section of the firm's home page, this provides a number of different historical approaches to Telstra development.

15-032 Siemens
http://www.siemens-hearing.com/about/150years/150yrsmain.htm
Useful chronology of the German firm over 150 years.

D. Biography

15-033 * Edwin Howard Armstrong by Mike Datzdorn
http://www.erols.com/oldradio/
Based on the extensive Harry Houck collection, this includes a host of historical documents in reproduction.

15-034 ** Alexander Graham Bell Family Papers
http://memory.loc.gov/ammem/bellhtml/bellhome.html
Library of Congress site with papers from the family—will eventually have some 4,700 items and 38,000 images. Includes his key laboratory notebook pages.

15-035 ** Bell's Path to the Telephone by Mike Gorman
http://jefferson.village.virginia.edu/albell/homepage.html
Highly detailed site complete with a master map, tracing Bell's early lab notebooks and how they record Bell's progress.

15-036 * Thomas Edison Papers by Rutgers University
http://edison.rutgers.edu/
The Web version of the printed and microfilm record (see 5-106), this includes a vast amount of searchable material, detailed chronology, information on patents, and on Edison's companies.

15-037 The Farnsworth Chronicles
http://songs.com/philo/
Primarily a narrative with some links to other material on the television pioneer.

15-038 Marconi Centenary by General Electric Co. (of England)
http://www.gec.com/marconi/
Extensive chronology of the man and his inventions with a bibliography; some of the items can be downloaded.

15-039 Samuel F.B. Morse Papers from the Library of Congress
http://memory.loc.gov/ammem/atthtml/mrshome.html
Not yet fully online as this volume went to press, the site will soon contain extensive material from the inventor/painter.

15-040 Nikola Tesla
http://www.mercury.gr/tesla/introen.html
German site with information on his life, major inventions, and links to related material.

15-041 Nikola Tesla
http://www.neuronet.pitt.edu/~bogdan/tesla/
Biography, collection of comments on his work, links, and more.

E. Telegraphy

15-042 Canadian Railway Telegraph History
http://web.idirect.com/~rburnet/
Offers a book and many useful pictures and information on the expansion of the telegraph parallel to Canadian railway growth.

15-043 * Commercial Telegraph Code Books by Jim Reeds
http://www.research.att.com/~reeds/codebooks.html
Includes a detailed essay on the hundreds of such books published. It also allows a link to the author's detailed list of such books.

15-044 Internet On-Line Telegraph & Scientific Instrument Cyber Museum by Tom Perera
http://www.chss.montclair.edu/~pererat/telegraph.html
Includes brief chronology and bibliography

15-045 Modern Practice of the Electric Telegraph by Franklin L. Pope
http://sd.znet.com/~cdk14568/mpet/contents.html
Offers the full text of an 1881 book; see also 6-047.

15-046 Telegram History [of Australia] by Larry Rice
http://www.omen.com.au/~larry/mfwahist.html
Offers good information on the rise and development of telegraph services down under including present-day hobbyists.

15-047 Telegraph Lore
http://www.cris.com/~Gsraven/index.shtml
History and information on Morse telegraphy with further links to other sites.

15-048 ** The Telegraph Office: A Tribute to Morse Telegraphy and Resources for Wire and Wireless Telegraph Key Collectors and Historians by Neal McEwen
http://fohnix.metronet.com/~nmcewen/tel_off.html
Very useful site with considerable reference value including lists of books and articles of use to key collectors.

F. Telephony

15-048a * Condensed History of Telecommunication by William von Alven
http://www.cclab.com/billhist.htm
An FCC official's private page offers a quite detailed chronology of (primarily) telephone development, including policy events of the past several decades that reflect on technological changes.

15-049 ** History of the Telephone Set by NTT
http://hannover.park.org/Japan/NTT/DM/html_f2/F2_menu_010_e.html
Colorful time line of telephone receivers and related equipment, "clickable" for further information.

15-050 The Strowger Telecomms Page by Michael Spalter
http://www.seg.co.uk/telecomm/
Information on electromechanical telephone switches, focusing on Britain.

15-051 * Telephone History Website by Chuck Eby
http://www.cybercom.com/~chuck/phones.html
Provides a host of links to many other telephone sites (U.S. and foreign companies, associations, telephone shows, individual pages, books) and offers a basic history.

15-051a ** Tribute to the Telephone by David Massey
http://hyperarchive.lcs.mit.edu/telecom-archives/tribute/students.html
A very extensive site with information on all types of telephones, plus research resources, organizations, and a host of links.

G. Radio

15-052 Antique Radio Page by D.J. Adamson
http://members.aol.com/djadamson/arp.html
Designed for those who collect old radios, this includes books, articles, links, classified ads, and more.

15-053 Antique Wireless Association Electronic Communication Museum
http://www.antiquewireless.org
Details on the collection and its accessibility.

15-054 The Broadcast Archive by Barry Mishkind
http://www.oldradio.com/
Includes equipment and programming sections and links, plus information about the FCC, old stations, and links to other archives and organizations.

15-054a On The Shortwaves by Jerry Berg and John C. Herkimer
http://www.ontheshortwaves.com/
All aspects of shortwave history and development, including both seconding and receiving, plus articles, reviews, and many links.

15-054b 100 Years of Radio: 1895–1995
http://monviso2.alpcom.it/hamradio/

Covers a century of radio, beginning with and focusing on Marconi's work. Available in English or Italian.

15-055 The RMS Titanic Radio Page
http://www.netinfo.com.au/anars/
Quite detailed page with information on the radio equipment used at the time (1912), the operators, and the actual messages sent and received.

15-056 * Surfing the Aether
http://www.northwinds.net/bchris/index.htm
Extensive site with chronology arranged by decade and incorporating many links to people and developments.

15-057 * United States Early Radio History** by Thomas H. White
http://www.ipass.net/~whitetho/index.html
A wonderfully useful site which offers full copy of a variety of pre-1920 articles and documents plus the author's valuable own research on early radio station list publications, call-letter policies, and the like.

15-058 World of Wireless
http://home.luna.nl/~arjan-muil/radio/history.html
A Dutch site (in both English and Dutch) takes the story through World War II and includes details of the owners' own collection.

H. Electroacoustics and Recording

15-059 Early Recorded Sound and Wax Cylinder Page
http://www.tinfoil.com
Primarily concerned with actual recordings themselves—and includes ability to hear some historical cylinder recordings.

15-060 * History of Recording Technology by Steve Schoenherr
http://ac.acusd.edu/History/recording/notes.html
Includes phonographs, tape recorders, and even musical jukeboxes. Offers a 16-part chronology with pictures and links.

15-061 The Online Antique Phonograph Gallery
http://www.inkyfingers.com/~inky/Record.html
A window onto a number of private collections of early hardware.

15-062 * The World's Earliest Television Recordings by Don McLean
http://www.dfm.dircon.co.uk/articles.htm
Disc recordings made by John Logie Baird in the 1920s in Britain are discussed and demonstrated.

I. Electron Tubes and Solid-State Devices

15-063 Fifty Years of the Transistor by Lucent Technologies
http://www.lucent.com/ideas/heritage/transistor
Sections cover inventors, history, applications, and current roles.

15-064 Transistor References by Donald Darrow
http://www.uni.edu/darrow/transistor.html
A host of historical links concerning the transistor.

15-065 Tubes versus Transistors
http://gprime.com/proaudio/tubes/tubes.htm
Reprints a 1973 article from *Audio Engineering Society Journal* as to whether there was an audible difference in amplification with the two means.

15-066 Vacuum Tube Links by Michelle Troutman
http://www.acadia.net/michelle/tubes.htm
Just that.

15-067 Vacuum Tube Valley by Charles Kittleson
http://www.vacuumtube.com/cooltube.htm
Links to all aspects of vacuum tubes past and present as assembled by a history quarterly of this name.

J. Television

15-068 ** History of Color Television by Edwin H. Reitan Jr.
http://www.novia.net/~ereitan/
Very useful site with considerable detail on both CBS and RCA systems and the debate over standards in the 1940s and 1950s.

15-069 Mark's Index to the History of TV Technology
http://members.aol.com/aj2x/oldtv.html
Offers links to museums, archives, and other sites emphasizing television's technology.

15-070 The Museum of Television
http://www.mztv.com/gallery.html
The Toronto-based museum offers a virtual tour of its galleries on the pioneering years of television (e.g., mechanical era, television in 1939) and of TV inventors and inventions.

15-071 * A U.S. Television Chronology, 1875–1970 by Jeff Miller
http://members.aol.com/jeff560/chronotv.html
Just that, and quite extensive.

K. Newer Media Technologies

15-072 * HDTV—An Historical Perspective by Corey Carbonara
http://web-star.com/hdtv/perspective.html
Primarily text, divided into extensive chapters carrying the story back to the 1930s.

15-073 National Cable Television Center
http://www.cablecenter.org
The industry-sponsored museum and archive of CATV history.

15-074 Soundsite
http://www.soundsite.com/index.html
Reviews all aspects of current home entertainment audio and video technology, including data from manufacturers and a brief chronology.

L. Transmission: Mobile, Satellite, Fiber Optic, and the Internet

15-075 A Fiber-Optic Chronology by Jeff Hecht
http://www.sff.net/people/Jeff.Hecht/Chron.html
Useful chronology (as of 1996) with a link to a brief narrative historical survey as well. See also author's book, 14-077.

15-076 ** Hobbes' Internet Timeline by Robert Hobbes Zakon (Mitre Corp.)
http://info.isco.org/guest/zakon/Internet/History/HIT.html
Useful coverage in a widely cited Web site. Includes several trend charts and tables of historical statistics.

15-077 Internet History Timeline by Steve Harris
http://www.njin.net/~srharris/inter/history/history.html
Very extensive chronology with many places where one can click for further information.

15-078 Low-Earth Orbit Satellite Networks by University of Iowa Business School
http://www.biz.uiowa.edu/class/6K180_park/Student-Reports/kburrows/index.html
Extensive site reviewing Iridium, Teledesic, and other related LEO projects as of April 1998.

15-079 ** NASA Experimental Communications Satellites
http://sulu.lerc.nasa.gov/dglover/satcom1.html
Extensive site covering the past four decades and many specific projects.

15-080 Satellite and Telecommunications History Trivia Quiz by Didactec
http://www.access.digex.net/~mrc/trivquiz.htm
Nineteen brain twisters concerning the development of telecommunications/satellite technology!

Name Index

References are to individual or corporate authors or (only for those entries lacking an author) to an entry's title in CAPS (for books) or "in quotes" (for articles and chapters). Web site titles (all are in chapter 15) are abbreviated in many cases. Initial words such as "The, A, An" and the like are ignored in titles, many of which are abbreviated. Numbers indicate chapter-entries (e.g., 4-028 is the 28th entry in chapter four). Coauthors are indexed separately from main authors, but coauthor teams with same last name are shown only under first author. Translators are not indexed.

Abbate, J. 14-083
Abbot, C. 3-015
Abbott, A. 7-038
Abbott, D. 5-001
Abbott, H. 7-043
Abernathy, J. 6-069
Abramson, A. 5-109, 5-225, 5-226, 10-113, 10-114, 12-110, 12-111
Adam, J. 14-084
Adams, G. 6-141
Adams, M. 3-108
Adams, S. 4-167, 8-026
Adley, C. 6-022, 6-142
Affel, H. 7-145
Ahvenainen, J. 6-125
Aitken, H. 8-001, 8-038, 9-090
Aitken, W. 7-044, 7-045
Alexander, W. 6-011
Alexanderson, E. 5-038, 9-048
Algar, J. 4-205
Allen, T. 6-170, 6-173, 6-183
Allsop, F. 7-024
Alper, J. 4-019
Alth, M. 9-157
Alvarado, M. 10-131
Amos, D. 9-122
Ampex, 15-019
Anderson, A. 4-211
Anderson, J. 6-048

Anderston, L. 5-213, 5-220
Andrew, W. 6-144
Andrews, F. 10-034
Anglin, R. 13-088
Angus, R. 10-047, 10-048
Angwin, A. 7-046
Ansted, D. 6-186
Antébi, E. 3-001
Appleby, T. 5-152
Appleyard, R. 4-090, 5-016
Archer, C. 6-013
Archer, G. 9-091
Ardenne, M. 11-038, 11-039, 12-035
Ardis, S. 2-001
Armagnat, H. 13-001
Arlen, G. 13-038
Armstrong, E. 5-039–043, 5-155, 8-113, 9-130–132, 9-140–143, 11-012, 11-013, 15-033
Arnold, E. 12-112
Arvin, W. 5-091
Ashbridge, N. 12-047, 12-077
Ashley, C. 9-046
Asimov, I. 5-003
Asmann, E. 3-024
Astley, C. 6-013
Atkinson, H. 6-145
Attman, A. 4-202
Auger, C. 3-016

Ault, P. 6-102
Austin, B. 3-109

Baark, E. 6-126
Baarslag, K. 3-149
Babani, B. 11-000
Bachman, W. 10-016
Baer, W. 13-069
Bagdikian, B. 13-039
Baglehole, K. 4-199
Bahtt, S. 14-068
Bailey, M. 5-004
Bain, A. 6-002
Baird, D. 5-138
Baird, J. 5-044–058, 10-123
Baird, M. 5-045
Baker, D. 3-109
Baker, E. 5-198
Baker, R. 3-017
Baker, T. 13-002–005
Baker, W. 4-215, 12-025
Balanis, C. 9-093
Balderson, W. 4-151
Baldwin, F. 7-128
Baldwin, J. 10-115
Baldwin, N. 5-097
Baldwin, T. 7-034, 13-057
Ball, J. 14-069
Ballantyne, R. 6-215
Bamber, E. 4-233
Bangay, R. 3-110
Bargellini, P. 5-184
Barnes, C. 6-221
Barnum, F. 4-151a
Bartlett, E. 13-058
Barty-King, J. 4-200
Bates, D. 3-131
Batten, J. 10-017
Bauer, R. 10-001
Beare, R. 9-047
Beauchamp, K. 3-002
Beck, A. 3-025
Bedi, J. 1-037
Beechey, F. 6-055
Begun, S. 10-049
Beitman, M. 12-052
Belitsos, B. 14-001
Bell, A. 5-037, 5-059–064, 5-129, 7-047, 7-069, 15-034, 15-035
Bell, F. 7-027
Bell, J. 6-088, 7-026
Bell, M. 7-140
Bellamy, R. 13-082, 13-083
Bello, F. 4-115, 11-069, 12-131
Beltz, C. 13-105
Bendix, B. 7-096
Beniger, J. 3-018
Bennett, A. 7-115
Bennett, R. 6-146
Bennett-Levy, M. 12-145, 12-146

Benson, D. 7-129
Benson, K. 13-106
Benson, T. 12-010
Benton, M. 8-014
Berg, J. 9-133
Bergen, J. 3-133
Bergman, L. 10-002
Berliner, E. 5-065
Berners-Lee, T. 14-085, 14-085a
Bernstein, J. 4-116
Besen, S. 3-060
Beverage, H. 8-027
Bhattacharyya, P. 5-066
Bidwell, S. 13-006
Bilby, K. 4-152
Birkenshaw, D. 12-042, 12-097
Birr, K. 4-128
Bishop, H. 4-191
Bitting, R. 12-078
Black, G. 10-094
Black, S. 13-059
Blaine, R. 9-011
Blake, G. 9-144
Blanchard, J. 5-139
Black, R. 3-026
Blackwell, D. 6-147
Blackwell, O. 3-028
Blondheim, M. 6-103
Bloom, L. 13-075
Blumlein, A. 12-079
Blundell, J. 6-208
Boehm, G. 10-096
Boettinger, H. 7-059
Bohringer, A. 12-001
Bolton, F. 6-060
Bolton, H. 2-002
Bond, D. 13-007
Bond, R. 6-035
Bondynpadhay, P. 5-067, 5-068
Boot, H. 11-055
Booth, J. 3-150
Borchardt, K. 4-052
Borkin, J. 12-051
Bose, J. 5-066–072
Botone, S. 10-018
Bottone, S. 9-006
Bouman, P. 4-230
Bourton, K. 12-132
Bouwman, H. 13-076
Bowers, R. 7-159, 14-011
Bowkers, B. 5-080
Bown, R. 11-070
Boyd, D. 10-132
Boylan, J. 7-097
Boyle, R. 10-003
Boyrer, W. 7-037
Braband, K. 4-123
Bracken, J. 1-001, 1-021
Bragaw, C. 4-032
Bramwell, F. 7-013

Name Index

Branley, E. 9-145
Brattain, W. 11-078
Braun, E. 11-071
Braun, K. 5-073–075
Braun, M. 10-097
Bray, J. 5-018
Bray, W. 12-061
Brett, J. 6-174
Bretz, R. 13-040, 13-084
Brewer, A. 4-173
Bridgewater, T. 5-046, 5-209, 12-062
Briggs, A. 4-192
Briggs, C. 6-176
Bright, B. 6-204
Bright, C. 3-111, 5-076, 6-148, 6-200, 6-218, 6-219
Bright, E. 5-076
Brightbill, G. 1-002
Brinkley, J. 13-108
Brinkman, W. 11-072
Brittain, J. 3-003, 4-087, 4-088, 5-038, 7-070, 9-048, 11-056
Brock, G. 4-053
Brodsky, I. 14-013
Brokmeyer, E. 5-108
Brooks, C. 6-177
Brooks, J. 4-097
Brouwer, 7-072
Brown, C. 6-032
Brown, F. 3-067
Brown, J. 3-134
Brown, M. 2-003, 14-025
Brown, R. 14-001a
Browne, D. 3-099
Bruce, R. 5-059
Bryant, J. 4-182, 8-028, 9-158, 13-053
Bubbers, J. 10-050
Bucciante, G. 5-156
Bucher, E. 9-040, 11-014
Buchwald, J. 5-140
Buckley, O. 5-144, 7-146
Bunch, B. 1-043
Bunis, M. 9-159, 9-160
Burlingame, R. 3-019
Burnett, W. 6-033
Burns, R. 5-047, 5-048, 12-063, 12-080, 12-081, 12-113, 12-114
Burt, L. 10-045
Bussey, G. 4-054, 4-059, 9-161
Butler, O. 4-167
Butrica, A. 6-127, 14-026
Button, K. 8-029
Byng, M. 7-027

Calhoun, G. 14-014, 14-015
Callahan, J. 13-008
Cameron, E. 10-079
Camm, F. 1-024, 12-022
Campbell, G. 7-070, 10-051
Campbell, L. 5-169

Campbell, W. 7-065
Campbell Swinton, A. 5-210
Cannell, J. 12-036
Cantelon, P. 4-147, 8-030
Caprara, G. 14-027
Carlson, W. 4-098
Carlton, N. 4-175
Carneal, G. 5-092
Carson, D. 7-158
Carson, J. 8-015
Carter, H. 11-001
Carter, L. 14-028
Carter, S. 5-120
Carty, J. 5-037, 5-077–079
CasaBianca, L. 13-109
Case, J. 4-129
Cashman, T. 7-130
Casson, H. 7-048
Cerver, F. 3-068
Chaffee, C. 14-075
Chanan, M. 10-052
Chanecka, S. 7-097a
Channing, I. 4-011
Channing, W. 6-014, 6-025
Chapple, H. 12-012, 12-023, 12-026, 12-049
Chapuis, R. 7-116, 7-117
Chayes, A. 14-071
Cheney, M. 5-214
Cherry, C. 3-085
Chew, V. 10-019
Chin, F. 13-061a
Chipman, R. 11-015
Christofferson, M. 13-076
Churchill, C. 6-052
Clark, G. 5-205
Clark, J. 4-198, 6-188
Clark, K. 3-086
Clark, M. 10-053, 10-054
Clark, P. 7-165
Clark, R. 5-098
Clarke, A. 3-027, 6-222, 14-029
Clarke, B. 12-064, 12-082–084
Clarke, L. 5-081
Clarkson, R. 9-095
Clarricoats, J. 4-094
Clayton, H. 6-223
Clayton, R. 4-205
Clement, E. 7-098
Clements, H. 10-020
Clifford, G. 4-081
Clune, F. 6-149
Coase, R. 4-033
Coates, A. 3-076
Coates, V. 6-224
Codding, G. 4-012, 4-013, 4-014
Coe, D. 5-157
Coe, L. 6-104, 7-073, 9-096
Cohen, R. 13-110
Cole, A. 13-009

292 Name Index

Cole, W. 9-063
Coleman, A. 9-050
Coll, D. 3-080, 13-062
Collette, R. 14-030
Collier, D. 5-162
Collins, A. 7-074, 9-027, 9-029, 9-030, 12-018
Collins, F. 9-051
Collins, P. 6-128, 12-147
Collins, R. 3-077
Colpitts, E. 3-028
Cones, H. 4-182, 9-158
Connell, S. 13-085
Conot, R. 5-099
Constable, A. 9-097, 9-162
Cooke, T. 5-082
Cooke, W. 5-079a–090
Coon, H. 4-098a
Cooper, D. 7-099
Cooper, I. 1-003
Coopersmith, J. 13-010, 13-011
Cope, S. 4-001
Copplestone, B. 3-151
Corbishley, H. 12-053
Cornell, A. 6-084
Cornell, E. 4-175
Cornell, W., 4-060
Cortada, J. 1-004, 1-005, 4-055, 5-019
Costigan, D. 13-012, 13-013
Coughlan, S. 3-150
Coulson, T. 5-131a, 9-052
Coursey, P. 9-053, 9-054
Cousino, B. 10-055
Cowan, L. 10-080
Crafton, D. 10-081
Craig, D. 6-066
Crane, R. 12-135
Crawley, C. 3-029
Crombie, D. 14-002
Cromwell, J. 7-100
Crookes, W. 9-001
Crowhurst, N. 10-098, 10-099
Culley, R. 6-036
Cunningham, J. 10-008
Curtis, P. 3-062
Cusomano, M. 10-133, 10-142
Czitrom, D. 3-030

Dachis, C. 9-163
Dahl, T. 4-228
Dalton, W. 9-098
Daniel, E. 10-056
Danielian, N. 4-099
Dantieth, J. 5-006
Darrow, K. 5-144
Dasgupta, S. 5-069
David, R. 5-200
Davidson, G. 5-100
Davis, C. 6-056
Davis, D. 6-012

Davis, H. 1-047
Davis, M. 6-076
Davis, W. 13-014
Dawson, K. 6-105
Day, F. 14-016
de Arcangelis, M. 3-112
de Boer, K. 10-100, 10-101
de Bruin, R. 13-111
de Cogan, D. 6-128a, 6-225
de Forest, L. 5-091–095, 9-055, 11-015–017, 11-025–026, 12-085
de Giuli, I. 6-226
de Tunzelmann, G. 8-008
de Villepin, M. 14-019
Dearling, R. 10-057
DeBlois, D. 6-111, 6-248
Deloraine, E. 8-016
Deloraine, M. 4-142
Denman, R. 3-031
DeSonne, M. 13-089
DeSoto, C. 4-076, 4-077
Devereux, T. 3-113
Dewalt, B. 14-003
Dholakia, R. 13-042
D'Humy, F. 6-089
Dibner, B. 5-020, 5-170, 6-227
Dickerson, E. 5-132
Dickson, E. 7-159
Dinsdale, A. 12-002, 12-019, 12-065
Dobbs, A. 7-029, 7-031
Dobrow, J. 10-134
Dodd, G. 6-040
Dodd, M. 14-073
Dodwell, R. 6-034
Dolbear, A. 7-001
Dolby, R. 13-091
Domb, C. 5-171
Donaldson, F. 4-217
Donow, K. 13-110
Dooner, K. 7-166, 7-167
Douglas, A. 4-056
Douglas, J. 6-053
Douglas, S. 3-162, 9-099
Dowding, G. 12-027
Dowling, M. 4-057
Dowsett, H. 9-056
Draper, W. 5-215
Drawbaugh, D. 5-096
Dreher, C. 4-153
Dubbe, R. 10-141
Duboff, R. 6-106–108
Dubreuil, S. 3-152
Duke, V. 13-007
Dummer, G. 1-048
Du Moncel, T. 7-010
Duncan, J. 6-090
Dunlap, O. 1-049, 3-032, 5-021, 5-158, 9-100, 12-020, 12-087
Dupagne, M. 13-112
Durham, J. 6-150

Name Index

Dyer, F. 5-101

Earl, R. 7-131, 7-176
Easton, K. 13-063
Eastwood, E. 9-101
Eccles, W. 9-057, 9-102
Eckhardt, G. 12-037
Edelson, B. 4-022
Edison, T. 5-097–107, 10-021–022, 10-025, 10-029, 11-035, 15-036
Edward, W. 10-088a
Edwards, S. 7-031
Eichhorn, G. 9-031
Einspruch, N. 11-086
Elder, D. 14-032
Elliott, C. 5-008
Elton, M. 7-160
Elwell, C. 9-059
Elwood, T. 6-228
Emerson, A. 7-168, 7-169
Emery, W. 4-039
Emmerson, A. 12-148
Engineer, M. 5-066
England, R. 2-004
Enochs, H. 3-033
Epstein, D. 12-048
Erickson, D. 5-039
Erlang, A. 5-108
Ernst, M. 9-012
Erskine-Murray, J. 3-153, 9-030
Espenschied, L. 9-103
Evans-Cameron, J. 10-079
Everitt, C. 5-172
Everson, G. 5-111
Eward, R. 3-087
Exwood, M. 5-049
Eyman, S. 10-082

Fabrizio, T. 10-023
Fackelman, M. 1-006
Fahie, J. 6-072, 9-004
Falk, H. 7-161
Faltas, S. 14-076
Faraday, M. 5-022, 5-028, 5-183, 6-237
Farnsworth, E. 5-112
Farnsworth, P. 5-109–115, 12-088, 15-037
Faulkner, B. 3-154
Fawcett, W. 5-206
Fay, H. 4-162
Fay, K. 4-079
Fedida, S. 13-077
Feldenkirchen, W. 4-235
Felix, E. 12-017
Fessenden, H. 5-116
Fessenden, R. 5-116–119, 9-060
Field, A. 6-109, 6-129
Field, C. 5-120–123, 6-212
Field, H. 6-201
Field, K. 7-003
Fielding, R. 12-115

Finholt, T. 6-122
Fink, D. 4-086, 11-028, 12-089–091, 12-136, 13-106
Finlaison, J. 6-002
Finn, B. 1-007, 6-224, 6-229, 6-230
Finn, W. 6-074
Fischer, C. 7-076-079
Fisher, D. 12-116
Fisk, J. 12-117
Fitch, R. 3-114
Flannery, G. 4-041
Flegel, G. 6-045
Fleming, J. 5-124–126, 9-013, 9-032, 9-041, 9-061–062, 11-003, 11-018–020
Fletcher, H. 10-103
Flichy, P. 3-034, 6-130
Forbes, W., 4-105
Ford, P. 10-059
Foreman-Peck, J. 7-133
Fortner, R. 3-088
Fowle, F., 6-091
Fox, B. 3-115
Franklin, H. 10-083
Fransman, M. 3-069
Fraser, D. 3-116
Frayne, J. 10-024, 10-084, 10-104
Frederick, H. 10-004
Freeman, J. 13-113
Freeze, K. 4-109
French, T. 6-246
Friedlander, A. 3-035
Friedman, H. 7-080
Friedman, J. 4-096
Froehlich, F. 1-026
Frow, G. 10-025

Gabel, D. 7-081
Gable, R. 9-135
Gabler, E. 6-110
Gabor, D. 12-137
Gaggioni, H. 12-118
Galambos, L. 4-100
Galbraith, I. 13-085
Galloway, J. 14-033
Galvin, P. 4-150
Gander, M. 10-005
Ganley, G. 10-135, 10-136
Ganley, O. 10-136
Gannett, E. 4-083
Garard, G. 14-017
Garbit, F. 10-026
Gardner, R. 1-027
Garner, L. 10-105
Garner, S. 7-004
Garnet, R. 4-101
Garnett, W. 5-169
Garnham, S. 6-231
Garratt, G. 3-159, 5-022, 6-232, 12-043, 12-119
Gatland, K. 14-034

Gavaghan, H. 14-035
Gebhard, L. 3-163
Geddes, K. 4-059, 4-193, 5-159
Geddes, P. 5-070
Geduld, H. 10-085
Gelatt, R. 10-027
Gerbner, G. 13-043
Gernsback, S. 1-028
Ghose, S. 6-151
Gibson, C. 9-063
Giles, A. 6-152
Gillett, W. 10-028
Ginsburg, C. 10-118
Gittman, L. 4-149
Glatzer, H. 8-039
Glazebrook, R. 5-173
Gleave, M. 13-092
Glover, J. 4-060
Gluckman, A. 8-002
Godfrey, D. 5-114
Godwin, G. 4-218
Goetzeler, H. 4-243
Goodman, D. 6-122
Goldman, M. 5-174
Goldmark, P. 5-127, 12-138
Goldsmid, F. 6-153
Goldsmith, A. 3-036, 9-064, 13-016
Gomery, D. 10-086, 12-092
Goodman, R. 6-238
Goodwin, W. 3-155
Gorham, J. 11-029
Gori, V. 5-033
Gorman, M. 5-060, 6-154
Gorman, P. 4-168
Gorokhov, P. 12-120
Gosling, B. 11-030
Gould, R. 14-036
Gouriet, G. 12-139
Graham, M. 10-137
Grant, A. 3-037, 13-044
Grant, E. 4-174
Graves, E. 6-155
Gray, E. 5-128–131, 7-005
Gray, G. 4-029, 7-049
Gray, T. 5-023
Green, F. 10-087
Green, V. 7-082
Greenwood, T. 8-017
Gregor, A. 4-117
Grey, V. 14-086
Gribble, P. 6-156
Griffith, T. 7-101a
Griscom, G. 6-205
Grisewood, E., 12-008
Grivet, P. 3-007
Gross, G. 4-042
Grosvenor, E. 5-061
Grotz, T. 5-221
Gunston, D. 5-161
Gunther, F. 8-018

Gust, F. 7-050

Haantjes, J. 11-001
Habermann, W. 13-114
Hackenburg, H. 7-102
Hadfield, R. 6-231
Hafner, K. 14-087
Haglund, H. 6-233
Haigh, K. 6-234
Hale, W. 8-040
Hall, J. 7-102a
Hallett, M. 5-050
Halloran, A. 12-038
Hamel, J. 6-157
Hammar, P. 10-060, 10-119
Hammond, J. 4-131, 9-117
Hancock, H. 3-156
Hancock, K. 13-062
Hanscom, C. 1-052
Harcourt, E. 4-229
Harder, W. 5-096
Harding, C. 12-054
Harding, R. 12-093
Harlow, A. 3-038, 3-135
Harmon, P. 5-175
Harpur, P. 1-053
Harriets, J. 9-050
Harris, L. 3-117
Harris, R. 6-111, 6-247, 6-248
Harris, W. 4-061
Harrison, A. 9-147
Harrison, C. 9-148
Harrison, H. 6-092
Harvey, A. 11-057
Harvey, F. 10-088
Harvith, J. 10-029
Haskin, E. 7-103
Hatcher, W. 6-004
Hathaway, K. 12-021
Hauben, M. 14-088
Hawes, R. 9-165, 9-166
Hawkins, D. 13-102
Hawkins, L. 4-132
Hawks, E. 3-008, 5-024
Hayward, C. 9-046
Hazewindus, N. 11-073
He, J. 14-089
Head, F. 6-158
Headrick, D. 3-089, 3-090
Hecht, J. 14-077
Heerding, A. 4-231
Heightman, D. 12-149
Hellemans, A. 1-043
Hempstead, C. 6-235, 6-236
Henderson, K. 13-093
Hendry, J. 5-176
Henley, W. 4-211
Henry, J. 5-131a–137, 6-027
Herbert, R. 5-051, 5-052
Herbert, T. 6-159, 7-134

Name Index

Herold, E. 11-040, 11-041, 11-074
Herring, J. 4-042
Hertz, H. 5-031, 5-138–143, 8-008, 8-010
Hertz, J. 5-141a
Hezlet, A. 3-157
Higgens, G. 4-194
Higgins, T. 1-008, 1-009, 5-009, 5-010
Highton, E. 6-015, 6-019
Hijiya, J. 5-094
Hill, J. 9-167-169
Hilliard, J. 10-106
Hillis, G. 6-093
Hills, L. 13-017
Hills, R. 12-121
Hines, B. 10-107
Hippisley, R. 3-118
Hirsch, E. 13-054
Hitchcock, H. 10-030
Hoddeson, L. 4-102, 11-083
Hodge, J. 6-131
Hofer, S. 5-115
Hoffer, T. 5-207
Hogan, J. 12-066, 13-018, 13-019
Hollins, T. 13-065
Holmes, T. 6-160
Holonyak, N. 14-078
Holzmann, G. 6-132
Homans, J. 7-033
Hong, S. 11-002, 11-003
Hoover, C. 10-006
Hopkins, W. 7-019, 7-028
Horn, C. 9-118
Horton, L. 5-208
Horwitz, R. 4-043
Hoskiaer, O. 6-049, 6-061
Hounshell, D. 1-038, 5-128-130
House, R. 6-019
Houston, E. 1-029, 3-008a, 3-039, 6-094, 7-023
Howe, F. 7-104
Howe, G. 5-195
Howe, P. 6-089
Howeth, L. 3-164
Howgrave-Graham, R. 9-040
Huang, D. 14-018
Hubbard, G. 5-087
Hubbell, R. 12-123
Hudgins, J. 9-014
Hudson, H. 14-037
Hudson, R. 1-054
Hughes, T. 3-009
Hugill, P. 3-090a
Hunkin, T. 13-021
Hunt, B. 5-025, 6-237
Hunt, F. 10-007
Hunt, I. 5-215
Huntington, F. 6-054
Hutchinson, R. 12-011, 12-028
Huth, A. 9-123
Huurdeman, A. 14-005

Hylander, C. 12-093
Hyndman, H. 8-011

Iardella, A. 4-169
Ibuka, M. 4-244
Ingham, J. 4-002
Ingram, D. 6-249, 12-117
Ingram, S. 11-031
Inglis, A. 3-064
Ireland, N. 5-011
Irwin, M. 4-062
Isakson, D. 13-022
Israel, P. 5-102, 6-112, 6-113
Ives, H. 5-144, 12-067, 13-023

Jackson, C. 8-041
Jackson, G. 9-105
Jacobsen, K. 4-210
Jacot, B. 5-162
Jaffe, L. 14-039, 14-040
Jagoda, A. 14-019
Jahnke, D. 4-079
James, J. 6-133
Jamieson, A. 6-213
Jeans, R. 4-203, 14-020
Jeans, W. 5-026
Jehl, F. 5-103
Jenkins, C. 5-145, 12-007, 13-024
Jensen, A. 12-124
Jensen, P. 9-065
Jepson, T. 7-105
Jeremy, D. 4-004
Jewett, F. 5-077, 5078, 7-051, 7-052
Jewkes, J. 3-021
Joel, A. 7-120, 7-121
Johannessen, N. 7-135
Johnson, A. 9-020
Johnson, G. 6-220
Johnson, J. 6-135
Johnson, L. 3-060
Johnson, S. 6-063
Johnston, R. 6-161
Johnston, W. 6-058, 6-068
Jolly, W.P. 5-147, 5-163
Jones, A. 6-016
Jones, C. 13-025
Jones, F. 3-039a
Jones, G. 12-094, 13-116
Jones, W. 6-082
Jordan, D. 7-083
Jordan, R. 10-008
Jorysz, A. 10-061
Josephson, M. 5-104
Joy, G. 7-067
Judson, I. 5-121
Jurgen, R. 13-094

Kahaner, L. 4-148
Kahtri, L. 5-015
Kamen, I. 13-046

Kaplan, S. 11-042
Kellogg, E. 10-013, 10-031, 10-088a
Kelly, M. 7-148
Kelsey, E. 9-134
Kelvin, Lord. 5-028
Kemp, J. 8-031
Kempner, S. 1-030
Kendall, A. 5-187
Kennan, G. 6-136
Kennedy, T. 9-106
Kennelly, A. 6-094, 7-023, 9-033
Kent, A. 1-026, 4-111
Kent, B. 3-158
Kerby, P. 12-055
Kerr, R. 9-002
Kieve, J. 6-162
King, W. 3-040
Kingsbury, H. 6-238
Kingsbury, J. 7-053
Kirstein, P. 14-090
Kittross, J. 1-010, 3-065, 4-031, 8-044
Klaue, W. 1-039
Klein, M. 6-114
Klemperer, O. 12-056
Kline, R. 1-037, 4-084
Knapp, J. 12-069
Knappen, R. 7-170
Knauer, L. 4-027
Knox, W. 13-066
Kobb, B. 8-045
Koenigsberg, A. 10-032
Kogan, P. 11-004
Kohlhass, H. 4-144
Kompfer, R. 11-058, 11-059
Kompfner, R. 14-046
Korn, A. 5-146, 13-026
Korn, E. 5-146
Korn, T. 5-146
Kraft, J. 10-062
Kragh, H. 7-122
Kraus, J. 12-003
Krauter, D. 5-117, 9-149–9-150c
Krekel, K. 1-006
Kreuzer, J. 5-164
Krivocheev, M. 13-117, 13-118
Krull, A. 11-075
Kuhl, V. 13-129
Kuhn, M. 3-078
Kumin, D. 13-095
Kurtz, E. 12-070
Kurylo, F. 5-073

Lafferty, W. 10-063
Land, H. 13-067
Landry, R. 9-119
Lane, B. 10-064
Lane, D. 9-170
Lane, H. 12-005
Lane, R. 9-170
Langdale, J. 7-084

Langdon, W. 7-054, 7-171
Lansford, W. 3-119
Lardner, D. 6-026
Lardner, J. 10-138
Larkin, O. 5-188
Larmor, J. 5-178
Larner, E. 12-006
Larsen, J. 11-085
Lavin, P. 13-096
Lavine, A. 3-136
Law, H. 11-043, 12-140
Lawford, G. 4-251
Lawrence, G. 6-163
Lawson, R. 7-055
Layer, H. 12-125
Lean, T. 3-101
LeBow, I. 3-041
Lee, K. 4-015
Lee, R. 14-041
Leigh-Bennett, E. 4-220
Leiner, B. 14-091
Leinwoll, S. 9-107
Lent, 10-132
Leon, G. 7-085
Lescarboura, A. 9-066
Lessing, L. 5-040, 5-041, 14-074
Leutz, C. 9-135
Levin, H. 8-046
Levy, M. 10-139
Lewis, E. 1-031
Lewis, G. 5-074
Lewis, T. 5-027
Li, G. 3-092
Lim, J. 13-119
Limbacher, J. 10-089
Lincoln, A. 3-131
Lindley, L. 6-115
Lipartito, K. 6-116, 7-086, 7-106
Lipman, A. 14-042
Lipshitz, S. 10-009
Lister, W. 13-027
Livingston, K. 3-079
Lockwood, D. 6-147
Lockwood, T. 6-070, 7-014
Lodge, O. 1-032, 5-148-150, 8-010, 8-020–021, 9-067, 9-151
Lohr, L. 12-095
Longbridge, J. 6-177
Loomis, M. 5-152–154
Loring, A. 6-062
Loughlin, B. 12-096
Loughney, K. 2-005
Lovette, F. 4-170
Lowe, A. 9-068
Lubar, S. 3-042
Lubcke, H. 5-113
Lubell, S. 5-095
Lubrano, A. 6-117
Lubszynski, H. 11-044
Luff, P. 7-172

Name Index

Lum, Y. 14-035
Lynch, A. 6-239
Lynd, W. 6-073
Lyon, E. 11-060
Lyon, M. 14-087
Lyons, E. 4-154
Lyons, F. 2-006
Lyons, N. 4-245

Mabee, C. 5-189
Mabon, P. 4-119
MacArthur, E. 11-032
MacCafferty, M. 13-028
Macdonald, B. 4-195
MacDonald, D. 5-028
MacDonald, H. 8-012
MacDonald, S. 11-071
Macintosh, J. 6-202
Mackay, J. 5-062
Mackenzie, C. 5-063, 10-033
Maclaren, M. 4-064
MacLeod, M. 5-165
MacNamara, T. 12-097
MacPherson, A. 4-065
Maddox, B. 14-006
Mahoney, S. 13-047
Maier, J. 7-017
Malcolm, H. 7-149
Malek, M. 1-055
Malia, T. 4-044
Malik, R. 13-077
Maloff, I. 12-048
Malone, M. 11-076
Malter, L. 12-025
Mance, H. 3-093
Manley, H. 1-033
Mann, F. 4-127, 5-074
Mann, R. 6-171
Marco, G. 10-034
Marcoarti, A. 6-192
Marconi, D. 5-166
Marconi, G. 5-022, 5-037, 5-155–168a, 8-022, 8-032, 9-005, 9-007, 9-009, 9-015, 9-024, 9-028, 9-069, 9-071–074, 9-086, 9-152, 11-003, 15-038
Marland, E. 3-043
Marlow, E. 10-120
Marriott, R 9-075
Marsden, C. 13-120
Marshall, 3-137
Marsten, R. 14-043
Martin, D. 14-044
Martin, J. 13-078, 14-007–009
Martin, M. 13-029
Martin, S. 4-005
Martin, T. 5-101, 5-216
Martin, W. 7-173
Marty, D. 10-035
Marvin, C. 3-044
Masini, G. 5-167

Mason, R. 4-164
Massa, F. 10-010
Massie, W. 9-042
Masters, K. 7-107
Mattison, I. 6-050
Maver, W. 6-076, 6-080, 6-081, 6-095, 9-016, 9-021, 9-022
Maverick, A. 6-176
Maxwell, J. 5-028, 5-169–183, 8-005–007
Mayes, T. 4-063
Maynard, K. 1-011
Mazzotto, D. 9-034
McCabe, E. 10-140
McCarthur, T. 5-053
McCarthy, M. 3-079a
McCarthy, T. 4-140
McChesney, R. 4-045
McClenachan, C. 6-193
McConnell, K. 13-029a
McDonald, E. 4-181
McDonald, P. 5-122
McGee, J. 5-203, 5-210, 5-211, 11-044–047
McGee, W. 13-048
McGillem, C. 3-045
McGinn, R. 10-066
McGregor, J. 3-102
McGregor-Morris, J. 5-126, 11-048
McKay, H. 12-039
McKendrick, J. 10-036
McKinsey & Co. 4-171
Mclaurin, W. 9-108
McLauchlan, W. 3-045
McLean, D. 10-121–123
McMahon, A. 4-085
McMeal, H. 7-057
McMean, S. 7-058
McNicol, D. 1-056, 6-095, 6-096, 9-109
McPhail, T. 3-080
McProud, C. 10-067
McVoy, D. 13-057
Meador, J. 10-037
Meadows, J. 3-037
Megaw, E. 11-061
Mehr, L. 1-040
Meinel, H. 8-033
Melick, D. 7-136
Merritt, J. 6-240
Meucci, A. 5-184–5-186
Meulstee, L. 3-120
Meurling, J. 4-203, 14-020
Meyer, R. 6-164, 7-174
Michaelis, A. 4-016
Miehling, R. 10-090
Millard, A. 5-105, 10-038
Miller, D. 5-012
Miller, J. 4-133, 4-134
Miller, K. 7-030, 7-059
Mimms, F. 14-079
Misray, J. 14-001

Name Index

Mitchell, O. 10-039
Moir, J. 3-121
Molloy, E. 1-034, 12-029
Moncton, C. 9-043
Monforte, J. 13-098
Montjoy, R. 7-175
Mooney, M. 10-068
Moore, C. 1-011a
Moore, E. 9-023
Moore, R. 9-171, 9-172
Morat, A. 6-206
Moreau, L. 3-122
Morgan, J. 3-046
Morgan, P. 1-057
Morita, A. 4-246, 4-248
Morris, P. 11-077
Morris, R. 7-137
Morrisey, J. 5-041a
Morrison, P. 5-142
Morse, A. 9-076
Morse, E. 5-191
Morse, S. 5-026, 5-037, 5-187–194, 6-009, 6-019, 6-041, 6-044, 15-039
Morton, D. 3-009a, 10-069, 10-070
Morton, G. 12-034, 12-109
Morus, I. 6-118, 6-165
Mosco, V. 13-079
Moseley, S. 5-054, 12-012, 12-039, 12-049
Mothersole, P. 12-150
Motorola, 15-024
Mott, R 9-124
Mottelay, P. 1-012
Moudin, R. 7-060
Mouromtseff, I. 11-005, 11-033
Moyal, A. 3-081, 4-250
Moyer, A. 5-133
Moyer, C. 6-241
Mueller, M. 7-087
Mueser, R. 4-120
Mullaly, J. 6-169, 6-178
Mullard, S. 11-034
Muller-Fischer, E. 9-110
Muller-Romer, F. 13-099
Mulligan, J. 5-143
Mullin, J. 10-071
Mumford, A. 12-119
Munro, J. 5-029
Murray, J. 4-184, 6-097
Muscio, W. 9-125
Myers, L. 12-040
Mylonadis, Y. 10-133

Nakagawa, Y. 6-119
Nalder, R. 3-123, 3-124
Nasrallah, W. 4-007
Neal, W. 6-250
Nebeker, F. 5-030
Neering, R. 6-137
Nelson, M. 3-103
Nese, M. 5-185

Neuhauser, R. 11-049
Newman, D. 14-080
Newman, J. 13-049
Newman, P. 4-227
Newville, L. 10-040
Niblock, M.13-123
Nichols, R. 7-150
Nicholson, L. 4-251
Nickelson, R. 13-124
Nicol, D. 6-079
Nicolay, H. 5-193
Nicotra, F. 5-185
Nielsen, H. 10-054
Nier, K. 6-113
Niven, W. 5-180
Noam, E. 3-051
Noble, D. 14-021
Noll, A. 3-049, 4-121. 7-162
Norman, B. 12-126
Nye, D. 4-135

O'Brien, J. 3-139
O'Brien, R. 12-127
O'Conell, J. 10-072
O'Day, W. 9-111
O'Dell, T. 9-152
O'Hara, J. 5-031
O'Leary, T. 13-100
O'Meara, W. 7-151
O'Neill, J. 5-217
Occomore, D. 7-138
Ogle, E. 7-139
Olson, H. 10-011
Orr, J. 10-069
Osborn, E. 12-030
Oslin, G. 3-048, 4-176
Osman, D. 12-060
Ososke, D. 10-060
Osterman, F. 9-173
Oswald, A. 8-024
Owen, W. 7-039

Pacent, L. 9-087
Paetsch, M. 14-022
Page, A. 4-103
Paglin, M. 4-028
Paine, A. 4-104
Pangborn, J. 6-077
Park, D. 7-108
Parkinson, J. 6-207
Parr, G. 11-050, 11-051
Parsons, F. 4-177
Passer, H. 4-066
Patchett, 12-141
Patten, W. 7-140
Paul, F. 10-012
Paul, G. 10-023
Pawley, E. 4-196
Pearson, G. 11-078
Pederson, R. 7-074

Pehrson, B. 6-132
Pelton, J. 4-019, 14-010
Percy, J. 5-055
Perera, T. 6-251
Perkins, W. 6-098
Perucca, E. 5-033
Pestre, D. 8-047
Peterson, M. 13-030
Peterson, R. 4-207
Petrakis, H. 4-150
Philip, C. 12-013
Philips, A. 4-230
Philips, F. 4-232
Phillips, M. 13-068
Phillips, V. 5-071, 9-077
Philson, T, 6-005
Pick, J. 11-004
Pickard, G. 11-079
Pickworth, G. 9-078, 9-153, 9-154
Pier, A. 4-105
Pierce, G. 9-079
Pierce, J. 3-049, 6-206, 14-045, 14-046–048
Piggot, W. 6-194
Pilnick, C. 13-069
Pitkin, W. 10-091
Pleasance, C. 7-088
Plum, W. 3-140
Pocock, R. 3-159, 4-067
Poincare, H. 8-007
Poland, J. 2-001
Pole, W. 4-237
Pond, C. 6-038a
Pool, I. 3-050, 3-051, 3-105, 7-089, 7-090
Poole, J. 7-018
Pope, F. 5-134, 6-047
Popov, A. 5-195–197
Porter, R. 5-014
Poulson, V. 9-058
Povey, P. 7-176
Powell, J. 1-013
Pratte, A. 5-114
Preece, W. 5-198–199, 6-057, 6-166, 6-191, 7-017, 7-020
Prentiss, S. 13-125
Prescott, G. 6-031, 6-059, 7-006, 7-015
Preston, S. 12-099
Price, A. 3-141
Pricha, W. 5-031
Prime, S. 5-194
Pritchard, W. 14-049
Progress, P. 6-006
Prosise, M. 9-174
Pupin, M. 5-200–201, 7-070
Purington, E. 9-117

Queisser, H. 11-080

Raby, O. 5-118
Radovsky, M. 5-196

Rae, F. 6-056
Ragazzini, J. 5-042
Raines, R. 3-142
Ramirez, R. 9-174
Ramsay, J. 8-034
Randall, J. 11-055, 11-062
Randell, W. 3-052
Ranger, R. 13-031–033
Ratzlaff, J. 5-218–220
Raucher, E. 5-221
Raytheon, 15-026
Read, L. 5-168
Read, O. 10-043, 10-073
Reader, W. 4-091
Rees, D. 14-050
Reeves, A. 8-016
Reich, L. 4-136
Reid, J. 5-035
Reid, R. 14-092
Reid, T. 11-081
Reingold, N. 5-135
Reis, J. 5-202
Reiss, E. 10-044
Rettenmeyer, F. 1-014
Reynolds, K. 10-124a
Reyner, J. 12-024, 12-050
Reynolds, F. 13-034
Rhoads, S. 6-099
Rhodes, F. 5-079, 7-061
Rice, C. 10-013
Rice, J. 13-126
Rice, P. 10-141
Rice, R. 13-050
Richards, W. 6-027
Ridenour, L. 11-082
Rienzi, T. 3-144
Rigby, C. 3-104
Riordan, M. 11-083
Rippey, J. 7-109
Risdon, P. 12-031
Ritchie, M. 12-101
Roberts, A. 6-100
Robertson, A. 13-051
Robertson, J. 7-142
Robinson, E. 12-032
Robinson, J. 6-187
Robison, S. 9-035
Rockett, F. 11-084
Rodrique, G. 11-063
Roess, A. 4-008
Rogers, D. 4-069
Rogers, E. 11-085, 13-052
Rohwer, J. 3-125
Roizen, J. 10-125
Romnes, H. 7-062
Ronalds, F. 1-015, 6-001
Rosen, H. 14-051
Rosen, P. 4-046
Rosenbloom, R. 4-109, 10-133, 10-142
Rosenthal, E. 9-128

Rossi, J. 4-157
Rothenberg, M. 5-135
Routledge, R. 6-086
Rowett, W. 6-179, 6-196
Rowland, J. 5-056
Rowland, W. 3-053
Rowlands, P. 5-151
Roysdon, C. 5-015
Ruhmer, E. 9-044
Russell, R. 6-019
Russell, W. 6-197
Rutkowski, A. 4-013, 4-014
Rutland, D. 9-175
Rycart, G. 7-165
Ryder, J. 4-086

Sabine, R. 6-042, 6-075
Saffady, W. 13-127
St. John, T. 9-036
Salvaggio, J. 13-053
Samson, 6-063
Sanjak, R. 4-068
Sarnoff, D. 4-152. 4-153. 4-154, 4-158
Sauer, G. 6-138
Savage, J. 4-017
Saward, G. 6-189, 6-210
Sawyers, D. 3-020
Schaffner, T. 6-023, 6-029
Schairer, O. 4-159
Scheips, P. 3-126
Schenck, H. 6-242, 7-152
Schiavo, G. 5-186
Schiffer, M. 9-176, 9-177
Schmeckebier, L. 4-047
Schmidbauer, M. 13-084
Schmidt, S. 13-034a, 13-080a
Schneider, A. 10-126
Schneiderman, R. 14-023
Schoenbert, I. 5-203
Scholtz, R. 8-048
Schreiner, G. 3-094
Schroeder, P. 3-165
Schubert, P. 9-112
Schweitzer, E. 8-035
Schwendler, L. 6-051, 6-064
Schwock, J. 9-127
Scott, J. 4-204, 4-238
Scott, O. 4-163
Scott, R. 6-243
Scowan, F. 7-143
Scriven, G. 3-145
Scroggie, M. 12-033
Scudder, S. 2-007
Secor, H. 12-003
Secrest, J. 4-082
Secunda, E. 10-120
Seel, P. 13-112
Sefl, A. 10-025
Segaler, S. 14-093
Seifer, M. 5-222

Seitz, F. 3-010, 11-086
Semsel, C. 1-037
Sengupta, D. 5-072
Sewall, C. 9-017
Sexton, D. 1-016
Sharp, C. 11-035
Sharpe, B. 6-190
Shaw, T. 7-063
Shearer, S. 7-110
Sheldon, H. 12 008
Sheperd, F. 9-128
Shiers, G. 1-017, 1-018, 5-057, 5-075,
  6-120, 7-091, 9-080, 11-022,
  12-071–074, 12-128
Shima, S. 10-143
Shimura, S. 7-153
Shiraishi, Y. 10-144
Shore, H. 13-008
Shortelle, D. 1-027
Shridharani, K. 6-167
Sibley, L. 11-005a
Siegel, N. 12-058
Siemens, 15-032
Siemens, A. 1-065
Siemens, C. 4-233
Siemens, G. 4-239
Siemens, W. 4-235, 4-237, 4-240, 4-241,
  4-242
Siemens, We. 4-240, 4-241
Siemens, Wi. 4-242
Sievers, M. 9-178
Sigel, E. 10-145
Silverstone, R. 13-054
Simonds, W. 7-111
Simons, R. 5-168a
Simonton, J. 6-209
Simpson, T. 5-181
Sinclair, D. 4-138
Singleton, T. 12-102
Sivewright, J. 6-057, 6-060
Sivowitch, E. 9-113, 10-014
Skinner, W. 4-069
Slater, E. 4-212
Sleeper, M. 9-137
Sleeper, W. 12-059
Slide, A. 2-008
Sloan, T. 1-035
Slotten, H. 8-049, 8-050
Smale, J. 13-009
Smith, A. 7-064, 7-065
Smith, D. 14-052
Smith, E. 13-070
Smith, G. 4-106
Smith, J. 6-038
Smith, N. 9-179
Smith, R. 3-082, 13-071, 13-072
Smith, W. 6-211, 6-214, 6-216
Smith-Rose, R. 8-025, 11-023
Smits, J. 13-111
Snow, M. 4-025

Name Index

Snyder, R. 10-108
Snyder, T. 3-148
Snyder, W. 4-032
Sobel, R. 4-145, 4-160
Solomon, L. 7-154
Solomon, R. 1-019, 3-105
Soresini, F. 7-177
Southwick, T. 13-073
Southworth, G. 4-122
Sova, H. 2-009
Spencer, K. 1-011a, 10-074
Sponable, E. 10-092
Springer, K. 13-101
Spurge, L. 4-149
Squire, G. 3-054
Stamp, G. 7-178
Standage, T. 6-121
Stanton, J. 10-127
Steneck, N. 8-036
Sterling, C. 1-001, 1-020, 1-021, 1-066, 3-065,
Sterling, G. 9-120
Stevenson, O. 5-064
Still, A. 3-055
Stillerman, R. 3-020
Stokes, J. 11-007
Stokowski, L. 10-066
Stone, J. 5-205
Story, A. 9-025
Stow, R. 13-133
Stranger, R. 1-034, 12-015
Straubhauer, J. 10-132
Strowger Telecoms. Page, 15-050
Stubblefield, N. 5-206–208
Stubbs, A. 7-020
Stumpers, F. 3-070
Sturmey, S. 9-114
Sudalnik, J. 13-129
Sugaya, H. 10-128, 10-129
Sullivan, T. 13-017
Sumner, C. 6-067
Sunier, J. 10-109, 10-110
Surtees, L. 7-144
Susskind, C. 5-073, 5-197, 8-003
Swierstra, N. 9-081
Swift, J. 6-071, 12-103
Swinton, A. 5-209–212, 12-104–105

Tabbush, V. 4-149
Taggart, D. 1-022
Taltavall, J. 5-036
Tannenbaum, M. 3-013
Tarbell, I. 4-137
Tarr, J. 6-122
Taussig, C. 9-082
Taylor, A. 3-166
Taylor, L. 7-066
Taylor, P. 6-168
Taylor, W. 5-136
Tebo, J. 12-069

Tedeschi, A. 4-126
Teer, K. 12-106
Tegg, W. 6-065
Terman, F. 9-121
Tesla, N. 5-213–224, 15-040–041
Theiss, J. 7-067
Thiele, H. 10-075
Thiessen, A. 4-139
Thom, C. 6-082
Thompson, M. 11-030
Thompson, R. 6-101
Thompson, S. 5-202, 9-003
Thompson, W. 5-037, 5-178, 6-203
Thrower, K. 9-156, 11-009
Tibbetts, J. 10-094
Tiffany, W. 4-080
Tiltman, R. 5-058, 12-004
Tinkham, R. 10-111
Tirman, W. 13-130
Tissot, C. 9-037
Tolstoy, I. 5-182
Tomlinson, J. 4-018
Tooker, R. 11-073
Torick, E. 10-112
Tosiello, R. 4-107
Towers, W. 5-037
Townsend, F. 11-052
Tran, H. 14-016
Trevert, E. 9-010
Tribolet, L. 3-096
Tricker, R. 5-183
Tucker, D. 5-199, 7-157
Tucker, J. 5-204
Turnbull, L. 6-017, 6-019, 6-020
Turner, H. 3-127
Tuttlebee, W. 13-102
Tydeman, J. 13-081
Tyers, P. 12-044
Tyler, K. 12-143
Tyler, V. 7-112
Tyndall, 5-026
Tyne, G. 11-010

Udelson, J. 12-130
Uges, H. 7-125
Ulriksson, V. 6-123
Umemoto, M. 10-130a
Underhill, C. 9-042
Ungar, J. 14-055
Ungerer, H. 3-097

Vail, A. 6-003
Vail, J. 6-003
Vail, T. 4-100, 4-104, 7-069
Van Choate, S. 6-198
Van-den Ende, J. 12-075
Van der Bijl, H. 11-024
Van Deusen, E. 3-022
Van Horn, L. 14-063
Van Praag, P. 10-076

Van Rensselaer, C. 6-181
Varian, D. 4-166
Varian, R. 4-166, 11-064
Varian, S., 4-166
Varley, C. 6-043
Velie, L. 13-036
Verhulst, S. 13-120
Verrill, A. 9-084
Vigoureux, P. 10-015
Viterbi, A. 13-103
von Tunzelmann, G. 9-008
Vreeland, F. 8-007
Vyvyan, R. 9-115

Wachhorst, W. 5-107
Waddell, P. 5-053
Waldick, L. 7-152
Waldrop, F. 12-051
Walker, A. 10-095
Walker, C. 6-010
Walker, J. 13-082, 13-083
Walker, R. 3-023, 7-136
Walmsley, R. 3-014
Walsh, J. 5-034, 7-113
Walters, H. 7-011
Wander, T. 9-085
Ward, J. 3-105, 13-074
Ward, S. 12-151
Warner, P. 3-128
Wasserman, N. 7-095
Wassiczek, N. 13-132
Wathen, R. 11-066, 11-067
Watson, T. 7-068–069
Weaver, W. 1-023
Webb, F. 5-088
Webb, G. 10-046
Webb, H. 7-022, 7-040, 7-126
Webb, M. 5-119
Webbink, D. 8-062, 8-063
Wedlake, G. 9-116
Weiker, S. 4-243
Weimer, P. 11-053
Weiss, S. 13-133
Welch, W. 10-043, 10-045
Wenger, P. 13-036a
Wenstrom, W. 9-139
Werle, R. 13-034a, 13-080a
Wesson, M. 5-061
West, C. 6-184
West, S. 12-060
Whately, M. 9-086
Wheatstone, C. 5-026, 5-079a–090, 6-076
Whitaker, J. 13-008
White, W. 9-038, 11-011, 11-068
Whitehouse, E. 6-182, 6-225
Whittaker, E 8-004
Whittemore, L. 4-089
Whyte, A. 4-206
Wilbert, W. 10-079
Wilder, G. 7-041

Wile, F. 5-065
Will, T. 4-037
Wilkinson, H. 6-217
Williams, A. 3-056
Williams, F. 3-098, 13-056
Williams, J. 7-035
Williams, R. 4-110, 4-111
Williams, W. 6-078
Williamson, M. 14-064
Wilmotte, R. 13-134
Wilson, A. 3-083
Wilson, C. 7-114, 10-077, 10-078
Wilson, G. 6-018, 6-030, 6-139
Wilson, J. 5-151, 12-045
Wilson, K. 13-087
Wilson, P. 10-046
Wilson, S. 7-026
Wiltse, J. 8-037
Window, F. 6-024, 6-172, 6-185
Winston, B. 3-057, 3-058
Winters, S. 5-153
Witte, E. 4-073
Witts, A. 1-036
Wolff, L. 7-179
Wood, D. 13-114
Wood, J. 3-106. 3-107, 11-037, 14-065
Wood, K. 6-245
Wood, S. 8-180
Woodbury, D. 3-059, 4-180
Woods, D. 3-129, 3-161

Yanczer, P. 12-076
Yarotsky, A. 6-140
Yates, J. 6-124
Yates, R. 9-087, 12-009
Yeudall, B. 1-041
Young, O. 4-129, 4-137, 5-154
Young, P. 4-248, 7-127
Yves, A. 7-180

Zavada, R. 3-066
Zenneck, J. 9-088
Zimmerman, A. 9-129
Zobel, O. 10-051
Zworykin, V. 5-225–227, 11-054, 12-034, 12-107–109, 13-037

Title Index

A.B.C. of Wireless Television, The, 12-001
A.T. and T.: The Story of Industrial Conquest, 4-099
A.W.A. Review, 9-092
A.A. Campbell Swinton, 5-209
ABC of Television, 12-009
ABC of the Telephone, 7-033
ABC of Wireless Telegraphy, 9-010
Account of the Electric Telegraph, 6-004
Account of the Remarkable Applications of the Useful Arts by Mr. Alexander Bain, 6-002
Acharya J.C. Bose, 5-066
Achievement in Radio, 4-032
Acoustical Engineering, 10-011
Adoption of Self-Inductance by Telephony, 1886–1889, 7-083
Advances in Communications, 7-062
Adventure in Vision, 12-103
Adventure into the Unknown, 4-132
Aetheric or Wireless Telegraphy, 9-011
After 10 Years of Satellites, 14-067
Age of Innovation, 3-005
Air Force Communications Command, 3-148
Aircraft Communication in World War I, 3-119
Alexander Graham Bell, 5-063
Alexander Graham Bell, Elisha Gray and the Speaking Telegraph, 5-060
Alexander Graham Bell: The Life and Times, 5-061
Alexander Popov, Inventor of Radio, 5-196
Alexander S. Popov, 5-195
Alexanderson Transoceanic Radio System, 9-048
Alexanderson, 5-038
Alice on the Line, 6-147

Alice Springs Telegraph Station, 6-143
All about Television, 12-003
All about the Telephone and Phonograph, 7-002
All Red Line: Annals and Aims of the Pacific Cable, 6-220
All-American System? Business-Government Relations and the Radio Corporation of America, 1917–1932, 4-157
Allan's System of National Telegraphic Communications, 6-183
Allan's Systems of Inland and Submarine Telegraphy, 6-173
Allocating the Spectrum: The Origins of Radio Regulation, 8-038
Allocation of the Radio Spectrum, 8-041
AM Stereo and the FCC, 10-097
Amateur Radio: An International Resource, 4-080
America Calling: A Social History of the Telephone to 1940, 7-079
America on Record: A History of Recorded Sound, 10-038
American Electro Magnetic Telegraph, 6-003
American Engineers of the Nineteenth Century: A Biographical Guide, 5-015
American Fire-Alarm Telegraph, 6-025
American Leonardo: The Life of Samuel F.B. Morse, 5-189
American Men and Women of Science, 5-002
American Phonograph Design Patents, 1897–1916, 10-037
American Popular Music Business in the 20th Century, 4-068
American Radio Industry and Its Latin American Activities, 1900–1939, 9-127
American Tel and Tel, 4-098a

303

American Telegraph Engineering–Notes on History and Practice, 6-095
American Telegraph Practice, 6-096
American Telegrapher: A Social History, 6-110
American Telegraphy after 100 Years, 6-087, 6-089
American Telegraphy and Encyclopedia of the Telegraph, 6-081
American Telephone Historical Collection, and the Growth of the Historical Collection, 7-171
American Telephone Practice, 7-030
American Women in Science, 5-004
Ampex Corp and Video Innovation, 4-109
Analysis of Submarine Cable and Communication Satellite Systems Reliabilities, 3-092
Anatomy of a Business Strategy: Bell, Western Electric, and the Origins of the American Telephone Industry, 4-106
Anatomy of a Failure: Picturephone Revisited, 7-162
Anatomy of a Transformation, 1985–1995, 4-226
And End to Silence: The Building of the Overland Telegraph, 6-168
And Now—The BBC Presents Television to the World, 12-046
Anecdotal History of Stereophonic Recording, 10-111
Anecdotes of the Telegraph, 6-007a
Annual Report (FCC), 4-048
Annual Report (FRC), 4-050
Annual Report (INTELSAT), 4-020
Annual Survey of Communication Services, 1-070
Antennas: History, Development and Theory, 9-093
Antique Phonograph Gadgets, Gizmos and Gimmicks, 10-023
Anton Philips of Eindhoven, 4-230
Antonio Meucci, 1808–1889, 5-185
Antonio Meucci: Inventor of the Telephone, 5-186
Applied Science and Technology Index, 2-012
Architects of the Net of Nets, 14-084
Architects of the Web, 14-092
Armour Research Foundation and the Wire Recorder, 10-070
Armstrong of Radio, 5-040
Armstrong's Fight for FM Broadcasting, 5-039
Art of Sound Pictures, 10-091
Atlantic and South Atlantic Telegraphs, 6-185
Atlantic Bridgehead, 6-223
Atlantic Cable Mismanagement, 6-209
Atlantic Cable, 6-227

Atlantic Ocean Telegraph from Ireland to Newfoundland, 6-187
Atlantic Telegraph Cable Centennial, 6-233
Atlantic Telegraph Cable, 6-203
Atlantic Telegraph, 6-197
Atlantic Telegraph: A History of Preliminary Experimental Proceedings, 6-171
Atlantic Telegraph: Its History, 6-195, 6-199
Atlantic Telegraph: The Rise, Progress, and Development of Its Electrical Department, 6-182
Atlantic Telegraphs, 6-204
Atwater Kent Radio Development, 4-110
Atwater Kent, Master of Marketing, 4-111
Audio Pioneers, 5-017
Audio! Audio! The British Hi-Fi Spotter's Directory, 9-169
Audion: A New Receiver for Wireless Telegraphy, 11-016
Audion-Detector and Amplifier, 11-017
Australian Radio: The Technical Story, 1923–1983, 9-125
Authorship of the Practical Electric Telegraph of Great Britain, 5-082
Autobiographical and Other Writings, 5-210
Automatic Telephone Exchange, 7-143
Automatic Telephone Systems, 7-044
Automatic Telephony, 7-065

Baird and Television, 5-046
Baird of Television, 5-058
Battery and the Boiler, 6-215
Battle Talk, 3-132
Battlefronts of Industry: Westinghouse in World War II, 4-180
BBC Engineering, 1922–1972, 4-196
BBC Handbook or Yearbook, 4-186
BBC Sound Broadcasting: Its Engineering Development, 4-187
BBC Television Centre, 4-189
BBC Television Service: A Technical Description, 4-190
BBC Television: A British Engineering Achievement, 4-188
Beating the Transistor: Developments in UHF Tube Technology, 11-037
Beehives of Invention: Edison and His Laboratories, 5-100
Beginnings of Satellite Communication, 14-048
Beginnings of Telephony, 7-061
Behind the Front Panel: The Design and Development of 1920s Radios, 9-175
Behind the Tube: A History of Broadcasting Technology, 3-064
Behn Brothers, 4-141
Bell and Gray: Contrasts in Style, Politics, and Etiquette, 5-129
Bell Canada: A Case Study, 7-136

Title Index

Bell Laboratories Innovation in Telecommunications, 1925–1977, 4-120
Bell Laboratories: Inside the World's Largest Communications Center, 4-117
Bell Labs' First 50 Years, 4-113
Bell System and Regional Business: The Telephone in the South, 1877–1920, 7-106
Bell System R&D Activities: The Impact of Divestiture, 4-121
Bell Telephone System, 4-103
Bell Telephone: The Deposition of Alexander Graham Bell, 7-047
Bell: Alexander Graham Bell and the Conquest of Solitude, 5-059
Bell's Electric Speaking Telephone, 7-015
Between the Lines: A Personal History of the British Public Telephone and Telecommunications Service, 1870–1990, 7-137
Beyond Babel: New Directions in Communications, 14-006
Beyond Broadcasting: Into the Cable Age, 13-065
Beyond the Dial: Development of the Netherlands Telephone System, 7-125
Beyond the Ionosphere: Fifty Years of Satellite Communication, 14-026
Bibliographic Guide to the History of Computer Applications, 1950–1990, 1-005
Bibliographic Guide to the History of Computing, 1-004
Bibliographical History of Electricity and Magnetism, 1-012
Bibliography of Bibliographies in Electrical Engineering, 1918–1929, 1-011
Bibliography of Colour Television, 12-132
Bibliography of Magnetic Recording, 1900–1953, 10-061
Bibliography of Radio Wave Phenomena and Measurement of Radio Field Intensity, 8-023
Bibliography of the History of Electronics, 1-017
Bibliography of Theses and Dissertations in Broadcasting, 1-010
Bibliography on Educational Broadcasting, 1-003
Big Picture: HDTV and High-Resolution Systems, 13-131
Biographic Encyclopedia of Science and Technology, 5-003
Biographical Bibliography of Electrical Engineers and Electrophysicists, 5-009
Biographical Dictionary of American Business Leaders, 4-002
Biographical Dictionary of Scientists, 5-014
Biographical Dictionary of Scientists: Engineers and Inventors, 5-001
Biographical Encyclopedia of Scientists, 5-006
Biographical Index to American Science, 5-008
Biographical Memoirs, 5-013
Biographies of Engineers and Scientists, 5-010
Biography Index, 5-005
Birth and Babyhood of the Telephone, 7-068
Birth and Early Years of the Bell Telephone System, 1876–1880, 4-107
Birth of a High Definition Television System, 12-099
Birth of Broadcasting, 9-081
Birth of Helical Scan Videotape Recording, 10-119
Birth of Long Distance Communication: Semaphore Telegraphs in Europe (1790–1840), 6-130
Birth of the German Magnetophon Tape Recorder, 1928–1945, 10-060
Birth of the Talkies, 10-085
Birth of Video Recording, 10-118
Blind Telephonist: Technology and Organization of a Vocation for the Blind, 7-050
Blue Book and Ma Bell, 7-080
Book of Electrical Wonders, 3-008
Book of Practical Television, 12-027
Book of Radio, 9-082
Book of the Telegraph, 6-012
Boston Post Book on Television, 12-005
Bottom of the Atlantic and the First Laying of the Electric Telegraph Cable, 6-186
Boyhood of an Inventor, 5-145
Brass Pounders: Young Telegraphers of the Civil War, 3-135
Bringing Information to People, 14-012
Britain and Commonwealth Telecommunications, 3-075
British Broadcasting, 1922–1982: A Selected and Annotated Bibliography, 4-194
British Radio and Television Pioneers: A Patent Bibliography, 9-150
British State Telegraphs, 6-164
British Television: The Formative Years, 12-113
Broadcasting at 50: Can It Adapt? 1-042
Broadcasting from Space, 14-070
Broadcasting in Britain: 1922–1972–A Brief Account of Its Engineering Aspects, 4-193
Broadcasting in the United Kingdom: A Guide to Information Sources, 4-195
Broadcasting Network Service, 9-094
Broadcasting Technology–Past, Present & Future, 3-061
Broadcasting with Digital Audio, 13-094
Building a Digital Network: Data Communications and Digital Telephony, 1950–1990, 14-003

Title Index

Building Television Receivers at Home, 12-060
Business Demand the Development of the Telegraph in the United States, 1844–1860, 6-106
Business Telematics, 14-001

C&P Story: Service in Action–Washington, D.C., 7-100
Cable & Station Coverage Atlas, 1-078
Cable & Wireless: A History, 4-201
Cable and Wireless Communications of the World, 3-067
Cable and Wireless Communications of the World: Some Lectures and Papers on the Subject, 1924–1939, 3-084
Cable Communication, 13-057
Cable Operator's Handbook, 13-060
Cable Television, 13-066
Cable Television for Europe, 13-059
Cable Television Technology and Operations, 13-058
Cable Television: A Comprehensive Bibliography, 13-061a
Cable Television: A Guide to the Technology, 13-069
Cable: The First Forty Years, 13-061
Cables and Wireless, 3-094
Cableship Characteristics, 6-242
Cableships and Submarine Cables, 6-234
Calling CQ: Adventures of Short-Wave Radio Operators, 4-077
Calling the World: The First 100 Years of Alcatel Australia, 1895–1995, 4-184
Cambridge Dictionary of Scientists, 5-012
Campbell Swinton and Television, 5-211
Canadian Contributions to Telecommunications, 3-080
Carrier Current Telephony and Telegraphy, 3-028
Catalogue of Books and Papers Relating to Electricity, Magnetism, the Electric Telegraph, 1-015
Catalogue of Scientific and Technical Periodicals, 1665–1895, 2-002
Catalogue of Scientific Serials of All Countries, 2-007
Catalogue of the Wheeler Gift of Books, Pamphlets and Periodicals in the Library of the American Institute of Electrical Engineers, 1-023
Cathode Ray Revolution: Foundations of Electronics, 11-004
Cathode-Ray Tubes, 11-038
CATV: A History of Community Antenna Television, 13-068
Cavalcade, 4-216
Cavity Magnetron, 11-055, 11-062
CCIR 50th Anniversary Special, 4-010

Cellular Communications: Worldwide Market Development, 14-017
Centennial of the Semiconductor Diode Detector, 5-072, 9-146
Century of Microphones, 10-001
Century of Service: A Brief History of Cable & Wireless Ltd., 1868–1968, 4-199
Century of Wireless, 9-089
Century One...A Prologue, 4-168
Changes, Challenges and Opportunities in the New Electronic Media, 13-048
Changing Picture in Video Tape for 1959–1960, 10-117
Channels Field Guide, 13-041
Chip: How Two Americans Invented the Microchip and Launched a Revolution, 11-081
Chronological History of Electrical Communication–Telegraph, Telephone and Radio, 1-056
Chronological History of Electrical Development from 600 B.C., 1-044
Chronological Table of the History of Radio Engineering, 9-110
Chronology of Development of Wireless Communications and Electronics, 1-055
Chronology of Television, 1-046
Chronology: 25 Years of Radio, 1-045
Cinderella in the Sky, 14-074
Circuits of Victory, 3-136
City and the Telegraph: Urban Communications in the Pre-Telephone Era, 6-122
City of Light: The Story of Fiber Optics, 14-077
City of Sound, 4-220
Classic TVs, 12-151
Classification of Electron Tubes, 11-001
Classified Bibliography of Publications on the History and Development of Electrical Engineering and Electrophysics, 1-008
Clear Across Australia: A History of Telecommunications, 4-250
Clerk Maxwell and Modern Science: Six Commemorative Lectures, 5-171
Coaxial Transmission Lines, Related Two-conductor Transmission Lines, Connectors, and Components: A U.S. Historical Perspective, 8-028
Coherer-Based Radio, 9-154
Collecting Old Radios and Crystal Sets, 9-157
Collector's Guide to Antique Radios, 9-159
Collector's Guide to Transistor Radios, 9-160
Color Television, 12-138
Color Television Standards, 12-136
Color Television Tubes: A History but Not Yet an Obituary, 11-049
Color TV: Who'll Buy a Triumph?, 12-131

Title Index

Colorado Springs Notes and Commentary, 1899–1900, 5-224
Colour 69: An Advertiser's Guide to Colour Television, 12-133
Colour Television, 12-141
Colour TV, 12-137
Come Quick, Danger–A History of Marine Radio in Canada, 3-152
Coming of Sound: Technological Change in the American Film Industry, 10-086
Commander McDonald of Zenith, 4-181
Commissioners of the FCC, 1927–1994, 4-041
Communication Applications of Millimeter Waves: History, Present Status, and Future Trends, 8-033
Communication Booknotes Quarterly, 1-020
Communication in the Space Age: The Use of Satellites by the Mass Media, 14-072
Communication Satellite Systems, 14-036
Communication Satellites, 1958–1995, 14-044
Communication Satellites, 14-047
Communication Satellites: Technical, Economic, and International Developments: Staff Report, 14-060
Communication Satellites: Their Development and Impact, 14-037
Communication Serials, 1992–93 Edition, 2-009
Communication Technology and Social Policy, 13-043
Communication Technology Update, 13-044
Communication Technology: The New Media in Society, 13-052
Communication through the Ages, 3-055
Communication Towers: New Architecture, 3-068
Communication via Satellite, 14-052
Communication: Stories of Man's Achievements, 3-059
Communications and the United States Congress: A Selectively Annotated Bibliography of Committee Hearings, 1870–1976, 1-002
Communications for a Mobile Society, 14-011
Communications in Space, 14-040
Communications in Space: From Marconi to Man on the Moon, 3-032
Communications Media: Properties and Uses, 13-040
Communications Miracle: The Telecommunication Pioneers from Morse to the Information Superhighway, 5-018
Communications Outlook, 1-060
Communications Receivers: 1932–1981, 9-171
Communications Revolution, 13-056

Communications Satellite, 14-064
Communications Satellites, 14-039, 14-062
Communications Satellites: A Monitor's Guide, 14-063
Communications Technology Update, 3-037
Communications-Electronics, 1962–1970, 3-144
Comparative Analysis of the Early History of Southern and Northern Telegraph Systems, 6-116
Compatibility for UHF Television: Final Report, 8-061
Compatibility Standards, Competition, and Innovation in the Broadcasting Industry, 3-060
Compendium of Communication and Broadcast Satellites, 1958 to 1980, 14-025
Competition and Technical Change in the Television Industry, 12-112
Competition in a Network Industry: The Telephone Industry, 1894–1910, 7-081
Compleat Talking Machine: A Collector's Guide to Antique Phonographs, 10-044
Complete Encyclopedia of Space Satellites: Every Civil and Military Satellite of the World since 1957, 14-027
Complete Radio Book, 9-087
Complete Wireless, 1-034
Component Innovation: The Case of Automatic Telephone Switching, 1891–1920, 7-086
Computer-Based Analysis and Restoration of Baird 30-line Television Recordings, 10-121
Comsat at 15, 4-124
Comsat Guide to the INTELSAT Satellite System, 4-021
Connecticut Pioneers in Telephony, 7-113
Conqueror of Space: An Authorized Biography of the Life and Work of Lee de Forest, 5-092
Conquest of Distance by Wire Telephony, 7-063
Conquest of the Microchip, 11-080
Constitution Faces Technology: The Relationship of the National Government to the Telegraph, 1866–1884, 6-115
Consumer Digital Radio: From Concept to Reality, 13-102
Consumer Electronics Industry in Review, 1-061
Consuming Technologies: Media and Information in Domestic Spaces, 13-054
Contact at Sea: A History of Maritime Radio Communications, 3-165
Continental Dash: The Russian-American Telegraph, 6-137
Continuous Loop Magnetic Tape Cartridge, 10-055

Title Index

Continuous Wave: Technology and American Radio, 1900–1932, 9-090
Contribution of A.A. Campbell Swinton to Television, 5-212
Contributions of Faraday and Maxwell to Electrical Science, 5-183
Contributions of the Bell Telephone Laboratories to the Early Development of Television, 12-063
Control Revolution: Technological and Economic Origins of the Information Society, 3-018
Convergence in European Digital TV Regulation, 13-120
Cooke and Wheatstone and the Invention of the Electric Telegraph, 5-087
Craig's Manual of the Telegraph, 6-066
Creating an Industry, 12-078
Creating the Craft of Tape Recording, 10-071
Creation of Scientific Effects: Heinrich Hertz and Electric Waves, 5-140
Creative Ordeal: The Story of Raytheon, 4-163
Creativity in Radio: Contributions of Major Edwin H. Armstrong, 5-042
Crossed Wires and Missing Connections: Valdemar Poulson, the American Telegraphone Company and the Failure to Commercialize Magnetic Recording, 10-054
Cruise of the "Agamemnon" Extracted from the "Times" Newspaper, 6-175
Crystal Clear: Vintage American Crystal Sets, Crystal Detectors, and Crystals, 9-178
Crystal Fire: The Birth of the Information Age, 11-083
Cultural Imperatives and Product Development: the Case of the Shirt Pocket Radio, 9-177
Cumulative Index to Entire IEEE Group, 1951–1971, 2-013
Current Bibliography in the History of Technology, 2-014
Current Bibliography of the History of Science, 2-015
Current CCIR Activities in HDTV, 13-117
Current Contents: Engineering, Technology & Applied Sciences, 2-016
Cyclopedia of Electrical Engineering, 1-025
Cyclopedia of Television Facts, 12-052
Cyrus Field: Man of Two Worlds, 5-120
Cyrus W. Field, His Life and Work, 5-121

DAB Progress Report–1997, 13-099
DAB–Is It Already out of Date?, 13-096
Daniel Drawbaugh: The Edison of the Cumberland Valley, 5-096
Dates in American Telephone Technology, 1-052
David Sarnoff: A Biography, 4-154
Dawn of the Digital Age, 10-009
De Forest and the Triode Detector, 11-015
Decade in the History of the English Telegraph, 6-155
Decade of Electroacoustic Reproduction (1920–1930), 10-012
Decade of Progress in the Use of Electronic Tubes, 11-031
Dedication of the Statue of Philo T. Farnsworth, 5-110
Deep Sea Telegraphs: Their Past History and Future Progress, 6-189
Defining Vision: The Battle for the Future of Television, 13-108
Demon in the Aether: The Story of James Clerk Maxwell, 5-174
Description of the American Electro Magnetic Telegraph, 6-003
Descriptions of an Electric Telegraph, 6-001
Design and Implementation of the ATSC Demonstration of HDTV at NAB'97, 13-116
Desk Telephones of the Bell System, 7-179
Detailed Report of the Successful Laying of the Atlantic Telegraph, 6-193
Development and Diffusion of Telephone Technology in Britain, 1900–1940, 7-133
Development and Evolution of the Telephone Service, 7-119
Development of American Industries, 4-060
Development of Early Practical Electromagnetic Telegraphs and the Mechanization of Skilled Operation, 6-119
Development of Photo-Telegraphy, 13-027
Development of Submarine Cable Communications, 6-230
Development of Telephonic Communication in Copenhagen, 1881–1931, 7-123
Development of the First Optical Videodisc, 10-141
Development of the Microphone, 10-004
Development of the Phonograph at Alexander Graham Bell's Volta Laboratory, 10-040
Development of the Receiving Valve, 11-034
Development of the Telephone in Europe, 7-126
Development of the Telephone in Oxford, 1877–1977, 7-131
Development of Valves for Wireless, 11-030
Development of Wireless Telegraphy, 9-018
Development of Wireless to 1920, 9-080
Developments in Telephotography, 13-022
Diamond Jubilee Edition [of Electrical Engineering], 3-004
Diamond Jubilee: 1909–1984, Seventy-Fifth Anniversary, 7-075
Diary Kept by Rasmus Petersen Aboard the S.S. Great Northern, 4-207
Dictionary of Business Biography, 4-004

Title Index

Dictionary of Electrical Words, Terms, and Phrases, 1-029
Dictionary of Scientific Biography, 5-007
Did He Invent Radio? 5-208
Digital Audio Broadcasting, 13-090
Digital Audio Broadcasting: Status Report and Outlook, 13-089
Digital Audio Broadcasting: United States Technologies and Systems, Terrestrial and Satellite, 13-088
Digital Cellular Radio, 14-014
Digital Radio Takes Off, 13-092
Digital Reproduction of Sound, 13-098
Digital Television Here at Last, 13-119
Digital Television Recording, History and Background, 10-115
Digital Video Broadcasting, 13-111
Digital Video Recording, 10-130, 10-130a
Directory of Telephone Museums, 1-041
Discovery of the Oscillating Crystal, 11-079
Disk Recording and Reproduction, 10-016
Distant Electric Vision, 11-045
Distant Signals: How Cable TV Changed the World of Telecommunications, 13-073
Distant Vision, 5-112
Documents in American Telecommunications Policy, 4-031
Dr. E.O.W. Whitehouse and the 1858 Trans-Atlantic Cable, 6-225
Dr. Nikola Tesla Bibliography, 5-220
Dr. Nikola Tesla: Complete Patents, 5-218
Dr. Nikola Tesla: Selected Patent Wrappers from the National Archives, 5-219
Drake's Radio Cyclopedia, 1-033
Dream Come True: Satellite Broadcasting, 14-043
Dynamical Theory of the Electromagnetic Field, 8-006
Dynamics of Modern Communication: The Shaping and Impact of New Communication Technologies, 3-034

Early British Radio Industry, 4-067
Early Days of Television, 12-066
Early Days: Some Recollections, 5-227
Early Electrical Communication, 3-043
Early Experiences with the ARPANET and Internet in the United Kingdom, 14-090
Early History of Data Networks, 6-132
Early History of Radio from Faraday to Marconi, 5-022
Early History of Radio Interference, 9-105
Early History of Single-Sideband Transmission, 8-024
Early History of the Automatic Telephone, 7-064
Early History of the Electro-Magnetic Telegraph from Letters and Journals of Alfred Vail, 6-003
Early Radio Wave Detectors, 9-155
Early Radio: In Marconi's Footsteps, 1894 to 1920, 9-065
Early Schemes for Television, 12-071
Early Television: A Bibliographic Guide to 1940, 1-018
Early Wireless, 9-097, 9-162
Early Years of Oceanic Telegraphy: Technology, Science, and Politics, 6-235
Easy Lessons in Television, 12-011
Economic Development of Radio, 9-114
Edison and the Business of Innovation, 5-105
Edison Cylinder Phonographs, 10-025
Edison Effect and Its Modern Applications, 11-035
Edison, 5-104
Edison, Musicians, and the Phonograph, 10-029
Edison: A Life of Invention, 5-102
Edison: His Life and Inventions, 5-101
Edison: Inventing the Century, 5-097
Edison: The Man Who Made the Future, 5-098
EIA 50: Electronic Industries Association: The First Fifty Years, 4-082
1888–1988: A Hundred Years of Magnetic Sound Recording, 10-058
Electric Ariel: Telegraphy and Commercial Culture in Early Victorian England, 6-165
Electric B & O, 6-077
Electric Century, 3-009a
Electric Flashes; or, The Systems of Wireless Telegraphy..., 9-020
Electric Medium: A Pattern for the Early Development of the Electric Telegraph in the United States, 6-247
Electric Tables and Formulae for the Use of Telegraph Inspectors and Operators, 6-075
Electric Telegraph in the U.K.: A Social and Economic History, 6-162
Electric Telegraph Manipulation, 6-010
Electric Telegraph Popularized, 6-026
Electric Telegraph, 6-042
Electric Telegraph: An Historical Anthology, 6-120
Electric Telegraph: Comprising a Brief History, 6-006
Electric Telegraph: Its History and Progress, 6-015
Electric Telegraph: Its History, Theory, and Practical Applications, 6-022
Electric Telegraph: Was It Invented by Professor Wheatstone?, 5-084, 5-085
Electric Telegraph; and the Patented Improvement Thereon, 6-033
Electric Telegraphy, 6-094
Electric Telephone, 7-015, 7-023
Electric Transmission of Intelligence, 3-039

Electric Waves, 8-012
Electric Waves: Being Researches on Propagation, 5-141
Electric Word: The Rise of Radio, 9-112
Electrical and Electronic Technologies: A Chronology of Events and Inventors, 1-047
Electrical and Electronics Abstracts, 2-017
Electrical Cable, 6-205
Electrical Manufacturers, 1875–1900, 4-066
Electrical Recording, 10-051
Electrical Reproduction from Phonograph Records, 10-031
Electrical Transmission of Pictures and Images, 13-020
Electricity and the Electric Telegraph, 6-059
Electricity and the Electric Telegraph, Together with the Chemistry of the Stars, 6-018
Electricity in Every-Day Life, 3-008a
Electricity in the Service of Man, 3-014
Electricity, Magnetism, and Electric Telegraphy, 6-070
Electroacoustics: The Analysis of Transduction and Its Historical Background, 10-007
Electromagnetic Spectrum Utilization–The Silent Crisis, 8-059
Electro-Magnetic Telegraph, 6-020, 6-027
Electromagnetic Telegraphy: Early Ideas, Proposals, and Apparatus, 6-105
Electron Optics in Television, with Theory and Application of Television Cathode-Ray Tubes, 12-048
Electron Optics, 12-056
Electron Optics: An Introduction, 12-040
Electron Tubes, 11-027
Electron Tubes in World War II, 11-029
Electron Tubes, 1930 to 1950, 11-028
Electron, or the Pranks of the Modern Puck: A Telegraphic Epic for the Times, 6-028
Electronic Business, 4-061
Electronic Classics: Collecting, Restoration, and Repair, 7-169
Electronic Delivery of Documents and Graphics, 13-013
Electronic Epoch, 3-001
Electronic Genie: The Tangled History of Silicon, 11-086
Electronic Inventions and Discoveries, 1-048
Electronic Mail: A Revolution in Business Communications, 13-085
Electronic Market Data Book, 1-062
Electronic Media: A Guide to Trends in Broadcasting and Newer Technologies, 1920–1983, 1-066
Electronic Motion Pictures: A History of the Television Camera, 12-110
Electronic Switching: Central Office Systems of the World, 7-120

Electronic Switching: Digital Central Office Systems of the World, 7-121
Electronic Telephone, 7-172
Electronic Television, 12-037
Electronic Warfare, 3-112
Electronics and Communications Abstracts Journal, 2-018
Electronics and Development of Electronic Tubes, 11-005
Electronics and Sea Power, 3-157
Electronics Challenge: An Historical Perspective, 3-006
Electronics in the West: The First Fifty Years, 3-046
Electronics Research in the Space Age, 3-033
Electronics, Computers and Telephone Switching: A Book of Technological History, 7-117
Electronics–A Bibliographical Guide, 1-011a
Electro-Telegraphy, 6-055
Elementary Manual of Radiotelegraphy and Radiotelephony, 9-041
Elementary Treatise on Electrical Measurement, 6-045
Elementary Treatise on Electricity, 5-179
Elephant and the Blind Men: The Phenomenon of HDTV, 13-130
Elisha Gray and the Telephone: On the Disadvantages of Being an Expert, 5-128
Elma Pem Gardner Farnsworth: The Pioneering of Television, 5-114
Emergence of a Mass Market for Fax Machines, 13-030
Emergence of Basic Research in the Bell Telephone System, 1875–1915, 4-102
Emergence of CATV: A Look at the Evolution of a Revolution, 13-070
EMI Cathode-Ray Television Transmission Tubes, 11-044
Emile Berliner, Maker of the Microphone, 5-065
Empire of the Air: The Men Who Made Radio, 5-027
Empire Telegraph Communications, 6-245
Encyclopedia of Recorded Sound in the United States: 1870–1970, 10-034
Engineer's Review of Antonio Meucci's Pioneer Work in the Invention of the Telephone, 5-184
Engineering Index, 2-019
Engineers & Electrons: A Century of Electrical Progress, 4-086
Engines of Democracy: Inventions and Society in Mature America, 3-019
Entrepreneurship in the Early Development of the Telephone, 4-098
Ether and Reality, 8-021
Ether of Space, 8-020
European Perspectives in the Development of HDTV Standards, 13-132

Title Index

Events in Telecommunications History, 1-050, 1-051
Evolution of Circuit Design for AM Broadcast Receivers, 9-156
Evolution of Consumer VTRs–Technological Milestones, 10-143
Evolution of Digital Audio and Video Format Conversions, 10-124a
Evolution of Digital Wireless Technology, 13-103
Evolution of Electrical and Electronic Engineering and the Proceedings of the IRE, 4-087, 4-088
Evolution of Electronics, 11-011
Evolution of Frequency Modulation, 9-131
Evolution of High-Definition Television, 13-113
Evolution of Modern Television, 12-124
Evolution of Naval Radio-Electronics and Contributions of the Naval Research Laboratory, 3-163
Evolution of Picturephone Service, 7-158
Evolution of the Audio Recorder, 10-076
Evolution of the Cathode-Ray Tube: A Survey of Developments over Three Decades, 11-039
Evolution of the Domestic Television Receiver, 12-150
Evolution of the ITU, 4-014
Evolution of the Radio, 9-180
Evolution of the Thermionic Valve, 11-023
Evolution of the Vacuum Tube, 11-025
Evolution of Video Technologies, 12-118
Evolution of Wire Transmission, 7-049
Examination of the Telegraphic Apparatus and the Processes in Telegraphy, 6-044
Exhibiting Electricity, 3-002
Experimental Investigation of the Laws Which Govern the Propagation of the Electric Current in Long Submarine Cables, 6-188
Experimental Researches in Electro-Harmonic Telegraphy and Telephony, 1867–1878, 7-005
Experimental Television, 12-018
Experiments with Stereophonic Records, 10-101
Exploding Political Power of Personal Media, 10-135
Exploring in the Seventies and the Construction of the Overland Telegraph Line, 6-152
Exposition on the Most Improved Telegraph Cable, 6-206
Extel 100: The Centenary History of the Exchange Telegraph Company, 4-204
Extracts from the Private Letters of Sir William Fothergill Cooke, 5-088

Extra-Terrestrial Relays: Can Rocket Stations Give Worldwide Radio Coverage?, 14-029
Eyewitness to Early American Telegraphy, 6-003

Fabulous Phonograph, 1877–1977, 10-027
Face to Face with Technology: The Siemens Museum in Munich, 4-234
Facsimile and Its Future Uses, 13-018
Facsimile Picture Transmission, 13-037
Facsimile Telegraphy and Phototelegraphy, 13-015
Facsimile, 13-017, 13-025, 13-034a
Facsimile: What It Can Do Now, 13-019
Facsimile's False Start, 13-010
Facts and Figures Relative to Submarine Telegraphy, 6-202
Failure Is Not an Option—How MCI Invented Competition in Telecommunications, 4-149
Fall of the U.S. Consumer Electronics Industry, 3-062
Far Eastern Telegraphs, 6-125
Faraday, Maxwell and Kelvin, 5-028
Fast Forward: Hollywood, the Japanese and the VCR Wars, 10-138
Father of Radio, 5-093
Fax and Teletext, 13-028
Fax: Digital Facsimile Technology and Applications, 13-029a
Fax: The Principles and Practice of Facsimile Communication, 13-012
FCC Approves FM Stereophonic Broadcasting, 10-102
FCC Chronology and Leadership from 1934 to 1970, 4-039
Federal Communications Commission, 4-040
Federal Radio Commission: Its History, Activities and Organization, 4-047
Federal Research and Development for Satellite Communication, 14-031
Federal Telephone and Radio Corporation: An Historical Review, 1909–1946, 4-127
Ferdinand Braun and the Cathode-Ray Tube, 5-075
Ferdinand Braun, 5-073
Ferdinand Braun–Inventor of the Cathode-Ray Tube, 5-074
Fessenden–Builder of Tomorrows, 5-116
Fiftieth Anniversary Golden Yearbook [of the Radio Club of America], 4-092
Fiftieth Anniversary Issue [of the IRE], 1912–1962, 3-036
Fiftieth Anniversary of the AIEE, 4-074
50th Anniversary [of the FCC], 1934–1984: A Chronological History, 4-049
Fifty Years of A.R.R.L., 4-078

Fifty Years of Electricity: The Memories of an Electrical Engineer, 5-124
50 Years of Fernseh, 1929–1979, 4-185
Fifty Years of High Definition Television Transmission, 12-121
50 Years of Japanese Broadcasting, 4-224
Fifty Years of Loudspeaker Developments as Viewed Through the Perspective of the 50 Years of Television, 12-117
Audio Engineering Society, 10-005
Fifty Years of Telecommunications Reports, 4-044
Fifty Years Research in Radio Wave Propagation, 8-025
Film Finds Its Tongue, 10-087
Film, Television and Video Periodicals: A Comprehensive Annotated List, 2-005
Final Report and Recommendations, 13-104
Final Report on the Relay 1 Program, 14-061
Final Report [of the Task Force on Communications Policy], 4-035
Fine-Tuning of a Golden Ear: High-End Audio and the Evolutionary Model of Technology, 10-072
First 50 Years: A History of Collins Radio Company, 4-123
First 80 Years [of NEC], 4-222
First Century of Lightwave Communications, 14-079
First Century of Service: Bell 1880–1980, 7-132
First Cross-Channel Telephone Cable: The London-Paris Telephone Links of 1891, 7-157
First Demonstration of Television, 5-047
First Electron Tube, 11-022
First Principles of Television, 12-019
First Scientific Exploration of Russian America and the Purchase of Alaska, 6-133
First Transatlantic Telephone Cable, 7-147
First Twenty Years of HDTV, 13-118
Flight of Speech, 7-101
FM Broadcast Coverage of the Coterminous United States, 1-082
FM Multiplexing: A New Approach to Stereophonic Sound, 10-107
FM Radio Handbook, 9-137
FM Stereo Multiplexing, 10-099
Forbes: Telephone Pioneer, 4-105
Forces at Work Behind the NTSC Standards, 12-084
Forecasting the Telephone: A Retrospective Technology Assessment of the Telephone, 7-090
Forgotten History: Sir Jagadish Bose and the Origin of Radio, 5-069
45 Years with Philips: An Industrialist's Life, 4-232

Forty Years of Electrical Progress: The Story of the G.E.C., 4-206
Forty Years of Radio Research, 4-122
40 Years of the Yukon Telegraph, 6-163
40 Years of Waveguides: A Glimpse at History, 8-035
4000 Years of Television, 12-123
Founder's Touch: The Life of Paul Galvin of Motorola, 4-150
Four Aspects of the Film, 10-089
Four Pioneer Deep-Sea Cables, 6-239
French Optical Telegraphy, 1793–1855: Hardware, Software, Administration, 6-129
Frequency Assignment Methodology: An Annotated Bibliography, 8-040
Frequency Modulation, 9-133a
Frequency Modulation–1922 and 1948, 8-013
Frequency Spectrum Deregulation Alternatives, 8-063
Froehlich/Kent Encyclopedia of Telecommunications, 1-026
From Attics to Ignominy, 12-094
From Coherer to Spacistor, 9-106
From Dots and Dashes to Tele and Datacommunications, 4-208
From Effect to Artifact (II): The Case of the Thermionic Valve, 11-002
From Immigrant to Inventor, 5-201
From Lighthouses to Laserbeams: A History of the U.S. Department of Commerce, 1913–1988, 4-030
From Machine Shop to Industrial Laboratory: Telegraphy and the Changing Context of American Invention, 1830–1920, 6-112
From Radiovision to Video....And Television Between, 12-125
From Semaphore to Satellite, 4-016
From Spark to Satellite: A History of Radio Communication, 9-107
From Spark to Space: A Pictorial Journey through 75 Years of Amateur Radio, 4-079
From Talking Drums to the Internet: An Encyclopedia of Communications Technology, 1-027
From Telegraphy to Television: The Story of Electrical Communications, 3-029
From Television to Home Computer: The Future of Consumer Electronics, 13-051
From Tin Foil to Stereo: Evolution of the Phonograph, 10-043
From Tinfoil to Stereo: The Acoustic Years of the Recording Industry, 1877–1929, 10-045
Fundamentals of Television, 12-010
Future Also Has a Past: The Telefax, a Young 150-Year Old Service, 13-036a

Title Index

Future Developments in Telecommunications, 14-008
Future for HDTV in Europe, 13-123
Future of Communications Technology, 14-004
Future of Television, 12-087
Future of Transoceanic Telephony, 7-146

GEC Research Laboratories, 1919–1984, 4-205
General Electric Story–A Photo History, 4-130
General Radio Company, 1915–1965, 4-138
General: David Sarnoff and the Rise of the Communications Industry, 4-152
Generation of Microwave Ferrite Devices, 11-063
Genesis of a Generator: The Early History of the Magnetron, 11-066
Gentlemen on Imperial Service: A Story of the Trans-Pacific Telecommunications Cable, 6-243
Geographical Areas Serviced by Bell and Independent Telephone Companies in the United States, 1-081
Germany's Imperial Wireless System, 9-078
Getting the Message Through: A Branch History of the U.S. Army Signal Corps, 3-142
Girdle Round the Earth: The Story of Cable and Wireless, 4-200
Global Communications since 1844, 3-090a
Global Players in Telecommunications, 4-073
Global Political Fallout: The First Decade of the VCR, 1976–1985, 10-136
Global Satellite Communications, 4-022
Global Telecommunications Regulation, 4-015
Glossary of Abbreviations for Names of Technical, Scientific, Industrial and Professional Organizations, 4-001
Golden Age of Televisions, 12-147
Good Connections: A Century of Service by the Men & Women of Southwestern Bell, 7-108
Goodbye Central: Automation and the Decline of "Personal Service," 7-082
Goodbye, Central: Hello, World: A Centennial History of Northwestern Bell, 7-109
Government by Commission, 4-040
Government Operations in Space (Analysis of Civil-Military Roles and Relationships), 14-057
Government Ownership of Electrical Means of Communication, 4-051
Government Use of Satellite Communications, 14-058

Graham Bell–Pioneer: An Era of Outstanding Developments in World Communications, 7-046
Gramophone: Its Past; Its Present; Its Future, 10-033
Gramophones and Phonographs, 10-020
Great Inventions, 3-015
Great Northern Company: An Outline of the Company's History 1869–1969, 4-209
Great Northern Telegraph Company: A Danish Company in the Service of Globalisation since 1869, 4-210
Great Television Race: A History of the American Television Industry, 1925–1941, 12-130
Growth of an Enterprise: The Life of Anton Philips, 4-230
Growth of Long-Distance Telephony in the Bell System: 1875–1907, 7-084
Guglielmo Marconi and Early Systems of Wireless Communication, 5-168a
Guglielmo Marconi, 5-159, 5-160, 5-161
Guide for the Electric Testing of Telegraph Cables, 6-049
Guide to Sources in Educational Media and Technology, 1-022
Guide to Telecommunications Transmission Systems, 14-005
Guide to the Electric Telegraph, 6-013
Guide to the Literature of Electrical and Electronic Engineering, 2-001
Guinness Book of Recorded Sound, 10-057

Half a Century of "Electrification" in Telephony Systems, 7-072
Handbook of Electrical Diagrams and Connections, 6-056
Handbook of Practical Telegraphy, 6-036
Handbook of the Electric Telegraph, 6-007
Handbook of the Electric Telegraph: Designed for the Use of Students and Operators, 6-050
Hand-Book of the Electro-Magnetic Telegraph, 6-062
Handbook of the Telegraph, 6-035
Handbook of Wireless Telegraphy, 9-039
Harmsworth's Wireless Encyclopedia, 1-032
Harper's Wireless Book, 9-084
Hawaiian Telephone Story, 7-111
HDTV Proceedings, 13-115
HDTV Standards–Understanding the Issues, 13-124
HDTV: Advanced Television for the 1990s, 13-106
HDTV: High-Definition Television, 13-125
HDTV: The Politics, Policies, and Economics of Tomorrow's Television, 13-126
Heinrich Hertz, 5-142
Heinrich Hertz: Classical Physicist, Modern Philosopher, 5-138

Heinrich Rudolph Hertz: A Collection of Articles and Addresses, 5-143
Hello Texas: A History of Telephony in the Lone Star State, 7-102a
Hello?! The History of the Telephone in Norway, 4-228
Herbert Eugene Ives, 5-144
Here's Looking at You: The Story of British Television, 1908–1939, 12-126
Hermes Bound: The Policy and Technology of Telecommunications, 3-045
Heroes of the Telegraph, 5-029
Hertz and the Maxwellians, 5-031
Hertz, the Discoverer of Electric Waves, 5-139
Hertz's Researches on Electrical Oscillations, 8-008
Hertzian Wave Wireless Telegraphy, 9-013
High Definition Television: A Bibliography, 13-127
High Definition Television: HII-Vision Technology, 13-122
High Fidelity: A Bibliography of Sound Reproduction, 10-074
High Frequency Oscillator and Amplifier, 11-064
High-Definition Television: A Global Perspective, 13-112
High-Definition Television: An Annotated Multidisciplinary Bibliography, 1981–1992, 13-129
Highlights in the History of Multichannel Sound, 10-112
Highlights in the History of Telecommunications, 1-057
High-Power Pulsed Magnetron: A Review of Early Developments, 11-061
High-Tech Maneuvers: Industrial Policy Lessons of HDTV, 13-105
His Master's Voice, 4-151a
Historic Televisions and Video Recorders, 12-145
Historical Account of the Introduction of the Galvanic and Electro-Magnetic Telegraph into England, 6-157
Historical Development of Sound Films, 10-092
Historical Dictionary of Data Processing Biographies, 5-019
Historical Dictionary of Data Processing Organizations, 4-055
Historical Evolution of U.S. Telecommunications, 3-013
Historical Notes on Television Before 1900, 12-072
Historical Papers–Television, 12-115
Historical Review of the Cathode-Ray Tube As a Television Display Device, 11-052
Historical Review of the Development of Television Pickup Devices (1930–1976), 11-053
Historical Review of the East Bay Exchange, 7-107
Historical Review of Ultra-Short-Wave Progress, 9-139
Historical Sketch of Ferrites and Their Microwave Applications, 8-029
Historical Sketch of Henry's Contribution to the Electro-Magnetic Telegraph, 5-136
Historical Sketch of the Electric Telegraph, 6-016
History and Development of Stereophonic Sound Recording, 10-108
History and Development of the Cathode-Ray Tube, 11-048
History and Development of the Color Picture Tube, 11-040
History and Development of the United Fruit Company's Radio Telegraph System, 4-164
History and Future of Commercial Satellite Communications, 14-049
History and Identification of Old Telephones, 7-170
History of Automatic Telephony, 7-055
History of Bell's Telephone, 7-003
History of Binaural Sound, 10-109
History of British Army Signals in the Second World War, 3-123
History of Broadcasting in Japan, 4-223
History of Broadcasting in the United Kingdom, 4-192
History of Color Picture Tubes and Some Future Projections, 11-042
History of Color Television Displays, 11-041
History of Communications-Electronics in the United States Navy, 3-164
History of Community Antenna Television, 13-064
History of Consumer Electronics: Commemorating a Century of Electrical Progress, 1884–1984, 3-063
History of Disc Recording, 10-024
History of Electric Telegraphy to the Year 1837, 6-072
History of Electric Wires and Cables, 3-026
History of Electrical Communication, 3-031
History of Electrical Technology: An Annotated Bibliography, 1-007
History of Electronic Imaging, 11-047
History of Engineering and Science in the Bell System, 4-112
History of GTE, 4-140
History of Home Videotape Recorder Development, 10-144
History of International Broadcasting, 3-106, 3-107

Title Index

History of Land-Mobile Radio Communications, 14-021
History of Magnetic Recording, 10-048, 10-068
History of MCI: The Early Years, 1968–1988, 4-147
History of Millimeter and Sub-Millimeter Waves, 8-037
History of N.V. Philips' Gloeilampenfabrieken, 4-231
History of Radio Telegraphy and Telephony, 9-144
History of Radio to 1926, 9-091
History of Semiconductor Research, 11-078
History of Some Foundations of Modern Radio-Electronic Technology, 9-117
History of Sound Motion Pictures, 10-088a
History of Sound Recording, 10-059
History of Stereophonic Sound Reproduction, 10-106
History of Tactical Communication Techniques, 3-129
History of Telecommunications in Australia, 1854–1930, 3-081
History of Television, 12-119
History of Television for Public Showing in Cinemas in the United Kingdom, 12-080
History of Television from Early Days to the Present, 12-122
History of Television, 1880 to 1941, 12-111
History of Television, Part I: Mechanical and Semi-Mechanical Systems, 12-069
History of the Atlantic Telegraph, 6-201
History of the British Radio Valve to 1940, 11-009
History of the Cathode-Ray Tube, 11-050
History of the Coherer Principle, 9-151
History of the Defense Satellite Communications System (1964–1986), 14-040
History of the Electro-Magnetic Telegraph, 6-084
History of the General Radio Company, 4-139
History of the House of Siemens, 4-239
History of the Institution of Electrical Engineers, 4-090, 4-091
History of the Invention of the Electric Telegraph, 6-019
History of the Invention of the Transistor and Where It Will Lead Us, 11-072
History of the Marconi Company, 4-215
History of the Roseville Telephone Company, 7-097a
History of the Royal Canadian Corps of Signals, 1903–1961, 3-121
History of the Telegraph Operations during the War in South Africa, 1899–1902, 3-118
History of the Telephone in the United Kingdom, 7-128

History of the Telephone, 7-047, 7-048
History of the Theories of Aether and Electricity, 8-004
History of the United States Signal Corps, 3-130
History of the World Semiconductor Industry, 11-077
History of U.S. Electronic Warfare, 3-141
History of Videotape Recording, 10-125
History of Wireless Telegraphy, 1838–1899, 9-004
History, Development and Future of Telecommunications in Europe, 3-070
History, Theory, and Practice of the Electric Telegraph, 6-031
Home Recording for the Digital Millennium, 13-095
Home Video & Cable Yearbook, 13-045
Hoosier Connections: The History of the Indiana Telephone Industry, 7-110
How I Put the Electrons to Work in the Radio Bottle, 11-020
How Sony Developed Electronics for the World Market, 4-244
How the Radio Tube Grew Up, 11-026
How the World Was One: Beyond the Global Village, 3-027
Hysterical Background of Radio, 9-095

I.T.& T.'s Nine Lives, 4-143
I.T.& T. Ends a Brilliant Decade, 4-141
Iconoscope–A Modern Version of the Electric Eye, 11-054
Illustrated Handbook to the Electric Telegraph, 6-034
Illustrated History of Phonographs, 10-035
Illustrations and Description of Telegraphic Apparatus, 6-074
Image Worlds: Corporate Identities at General Electric, 1890–1930, 4-135
Images of the Future: The EBU's Part to Date in HDTV System Standardization, 13-114
Impact: A Compilation of Bell System Innovations in Science and Engineering..., 4-118
Imperial Telegraphic Communication, 6-148
Important Dates in Radiotelegraphy, 1-059
In One Man's Life; Being Chapters from the Personal and Business Career of Theodore N. Vail, 4-104
In the Shadow of the Shield: The Development of Wireless Telegraphy and Radio Broadcasting in Kingston and at Queen's University, 1902–1957, 9-129
Independent Telephony in New England: A History, 1876–1976, 7-103
Indexed Bibliography of Electron Tubes and Their Applications, 11-032
Index to IEEE Publications, 2-020

Index to Radio and Electronics Patents: A Keyword Index, 9-150b
Index to Scientists of the World from Ancient to Modern Times: Biographies and Portraits, 5-011
Infoculture: The Smithsonian Book of Information Age Inventions, 3-042
Information Highways & Byways: From the Telegraph to the 21st Century, 3-041
Information Machines: Their Impact on Men and the Media, 13-039
Information Sources in Patents, 3-016
Inspector and the Trouble Man, 7-031
Institute of Radio Engineers–Forty-Five Years of Service, 4-089
Instructions for Testing Telegraph Lines, 6-064
Instructions for the Use of Wireless-Telegraph Apparatus, 9-014
INTELSAT Global Satellite System, 4-019
INTELSAT Memoirs, 4-023
INTELSAT Research and Development Program, 4-024
Interactive Media–Special Report, 13-086
Interactive Videotex, 13-080a
Interdepartment Radio Advisory Committee, 4-033
Interdepartment Radio Advisory Committee: 50 Years of Service, 4-036
International Aspects of Electrical Communications in the Pacific Area, 3-096
International Audio Broadcasting for the Twenty-First Century, 3-100
International Cable Telegraph and the Austro-Hungarian Monarchy, 6-128a
International Communication, 3-088
International Communications: The American Attitude, 3-086
International Conference on 100 Years of Radio, 9-104
International Conference on Wireless Telegraphy: Preliminary Conference at Berlin, August 1903, 9-019
International Control of Radiocommunications, 4-018
International Directory of Company Histories, 4-003
International Facilities Study, 3-091
International Film, Radio, and Television Journals, 2-008
International Radio Broadcasting, 3-099
International Radio Tube Encyclopedia, 11-000
International Submarine Cable Systems, 7-153
International Telecommunication Union in a Changing World, 4-013
International Telecommunication Union: 130 Years, 1865–1995, 4-011
International Telecommunication Union: An Experiment in International Cooperation, 4-012
International Telecommunications, 3-093
International Telecommunications Bibliography, 1-006
International Telecommunications Satellite Organization (INTELSAT), 4-025
International Telecommunications Standards Organizations, 4-065
International Telephone Statistics, 1-065
Internet and the Meltdown of Audiotext and Videotext in Europe, 13-076
Introduction of the Internet and World Wide Web, 14-089
Introduction of the Loading Coil: George A. Campbell and Michael I. Pupin, 7-070
Introduction to Colour Television, 12-139
Introduction to Key Collecting, 6-246
Introduction to Television, 12-093
Introduction to the Photoplay, 10-094
Inventing American Broadcasting, 1899–1922, 9-099
Inventing the Internet, 14-083
Invention and Development of the Klystron, 11-065
Invention and Evolution of the Electrotechnology to Transmit Electrical Signals Without Wires: An Annotated Bibliography, 8-002
Invention and Innovation in the Radio Industry, 9-108
Invention of Fiber-Optic Communications, 14-076
Invention of the Electric Telegraph, 5-086
Invention of the Traveling Wave Tube, 11-059
Inventions, Researches and Writings of Nikola Tesla, 5-216
Inventor and Entrepreneur: Recollections of Werner von Siemens, 4-240
Inventor and the Pilot: Russell and Sigurd Varian, 4-166
Inventor in Eclipse, 3-022
Inventor of the Valve: A Biography of Sir Ambrose Fleming, 5-126
Inventor's Handbook of the Phonograph, 10-022
Inventors of the Telegraph and Telephone, 5-023
Investigation of the Telephone Industry in the United States, 4-108
Invisible Resource: Use and Regulation of the Radio Spectrum, 8-046
Invisible Weapon: Telecommunications and International Politics, 1851–1945, 3-089
Irony of Regulatory Reform: The Deregulation of American Telecommunications, 4-043

Title Index

Italian Navy "Coherer" Scandal Revisited, 5-068
Italian Navy Coherer Affair: A Turn-of-the-Century Scandal, 5-071
ITT: The Management of Opportunity, 4-145

J.L. Baird: Success and Failure, 5-048
J.L. Baird's Colour Television, 1937–1946, 5-051
James Clerk Maxwell, 5-170
James Clerk Maxwell: A Biography, 5-182
James Clerk Maxwell: A Commemoration Volume: 1831–1931, 5-177
James Clerk Maxwell and Modern Physics, 5-173
James Clerk Maxwell and the Theory of the Electromagnetic Field, 5-176
James Clerk Maxwell: Physicist and Natural Philosopher, 5-172
Jamming of Western Radio Broadcasts to the Soviet Union and Eastern Europe, 3-102
Japan's Computer and Communications Industry, 3-069
Japan's Progress in International Telecommunications, 4-213
Joe Batten's Book: The Story of Sound Recording, 10-017
John Baird: The Romance and Tragedy of the Pioneer of Television, 5-054
John J. Carty, Telephone Engineer, 5-077
John J. Carty–An Appreciation, 5-079
John Joseph Carty, 1861–1932, 5-078
John L. Baird: The Founder of British Television, 5-055
John Logie Baird and Television, 5-050
John Logie Baird: 50 Years of Television, 5-049
Joseph Henry and the Magnetic Telegraph: An Address..., 5-132
Joseph Henry: His Life and Work, 5-131a
Joseph Henry: The Rise of an American Scientist, 5-133
Journal of the SMPTE, Anniversary Issues, 4-095
Jubilee Year [of Marconi Co.], 4-219
Jump Cut! Memoirs of a Pioneer Television Editor, 10-126
Just a Few Lines..., 12-062
Just Give Me the Fax, 13-021

Keeping Pace with the New Television, 13-047
Keys, Keys, Keys, 6-249
Klystron, 4-165
Kolster Radio Corporation, 4-146
Krarup Cable: Invention and Early Development, 7-122

L.M. Ericsson: 100 Years, 4-202

Latest Advance in Wireless Telephony, 5-206
Laying and Repairing of Electric Telegraph Cables, 6-061
Laying and Repairing Submarine Telegraph Cables, 6-213
Laying of the Cable, 6-178
Leading Telegraph Patents, Including Original and Reissued Patents, of S.F.B. Morse, 5-192
Lectures on the Electro-Magnetic Telegraphy, 6-017
Lee de Forest and the Fatherhood of Radio, 5-094
Legacies of Edwin Howard Armstrong, 5-041a
Legislative History of the Communications Act of 1934, 4-028
Life and Work of Joseph Henry, 5-134
Life and Work of Lee de Forest, 5-091
Life and Work of Sir Isaac Shoenberg, 5-203
Life and Work of Sir Jagadis C. Bose, 5-070
Life and Works of A.K. Erlang, 5-108
Life of James Clerk Maxwell, 5-169
Life of John Stone Stone, 5-205
Life of Samuel F.B. Morse, 5-194
Life of Sir William Siemens, 4-237
Life on the Yukon 1865–1867, 6-141
Life Story of the Late Sir Charles Tilston Bright, 5-076
Lightning Flashes and Electric Dashes, 6-058
Lightning in His Hand: The Life Story of Nikola Tesla, 5-215
Lightning Wires: The Telegraph and China's Technological Modernization, 1860–1890, 6-126
Lincoln in the Telegraph Office, 3-131
Lines between the Rivers: A History of Telephony in Iowa, 7-101a
Literature Survey of Communication Satellite Systems and Technology, 14-055
Live via Satellite: The Story of Comsat, 4-126
Lives of the Electricians: Professors Tyndall, Wheatstone and Morse, 5-026
London Television Service, 12-097
London Television Station, 12-057
Long Distance Please: The Story of the Trans Canada Telephone System, 7-139
Longest Wire, 6-145
Looking Ahead: The Papers of David Sarnoff, 4-158
Looking Back at Distant Vision: Television Technology from 1936 to 1986, 12-106
Losing the Race: The British Post Office and Picture Telegraphy, 13-011
Low Voltage Cathode-Ray Tube and Its Applications, 11-051
Lowering Barriers to Telecommunications Growth, 14-002

318 Title Index

Made in Japan: Akio Morita and Sony, 4-246
Magnetic Recording, 10-049, 10-050
Magnetic Recording, 1888–1952, 10-078
Magnetic Recording, 1900–1949, 10-077
Magnetic Recording: The First 100 Years, 10-056
Magnetic Tape from the Earliest Days to the Present, 10-065
Magnetic Telegraph: Price and Quantity Data, and the New Management of Control, 6-109
Magnetron and the Beginnings of the Microwave Age, 11-056
Magnificent Failure, 5-095
Mahlon Loomis, Inventor of Radio, 5-152
Mahlon Loomis, the Discover and Inventor of Radio, 5-154
Makers of Electricity, 5-034
Making of a Profession: A Century of Electrical Engineering in America, 4-085
Making of American Industrial Research: Science and Business and GE and Bell, 1876–1926, 4-136
Man of High Fidelity: Edwin Howard Armstrong, 5-041
Managing the Moving Image–From an Engineering Point of View, 3-066
Manual of Rural Telephony, 7-035
Manual of Submarine Telegraph Companies, 6-208
Manual of Telegraph Construction, 6-053
Manual of Telegraphy, 6-078
Manual of Telegraphy, Designed for Beginners, 6-038a
Manual of Telephony, 7-020
Manual of the Various Electro-Magnetic Telegraphs at Present in Use, 6-005
Manual of Wireless Telegraphy and Telephony, 9-029
Manual of Wireless Telegraphy (Radio) for the Use of Naval Electricians, 9-035
Manufacturing Policy in the Electronics Industry, 4-069
Manufacturing the Future: A History of Western Electric, 4-167
Manuscripts in U.S. Depositories Relating to the History of Electrical Science and Technology, 1-038
Map of Coast Stations Open to Public Correspondence, 1-079
Marconi, 5-156, 5-163, 5-167
Marconi, 1939–1945: A War Record, 4-218
Marconi and His South Wellfleet Wireless, 9-069
Marconi and the Discovery of Wireless, 5-168
Marconi and Wireless, 9-115
Marconi Scandal, 4-217
Marconi Wireless on Cape Cod, 9-086
Marconi Wireless Telegraph System, 9-015

Marconi Wireless Telegraphy: A Short History, 9-074
Marconi, Pioneer of Radio, 5-157
Marconi, The Man and His Wireless, 5-158
Marconi's Magnetic Detector: 20th Century Technique Despite 19th Century Normal Science, 9-152
Marconi-EMI Television System, 12-079
Marconi–Master of Space, 5-162
Marconi–The Man and His Apparatus, 5-164
Mass Communications Research Resources: An Annotated Guide, 1-021
Mass Media: A Chronological Encyclopedia, 1-054
Masters of Space, 5-037
Maver's Wireless Telegraphy and Telephony, 9-022
Maver's Wireless Telegraphy: Theory and Practice, 9-022
Maverick Inventor: My Turbulent Years at CBS, 5-127
Maxwell on the Electromagnetic Field: A Guided Study, 5-181
Maxwell's Theory and Wireless Telegraphy, 8-007
Maxwellians, 5-025
Mechanical Developments of Facsimile Equipment, 13-033
Mechanics of Television: The Story of Mechanical Television, 12-076
Media and the American Mind from Morse to McLuhan, 3-030
Media for Interactive Communication, 13-084
Media Technology and Society: A History from the Telegraph to the Internet, 3-058
Media Use in the Information Age: Emerging Patterns of Adoption and Consumer Use, 13-053
Mementos of Early Photographic Sound Recording, 10-088
Memoir on the Euphrates Valley Route to India, 6-144
Memoirs, Letters, Diaries: Heinrich Hertz, 5-141a
Memorial of Joseph Henry, 5-137
Memorial of Samuel Finley Breese Morse Including Appropriate Ceremonies of Respect at the National Capitol, and Elsewhere, 5-190
Memories of a Scientific Life, 5-125
Men and Volts at War: The Story of General Electric in World War II, 4-133
Men and Volts: The Story of General Electric, 4-131
Menlo Park Reminiscences, 5-103
Messenger Gods of Battle: Radio, Radar, Sonar, the Story of Electronics in War, 3-113
Messengers for Mankind, 3-052

Title Index

Method of Reducing Disturbances in Radio Signaling by a System of Frequency Modulation, 9-130
Methods and Equipment in Cable Telegraphy, 6-238
Methods Employed for the Wireless Communication of Speech, 9-053
Michael Faraday, Cable Telegraphy, and the Rise of Field Theory, 6-237
Michael Idvorsky Pupin, 5-200
Micro Radio Waves, 8-032
Microprocessor: A Biography, 11-076
Microwave Antenna and Waveguide Techniques Before 1900, 8-034
Microwave Debate, 8-036
Microwave Instrumentation: An Historical Perspective, 8-026
Microwave Tubes: An Introductory Review with Bibliography, 11-057
Milestones in Motion Picture and Television Technology, 4-096
Milestones of Communication Progress, 4-144
Military Communications Explosion, 1914–1918, 3-122
Military Communications: A Test for Technology–The U.S. Army in Vietnam, 3-133
Military Signal Communications, 3-126
Military Telegraph during the Civil War in the United States, 3-140
Military Television: The First Public Demonstration of Combat Television, 3-138
Million Words a Minute, 13-036
Mission Communications: The Story of Bell Laboratories, 4-119
Misunderstanding Media, 3-057
Mobile Communications, 14-019
Mobile Communications in the U.S. and Europe: Regulation, Technology, and Markets, 14-022
Mobile Phone Book: The Invention of the Mobile Telephone Industry, 14-020
Modern Campaign; Or, War and Wireless Telegraphy in the Far East, 3-116
Modern Gramophones and Electrical Reproducers, 10-046
Modern Practice of the Electric Telegraph, 6-047
Modern Service of Commercial and Railway Telegraphy, 6-069
Modern Stentors: Radio Broadcasters and the Federal Government, 1920–1934, 4-046
Modern Telegraph Systems and Equipment, 6-098
Modern Telegraphy: Some Errors of Dates of Events and of Statements in the History of Telegraphy Exposed and Rectified, 6-041
Modern Wireless, 9-047
Moody's Industrial Manual, 4-006

Morse's Patent: Full Exposure of Dr. Chas. T. Jackson's Pretensions to the Invention of the American Electro-Magnetic Telegraph, 5-187
Motion Pictures, Television, and Radio: A Union Catalogue of Manuscript and Special Collections in the Western United States, 1-040
Municipal Electric Telegraph: Especially in Its Application to Fire Alarms, 6-014
Music Machines–American Style, 10-006
Musical Broadcasting in the 19th Century, 10-014
Muttering Machines to Laser Beams: A History of Mountain Bell, 7-102
My Father Marconi, 5-166
My Inventions: The Autobiography of Nikola Tesla, 5-223
My Philosophy, Representing My Views on the Many Functions of the Ether of Space, 5-150
Myths of Telephone History, 7-054

NAB Guide to Advanced Television Systems, 13-121
NAB Guide to HDTV Implementation Costs, 13-133
Nathan B. Stubblefield and His Wireless Telephone, 5-207
Natural Monopoly and Universal Service: Telephones and Telegraphs in the U.S., 1837–1940, 3-035
Nerds 2.0.1: A Brief History of the Internet, 14-093
Netizens: On the History and Impact of Usenet and the Internet, 14-088
Networks of Power: Electrification in Western Society, 1880–1930, 3-009
New & Improved: Inventors and Inventions That Have Changed the Modern World, 3-017
New Bibliography of Reginald A. Fessenden, 5-117
New Infotainment Technologies in the Home: Demand-Side Perspectives, 13-042
New Key West-Havana Carrier Telephone Cable, 7-145
New Media: Communication, Research, and Technology, 13-050
New Method for the Reception of Weak Signals at Short Wave Lengths, 9-140
New Submarine Telegraph Cable, 6-179
New Technologies Affecting Radio and Television Broadcasting, 13-048a
New Telephotograph System, 13-034
New TV: A Comprehensive Survey of High Definition Television, 13-109
Newcomen Papers, 4-175

320 Title Index

Newnes Television and Short-Wave Handbook, 12-022
News over the Wires: The Telegraph and the Flow of Public Information in America, 1844–1897, 6-103
Nikola Tesla: On His Work with Alternating Currents and Their Application to Wireless Telegraphy, Telephone, and Transmission of Power, 5-213
1929, A Contemporary Account of the Transition to Sound in Film, 10-094
1942–1967: Twenty-Five Years at RCA Labs, 4-155
1990 World's Submarine Telephone Cable Systems, 7-152
1992 World Administrative Radio Conference: Issues for U.S. International Spectrum Policy, 8-054
1992 World Administrative Radio Conference: Technology and Policy Implications, 8-055
Nortel: Past, Present, Future, 4-227
North Carolina Telephone Story: The First Ninety-Eight Years, 7-098
Notable Corporate Chronologies, 4-005
Notes on Military Telegraph Instruments with Diagrams of Connections, 3-127
Notes on the Development of a New Type of Hornless Loud Speaker, 10-013
Notes on the Theory of Modulation, 8-015
"Number Please!" A History of the Early London Telephone Exchanges from 1880 to 1912, 7-138
Numerical Index to Radio and Electronics Patents, 9-150c

Observations of Electromagnetic Wave Radiation Before Hertz, 8-003
Ocean Cable Lore, 6-241
Ocean Telegraph Cable, 6-196
Ocean Telegraph to India, 6-207
Ocean Telegraphing, 6-198
Ocean Telegraphy, 6-170
Ocean Telegraphy: The 25th Anniversary, 6-212
Old Telegraphs, 6-139
Old Telephones, 7-168
Old Television, 12-148
Old Time Telephones! Technology, Restoration and Repair, 7-174
Old Wires and New Waves: The History of the Telegraph, Telephone, and Wireless, 3-038
Oliver Lodge and the Invention of Radio, 5-151
On Methods Whereby the Radiation of Electric Waves May Be Mainly Confined to Certain Directions..., 8-022
On Submarine Electric Telegraphs, 6-172

On the Conversion of Electric Oscillations into Continuous Currents by Means of a Vacuum Valve, 11-018
On the Electric Telegraph and the Principal Improvements in its Construction, 6-024
On the Importance of Ocean Telegraphy, 6-194
On the Line: The Men of MCI–Who Took on AT&T, Risked Everything, and Won!, 4-148
On the Maintenance and Durability of Submarine Cables, 6-191
On the Origin and Progress of the Ocean Electric Telegraph, 6-174
On the Short Waves, 1923–1945: Broadcast Listening in the Pioneer Days of Radio, 9-133
On the Submerging of Telegraphic Cables, 6-177
On the Theory of Duplex Telegraphy–Conference Proceeding, 6-051
150 Years of Siemens: The Company from 1947 to 1997, 4-236
150th Anniversary of the Electromagnetic Telegraph, 6-140
100 Years of Bell Telephones, 7-175
100 Years of Communications Progress, 3-047
One Hundred Years of Maritime Radio, 3-155
One Hundred Years of Submarine Cables, 6-232
100 Years of Telephone Switching Part 1: Manual and Electromechanical Switching (1878–1960s), 7-116
101 Years of Television Technology, 12-127
One Hundred Years: The Story of W.T. Henley's Telegraph Works Company, 4-212
Operating Features of the Audion, 11-012
Orbital Radio Relays, 14-045
Origin or Basis of Wireless Communication, 9-067
Origins and Development of Radiotelephony, 9-103
Origins of Clerk Maxwell's Electric Ideas, as Described in Familiar Letters to W. Thomson, 5-178, 8-005
Origins of Maritime Radio: The Story of the Introduction of Wireless Telegraphy in the Royal Navy between 1896 and 1900, 3-159
Origins of Microwave Telephone–Waves of Change, 8-030
Origins of Modern Television, 12-120
Origins of Spread-Spectrum Communications, 8-048
Out from Behind the Eight-Ball: A History of Project Echo, 14-032
Outline of the History of Telegraphs in Japan, 6-134

Title Index

Outline of Theoretical Telegraphy, 6-038
Outlook for Television, 12-020
Overland Explorations in Siberia, 6-128
Overland Telegraph: The Story of a Great Australian Achievement, 6-149
Overview of Twenty-Five Years of Electrical and Electronic Engineering in the Proceedings of the IEEE, 1963–1987, 4-084
Owen D. Young and American Enterprise, 4-129
Owen D. Young: A New Type of Industrial Leader, 4-137

Papers of Joseph Henry, 5-135
Papers of Thomas A. Edison, 5-106
Past and Future History of the Internet, 14-091
Past Quarter-Century and the Next Decade of Videotape Recording, 10-129
Past Years, An Autobiography, 5-149
Patent History of the Phonograph, 1877–1912, 10-032
Patent Policies of Radio Corporation of America, 4-159
Patent Profiles, 3-021
Patented Telephony: A Review of the Patents Pertaining to Telephones and Telephone Apparatus, 7-025
Patents as Scientific and Technical Literature, 3-023
People Machine: An Illustrated History of the Telephone on the Central West Coast of Florida, 7-099
Person to Person: The International Impact of the Telephone, 7-127
Personnel of the Telegraph, 5-032
Perspectives on Television: The Role Played by the Two NTSC's in Preparing Television Service for the American Public, 12-083
Philco Radio: A Pictorial History of the World's Most Popular Radios, 1928–1942, 9-174
Philco: Autobiography of Progress, 4-151
Philipp Reis: Inventor of the Telephone, 5-202
Philo Farnsworth: Television's Pioneer, 5-115
Phone–An Appreciation, 7-165
Phonograph and How to Construct It, 10-028
Phonograph and Its Future; and the Aurophone and Its Future, 10-021
Phonograph and Its Inventor, Thomas Alva Edison, 10-026
Phonograph and Our Musical Life, 10-030
Phonograph and Sound Recording after One-Hundred Years: Centennial Issue, 10-041
Phonographs and Gramophones, 10-042

Photoradio Apparatus and Operating Technique Improvements, 13-008
Photoradio Developments, 13-032
Phototelegraphy, 13-005
Picturephone and Beyond, 7-161
Picturephone System, 7-164
Picturephone, 7-163
Pioneer Telegraphy in Chile, 1852–1876, 6-135
Pioneering Days: The First Live Transatlantic Link, 12-064
Pioneering Days: The Quest for High-Definition, 12-082
Pioneering in Educational Television, 1932–1939, 12-070
Pioneering in Industrial Research: The Story of the General Electric Research Laboratory, 4-128
Pioneering in Television: Prophecy and Fulfillment, 12-098
Pioneering the Cathode-Ray and Television Arts, 12-086
Pioneering the Telephone in Canada, 7-140
Pioneers in Electrical Communications, 5-033
Pioneers of Electrical Communications, 5-016
Pioneers of Television–Philo Taylor Farnsworth, 5-109
Pioneers of Television–Vladimir Kosma Zworykin, 5-226
Pioneers of Wireless, 5-024
Plain Talk about the Postal Telegraph, 6-067
Plan and Description of the Original Electro-Magnetic Telegraph, 6-011
Planning and Implementation of the Public Television Satellite Interconnection System, 14-069
Please Stand By: A Prehistory of Television, 12-101
Poles, Wires and Cables; or, Electric Telegraphs, 6-046
Politics and Technology of Satellite Communications, 14-033
Politics of International Standards: France and the Color TV War, 12-135
Politics of International Telecommunications Regulation, 4-017
Popov and the Beginnings of Radiotelegraphy, 5-197
Popular Explanation of the Electric Telegraph, 6-008
Popular Guide to Commercial and Domestic Telephony, 7-027
Popular Television, 12-026
Portable Radio in American Life, 9-176
Possibilities of Television, 12-105
Possibilities of the Iconoscope in Television, 12-034

Post Office Contributions to the Early History of the Development of Television in the United Kingdom, 12-061
Post Office Electrical Engineers' Journal, 7-141
Posts and Telegraphs, Past and Present; with an Account of the Telephone and Phonograph, 6-065
Poulson Arc Generator, 9-058
Power of Speech: A History of Standard Telephones and Cables, 1883–1983, 4-248
Practical Features of Telephone Work, 7-029
Practical Information for Telephonists, 7-014
Practical Systems of Electrical Telegraphy, 6-080
Practical Telegrapher, 6-071
Practical Telegraphist and Guide to the Telegraph Service, 6-073
Practical Telephone Hand Book and Guide to Telephonic Exchange, 7-034
Practical Telephone Handbook and Guide to the Telephone Exchange, The, 7-018
Practical Telephone: A Short Historical Sketch, 7-011
Practical Telephony, 7-026
Practical Television, 12-006
Practical Wireless Telegraphy, 9-049
Present and Probable CATV/Broadband-Communications Technology, 13-074
Present and Projected Business Utilization of International Telecommunications: 1985, 3-087
Present Status of Color Television: Report of the Advisory Committee..., 12-144
Presidential Address, 12-104
Press, Film, Radio, 3-071
Prestel 1980, 13-080
Principles of Electric Wave Telegraphy and Telephony, 9-032
Principles of Television Engineering, 12-089
Principles of Wireless Telegraphy, 9-079
Printing Telegraph Systems and Mechanisms, 6-092
Printing Telegraph Systems, 6-088
Proceedings of the IEEE: The First 75 Years, 4-083
Proceedings of the IRE, 8-051, 12-134
Process of Technological Change: New Technology and Social Choice in the Workplace, 4-198
Prodigal Genius: The Life of Nikola Tesla, 5-217
Progress in Wireless Telegraphy, 9-021
Progress of Electric Space Telegraphy, 9-009
Progress of the Telegraph, 6-030
Public Utilities: An Annotated Guide to Information Sources, 4-008
Publication Date: Early U.S. Radio Magazines, 2-006

Purdue University Experimental Television System, 12-054
Pushbutton Fantasies: Critical Perspectives on Videotex and Information Technology, 13-079

Quadruplex: With Chapters on the Dynamo-Electric Machine in Relation to the Quadruplex,, 6-076
Quarter Century of Electronics, 11-033
Quarter Century of Transcontinental Telephony, 7-052
Questions and Answers about Pay TV, 13-046
Quick Tidings of Hong Kong, 3-076

Radiation: An Elementary Treatise on Electromagnetic Radiation and on Röntgen and Cathode Rays, 8-011
Radio and Electronics Pioneers: A Patent Bibliography, 9-149, 9-150a
Radio and Television Almanac, 1-049
Radio Art, 9-165
Radio at Ultra-High Frequencies, 9-136
Radio Communication: Its History and Development, 9-111
Radio Corporation of America, 12-142
Radio Engineering, 9-121
Radio Engineers, the Federal Radio Commission and the Social Shaping of Broadcast Technology: Creating "Radio Paradise," 8-049
Radio Facsimile, 13-016
Radio for Everybody, 9-066
Radio for the Fireline: A History of Electronic Communication in the Forest Service, 4-029
Radio Frequency Spectrum: United States Use and Management, 8-060
Radio Magazines and the Development of Broadcasting: Radio Broadcast and Radio News, 1922–1930, 2-003
Radio Manual, 9-120
Radio Manufacturers of the 1920s, 4-056
Radio Reminiscences: A Half Century, 3-166
Radio Sound Effects: Who Did It, and How, in the Era of Live Broadcasting, 9-124
Radio Spectrum below 550 KHz, 8-017
Radio Spectrum Conservation: A Program of Conservation Based on Present Uses and Future Needs, 8-042
Radio Spectrum Utilization: A Program for the Administration of the Radio Spectrum, 8-043
Radio Stations of the United States, 9-083
Radio Stations of the World, 9-050
Radio Telegraphy, 3-167. 9-073
Radio Telephony, 9-064
Radio Today: The Present State of Broadcasting, 9-123

Title Index 323

Radio! Radio!, 9-168
Radio: Beam and Broadcast–Its Story and Patents, 9-076
Radio: Its Past, Present, and Future, 9-059
Radio's 100 Men of Science, 5-021
Radio's Conquest of Space: The Experimental Rise in Radio Communication, 9-109
Radio's First Voice: The Story of Reginald Fessenden, 5-118
Radio's Memorable Anniversary, 9-052
Radio-Electronic Bibliography, 1-014
Radiofrequency Use and Management: Impacts from the World Administrative Radio Conference of 1979, 8-053
Radiomovies, Radiovision, Television, 12-007
Radios by Hallicrafters, 9-163
Radio-Telegraphy, 9-043
Radiotelegraphy, 9-072
Radiotelegraphy: A Retrospect of Twenty Years, 9-062
Railway Telegraphs and the Application of Electricity to the Signaling and Working of Trains, 6-166
Railways, Steamers, and Telegraphs: A Glance at Their Recent Progress and Present State, 6-040
Rainbow in the Sky: FM Radio, Technical Superiority, and Regulatory Decision-Making, 8-050
RCA, 4-160
RCA and the Videodisc: The Business of Research, 10-137
RCA's Television, 12-100
RCA—An Historical Perspective, 4-156
Real Story of the Magnetron, 11-060
Rebirth in Digital Recording, 13-093
Recent Advances in Wireless Telegraphy, 9-028
Recent Developments in the Work of the Federal Telegraph Company, 9-055
Recording and Reproduction of Sound, 10-073
Recording Sound for Motion Pictures, 10-080
Recovery of Phonovision, 10-122
Reginald Fessenden: Radio's Forgotten Voice, 5-119
Regulation of Wireless Communications Systems, 14-016
Reminiscences of the Cape Government Telegraphs, 6-146
Remote Control in the New Age of Television, 13-082
Repeated Takes: A Short History of Recording and Its Effect on Music, 10-052
Reply to Mr. Cooke's Pamphlet, "The Electric Telegraph; Was it Invented by Professor Wheatstone?", A, 5-089

Report by the ITU on Telecommunication and the Peaceful Uses of Outer Space, 14-038
Report of the Chief Signal Officer to the Secretary of War, 1919, 3-143
Report of the FTC on the Radio Industry..., 4-072
Report of the Interdepartment Radio Advisory Committee, 8-058
Report on Communication Companies, 4-071
Report on the Condition of the Lines of the Western Union Telegraph Co., 6-043
Report to the President and the Congress, 4-125
Representations of Transatlantic Telegraphy, 6-236
Resonance of Antenna Systems in Wireless Telegraphy, 9-037
Restoration of Baird Mechanical Television Recordings, 10-123
Résumé of the Earlier Days of Electric Telegraphy, 6-214
Retrospective Technology Assessment: Submarine Telegraphy, 6-224
Review of Cable Television: The Urban Distribution of Broad-Band Visual Signals, 13-062
Review of Some Television Pick-Up Tubes, 11-046
Revolution and Evolution from Dot Sequential to NTSC, 12-096
Revolution in Electronics, 11-082
Revolution in Miniature: The History and Impact of Semiconductor Electronics, 11-071
Revolution in Radio, 5-043
Revolution in Rural Telephony, 1900–1920, 7-076
Rewiring of America: The Fiber Optics Revolution, 14-075
Rhodesia: A Postal History–Its Stamps, Posts and Telegraphs, 3-082
"Ring up Britain"—The Early Years of the Telephone in the United Kingdom, 7-135
Rise and Extension of Submarine Telegraphy, 6-216
Rise of Mechanical Television, 1901–1930, 12-074
Rise of the Electrical Industry during the Nineteenth Century, 4-064
Role of the Telegraph in the Consolidation and Expansion of the Argentine Republic, 6-131
Romance and History of the Electric Telegraph, 6-100
Romance and Reality of Television, 12-014
Romance of Television, 12-036

Royal Corps of Signals: A History of Its Antecedents and Development (Circa 1800–1955), 3-124
Rusty Ribbon: John Herbert Orr and the Making of the Magnetic Recording Industry, 1945–1960, 10-069

S. Gernsback's Radio Encyclopedia, 1-028
Saga of the Seas: The Story of Cyrus W. Field and the Laying of the First Atlantic Cable, 5-122
Saga of the Vacuum Tube, 11-010
Sam Johnson; The Experience and Observations of a Railroad Telegraph Operator, 6-063
Samuel F.B. Morse and American Democratic Art, 5-188
Samuel F.B. Morse, His Letters and Journals, 5-191
Sarnoff: An American Success, 4-153
Satellite Broadcasting, 14-071
Satellite Communications & DBS Systems, 14-065
Satellite Communications: Military-Civil Roles and Relationships, 14-056
Satellite Communications: The First Quarter Century of Service, 14-050
Satellite Invasion of India, 14-068
Satellites and Radio Broadcasting: Historical Review and Market Developments, 14-073
Satellites for World Communications, 14-059
Schoenberg, 5-204
Science Citation Index, 2-021
Scientific & Technical Papers of Werner von Siemens, 4-241
Scientific Letters and Papers of James Clerk Maxwell, 5-175
Scientific Papers of James Clerk Maxwell, 5-180
Scientific Papers of Sir Charles Wheatstone, 5-079a
Scientific Works of C. William Siemens, 4-233
Searching for Railway Telegraph Insulators, 6-250
Secret Life of John Logie Baird, 5-053
Seeing by Electricity, 12-081
Seeing by Radio, 13-014
Seeing by Wireless, 12-015
Seeing by Wireless: The Story of Baird Television, 5-052
Selective Guide to Literature on Telecommunications, 1-013
Selenium and Its Applications to the Photophone and Telephotography, 13-006
Semaphore: The Story of the Admiralty-to-Portsmouth Shutter Telegraph and Semaphore Lines, 1796 to 1847, 6-160

Semaphores to Short Waves: Proceedings of a Conference on the Technology and Impact of Early Telecommunications, 3-039a
Semiconductor Laser: A Thirty-Five Year Perspective, 14-078
Semiconductors and the Transistor, 11-074
Sermons, Soap and Television: Autobiographical Notes, 5-044
Serving Hawaii: The First 100 Years: The 100th Anniversary of the Hawaiian Telephone Company, 7-096
Setmakers: A History of the Radio and Television Industry, 4-059
Seventy-Fifth Anniversary Diamond Jubilee Yearbook [of the Radio Club of America], 1909–1984, 4-093
75 Years of Magnetic Recording, 10-047, 10-064
Seventy-Five Years of the Telephone: An Evolution in Technology, 7-173
70 Years of Radio Tubes and Valves: A Guide for Electronic Engineers, Historians, and Collectors, 11-007
Shadow Mask Color Picture Tube: How It Began–an Eyewitness Account of Its Early History, 11-043, 12-140
Shaffner's Telegraph Companion, 6-023
Shaping the Early Development of Television, 12-075
Shattered Silents: How the Talkies Came to Stay, 10-095
Shifting Time and Space: The Story of Videotape, 10-120
Ships of the Line: A History of Cableships, 6-228
Short History of Motion-Picture Sound Recording in the United States, 10-084
Short History of Television Recording, 10-113, 10-114
Short Review of Connecticut Telephony, 1878 to 1907, 7-112
Short Waves, 9-135
Shortwave Radio and Its Impact on International Telecommunications, 3-090
Short-Wave Receivers Past and Present, 1942–1997, 9-173
Show-Me State Story: A History of Telephony in Missouri, 7-107a
Siemens Brothers, 1858–1958: An Essay in the History of Industry, 4-238
Siemens Company–Its Historical Role in the Progress of Electrical Engineering, 1947–1980, 4-243
Signal Corps Centennial Issue, 3-146
Signal Corps, 3-147
Signal Corps, U.S.A. in the War of the Rebellion, 3-134
Signal Venture, 3-117

Title Index 325

Signal: A History of Signaling in the Royal Navy, 3-158
Signaling and Communicating at Sea, 3-161
Signaling: Marconi Wireless Telegraphy Explained, 9-024
Signals Across the Atlantic, 9-007
Signals from the Atlantic Cable, 6-181
Signals Intelligence in World War II, 1-016
Signals: The Science of Telecommunications, 3-049
Silicon Valley Fever: Growth of High-Technology Culture, 11-085
Simple Guide to Television, 12-049
Singing Wires: The Telephone in Alberta, 7-130
Single-Control Turning: An Analysis of an Invention, 9-147
Single-Sideband in Communication Systems: A Bibliography, 8-014
Sir Charles Wheatstone, 5-080
Sir Oliver Lodge: Physical Researcher and Scientist, 5-147
Sir William Fothergill Cooke, 5-081
Sir William O'Schaughnessy, Lord Dalhousie, and the Establishment of the Telegraph System in India, 6-154
Sir William Preece, 5-198, 5-199
Sir William Siemens: A Man of Vision, 4-242
60 Years of Electronics, 3-007
Sixty Years of PBX Development, 7-043
Sixty Years of Progress, 4-221
Social and Cultural Aspects of VCR Use, 10-134
Social Impact of the Telephone, 7-089
Some Aspects of the Genesis of Radio Engineering, 9-148
Some Personal Recollections of Early Experiences on the New Frontier of Electroacoustics during the Late 1920s and Early 1930s, 10-010
Some Phases of Railroad Telegraph and Telephone Engineering, 6-099
Some Possibilities of Electricity, 9-001
Some Recent Developments in the Audion Receiver, 11-013
Some Recent Developments in the Multiplexed Transmission of Frequency Modulated Broadcast Signals, 9-132
Something New under the Sun: Satellites and the Beginning of the Space Age, 14-035
Sony Vision, 4-245
SOS to the Rescue, 3-149
SOS: The Story of Radio Communication, 9-116
Sound and the Cinema: The Coming of Sound to American Film, 10-079
Sound Motion Pictures: From the Laboratory to Their Presentation, 10-083
Sound of High Fidelity, 10-008
Sound Projection, 10-090

Sound Recording in Europe up to 1945, 10-075
Sound Recording–Will There Be Progress Forever?, 13-091
Sounds out of Silence: A Life of Alexander Graham Bell, 5-062
Sources and Detectors for Optical Fibre Communications Applications: The First 20 Years, 14-080
Sources in Electrical History, 1-037
Sources of Invention, 3-020
Space Communications in Europe: How Did We Make it Happen?, 14-030
Spark That Gave Radio to the World, 9-153
Sparks of Genius: Portraits of Electrical Engineering Excellence, 5-030
Speaking Telephone, Electric Light, and Other Recent Inventions, The, 7-006
Special Issue on Two Centuries in Retrospect, 3-012
Special Issue: 50th Anniversary of the Transistor!, 11-087
Special Issue: Historical Notes on Important Tubes and Semiconductor Devices, 11-006
Special Issue: Telephone Centenary, 7-092
Special Issue: The History of Electrical Engineering, 3-011
Special Satellite Number, 14-053
Spectrum Guide: Radio Frequency Allocations in the United States, 30 MHz-300 GHz, 8-045
Spectrum Reallocation Final Report, 8-057
Speed of Sound: Hollywood and the Talkie Revolution, 1926–1930, 10-082
Spirit of Independent Telephony, 7-088
Spirit of the Web: The Age of Information from Telegraph to Internet, 3-053
Sprit of Discovery: An Appreciation of the World of Marconi, 5-155
Stage to Studio: Musicians and the Sound Revolution, 1890–1950, 10-062
Standard & Poor's Industrial Surveys, 4-009
Standard Broadcast Allocation Maps: 540 Kc to 1600 Kc, 1-080
Standard Electrical Dictionary, 1-035
Statistical Yearbook: Film, Television, Video and New Media in Europe, 1-063
Statistics of Communications Common Carriers, 1-073
Statistics of Telegraphy, 6-048
Statistics on Radio and Television, 1-068
Stay Tuned: A Concise History of American Broadcasting, 3-065
Stereo Goes to Market, 10-096
Stereo Then and Now, 10-105
Stereophonic Recording and Reproducing Equipment, 10-104
Stereophonic Sound Film System–General Theory, 10-103

Stereophonic Sound Production, 10-100
Stereophonic Sound, 10-098
Stokers and Pokers, 6-158
Stokowski and the Bell Telephone Laboratories: Collaboration in the Development of High-Fidelity Sound Reproduction, 10-066
Story of Cyrus Field: The Projector of the Atlantic Telegraph, 5-123
Story of Independent Telephony, 7-057
Story of Mahlon Loomis, Pioneer of Radio, 5-153
Story of My Life: By the Submarine Telegraph, 6-184
Story of Pye Wireless, 9-161
Story of Radio, 9-098, 9-100
Story of S.T.C. 1883–1958, 4-247
Story of Scophony, 12-102
Story of Stereo: 1881–, 10-110
Story of Telecommunications, 3-048
Story of Television: The Life of Philo T. Farnsworth, 5-111
Story of the Atlantic Cable, 6-219
Story of the Commonwealth Wireless Service, 9-122
Story of the First Trans-Atlantic Short Wave Message, 9-138
Story of the Indian Telegraphs: A Century of Progress, 6-167
Story of the Super-Heterodyne, 9-141
Story of the Telegraph, 6-097
Story of the Telegraph, and a History of the Great Atlantic Cable, 6-176
Story of the Telegraph in India, 6-142
Story of the Telephone...of Britain, 7-142
Story of the U.S. Army Signal Corps, 3-137
Story of Western Union, 4-178
Story of Wireless Telegraph, 9-025
Strategic Investments in Innovation: The Telecommuniations Equipment Industry, 1975–1986, 4-057
Strategic Maneuvering and Mass Market Dynamics: The Triumph of VHS over Beta, 10-133
Streak of Luck: The Life and Legend of Thomas Alva Edison, 5-099
Strowger Automatic Telephone Exchange, 7-060
Structure and Performance of the U.S. Communications Industry, 4-052
Studies of the Ionosphere and Forecasts for Radiocommunications: Physicists and Engineers, the Military, and National Laboratories in France (and Germany) after 1945, 8-047
Study of Future Directions for the Voice of America in the Changing World of International Broadcasting, 3-105

Study of International Telecommunications Policies, Technology, and Economics, 3-095
Study of the Operating Characteristics of the Ratio Detector and Its Place in Radio History, 9-143
Study of Western Electric's Performance, 4-171
Submarine and Land Telegraph Systems of the World, 1-072
Submarine Cable: The Story of the Submarine Telegraph Cable from Its Invention down to Modern Times, 6-231
Submarine Cable-Laying and Repairing, 6-217
Submarine Cables for Long-Distance Telephone Circuits, 7-151
Submarine Signal Log, 4-162
Submarine Telecommunication and Power Cables, 6-221
Submarine Telegraphs: Their History, Construction and Working, 6-218
Submarine Telegraphy, 6-226
Submarine Telegraphy: The Grand Victorian Technology, 6-229
Submarine Telephone Cables and International Communication, 7-150
Super-Heterodyne–Its Origin, Development and Some Recent Improvements, 9-142
Suppressing Innovation: Bell Laboratories and Magnetic Recording, 10-053
Survey of Television Progress in America, 12-025
Switch in Time: An Engineer's Tale, 4-203
Switchboard Problem: Scale, Signaling, and Organization in Manual Telephone Switching, 1877–1897, 7-087
Symposium on Communications Satellites, 14-028
Symposium on the Transatlantic Telephone Cable, 1957, 7-155
Synchronized Reproduction of Sound and Scene, 10-093
SYNCOM and Its Successors, 14-051
Syntony and Credibility: John Ambrose Fleming, Guglielmo Marconi, and the Maskelyne Affair, 11-003
Syntony and Spark: The Origins of Radio, 8-001

Tales of S.O.S. and T.T.T., 3-151
Tales of the Nineties, 7-097
Talkies: American Cinema's Transition to Sound, 1926–1931, 10-081
Talking by Electricity: Telephones..., 7-013
Talking Machine Industry, 10-039
Talking Machines and Records, 10-018
Talking Machines: 1877–1914, 10-019
Talking Wire: The Story of Alexander Graham Bell, 5-064

Title Index 327

Talks about Radio, 5-148
Taming the Tyrant: The First One Hundred Years of Australia's International Communication Services, 4-229
Tangled Prelude to the Age of Silicon Electronics, 3-010
Technical Description of Broadcasting House, 4-197
Technical Description of the Marconi-E.M.I. System of Television at the London Television Station, 12-042
Technical Development of Broadcasting in Asia-Pacific, 1964–1984, 9-128
Technical Development of Television, 12-128
Technological Boundaries of Television, 13-134
Technological Innovation and Organizational Change: The Navy's Adoption of Radio, 1899–1919, 3-162
Technological Pioneering and Competitive Advantage: The Birth of the VCR Industry, 10-142
Technological Survey of Broadcasting's Prehistory, 9-113
Technologies of Control: The New Interactive Media for the Home, 13-087
Technologies of Freedom: On Free Speech in an Electronic Age, 3-050
Technologies without Boundaries: On Telecommunications in a Global Age, 3-051
Technology and the Raj: Western Technology and Technical Transfers to India, 1700–1947, 6-151
Technology's Retreat: The Decline of Rural Telephony in the United States, 1920–1940, 7-077
Telcon Story: 1850–1950, 4-251
Tele: Retrospective Survey of the Last Twenty-Five Years, 4-249
Telecasting and Color, 12-143
Telecom History, 7-093
Telecommunications Act Handbook: A Complete Reference for Business, 4-027
Telecommunications and the Computer, 14-007
Telecommunications in Canada: A Century of Symbiotic Development, 3-078
Telecommunications in Europe: Free Choice for the User in Europe's 1992 Market, 3-097
Telecommunications in Japan, 4-214
Telecommunications Industry: Growth and Structural Change, 4-058
Telecommunications Industry: Integration vs Competition, 4-062
Telecommunications Industry: The Dynamics of Market Structure, 4-053

Telecommunications Policy, High Definition Television, and U.S. Competitiveness, 13-110
Telecommunications Research Resources: An Annotated Guide, 1-001
Telecommunications Satellites, 14-034
Telecommunications Structure and Management in the Executive Branch of Government, 1900–1970, 4-037
Telecommunications, Mass Media & Democracy: The Battle for the Control of U.S. Broadcasting, 1928–1935, 4-045
Telecommunications: A Program for Progress, 4-034
Telecommunications: Economics and Regulation, 4-042
Telecommunications: The Booming Technology, 14-001a
Telefoni/Telephone Sets, 7-177
Telegraph, 6-086
Telegraph and Its Proposed Acquisition by the Government, 6-161
Telegraph and Telephone Considered in Relation to Economy and Efficiency, 6-079
Telegraph and the Structure of Markets in the United States, 1845–1890, 6-107
Telegraph and the Telephone, 3-040
Telegraph and the Telephone: Their Development and Role in the Economic History of the United States–The First Century, 1844–1944, 3-024
Telegraph and Travel: A Narrative of the Formation and Development of Telegraphic Communication between England and India, 6-153
Telegraph Collector's Guide, 6-251
Telegraph Comes to Colorado: A New Technology and Its Consequences, 7-105
Telegraph in America: Its Founders, Promoters and Noted Men, 5-035
Telegraph in Europe, 6-138
Telegraph in Nineteenth-Century America: Technology and Monopoly, 6-108
Telegraph in World War I, 3-114
Telegraph Industry, 4-176
Telegraph Manual: A Complete History and Description, 6-029
Telegraph Monopoly, 4-177
Telegraph Pocket Book, 6-060
Telegraph Register of the Electro-Magnetic Telegraph Companies, 6-009
Telegraph Secrets: By a Stationmaster, 6-039
Telegraph to India and Its Extension to Australia and China, 6-200
Telegraph Transmission, 6-091
Telegraph: A History of Morse's Invention and Its Predecessors in the United States, 6-104

Telegraph: How Technological Innovation Caused Social Change, 6-117
Telegraph's Effects on Nineteenth Century Markets and Firms, 6-124
Telegrapher's Souvenir, 6-054
Telegraphers of To-Day: Descriptive, Historical, Biographical, 5-036
Telegraphers, Their Craft and Their Unions, 6-123
Telegraphers' Handbook, 6-085
Telegraphic Connections, 6-082
Telegraphic Railways, 5-083
Telegraphic Submarine Lines between Europe and America and the Atlantic and Pacific, 6-192
Telegraphic Tales and Telegraphic History, 6-068
Telegraphic Trail of the Transcontinental Telegraph, 6-032
Telegraphic Transmission of Photographs, 13-002
Telegraphing in Battle: Reminiscences of the Civil War, 3-139
Telegraphs in Victorian London, 6-150
Telegraphy, 6-057
Telegraphy: A Detailed Exposition of the Telegraph System of the British Post Office, 6-159
Telegraphy across Space, 9-003
Telegraphy, Aeronautics and War, 3-111
Telegraphy and Telephony, 3-056
Telegraphy and the Genesis of Electrical Engineering Institutions in France, 1845–1895, 6-127
Telegraphy and the Technology of Display: The Electricians and Samuel Morse, 6-118
Telegraphy in the Bell System, 6-090
Telegraphy of Photographs, Wireless and by Wire, 13-003
Telegraphy–Pony Express to Beam Radio, 6-093
Telephone, 7-001, 7-017
Telephone: An Historical Anthology, 7-091
Telephone and Automobile Diffusion in the United States, 1902–1937, 7-078
Telephone and Its Several Inventors: A History, 7-073
Telephone and Telephone Exchanges, 7-053
Telephone Book: Bell, Watson, Vail and American Life, 1876–1983, 7-069
Telephone Boxes, 7-178
Telephone Calendar, 1-058
Telephone Cards, 7-180
Telephone Cases...Adjudged in the Supreme Court, 7-016
Telephone Collecting: Seven Decades of Design, 7-167

Telephone Enterprise: The Evolution of the Bell System's Horizontal Structure, 1876–1909, 4-101
Telephone Handbook, 7-018
Telephone Hand-Book, 7-022
Telephone: Its History, Construction, Principles and Uses, 7-004
Telephone: Its Wonders and Lessons, 7-008
Telephone Lines and Methods of Constructing Them Overhead and Underground, 7-039
Telephone Lines and Their Properties, 7-019
Telephone: Outlines of the Development of Transmitters and Receivers, 7-028
Telephone Principles & Practice, 7-041
Telephone Service in 1882 in the United States and the World at Large, 7-012
Telephone Service: Its Past, Its Present, and Its Future, 7-040
Telephone Statistics, 1-067
Telephone Switchboard–Fifty Years of History, 7-051
Telephone System of the British Post Office: A Practical Handbook, 7-134
Telephone Systems of the Continent of Europe, 7-115
Telephone Theory & Practice, 7-059
Telephone Volume, 7-042
Telephone, the Microphone, and the Phonograph, 7-010
Telephone: The First Hundred Years, 4-097
Telephone's First Century–and Beyond, 7-094
Telephones and Telegraphs, 1-069
Telephones Antique to Modern, 7-166
Telephones: Their Construction and Fitting, 7-024
Telephony [1902], 7-036
Telephony [1905], 7-037
Telephony without Wires, 9-054
Telephony: A Complete and Detailed Exposition of the Theory and Practice of the Telephone Art, 7-058
Telephony: A Detailed Exposition of the Telephone System of the British Post Office, 7-134
Telephony: A Manual of the Design, Construction and Operation of Telephone Exchanges, 7-038
Teleports and the Intelligent City, 14-042
Teletext and Videotex in the United States, 13-081
Televiewing, 12-032
Television, 12-033, 12-041, 12-047, 12-067, 12-108
Television and Short-Wave World Practical Handbook, 12-053
Television and the Remote Control: Grazing on a Vast Wasteland, 13-083

Title Index

Television and the Wired City: A Study of the Implications of a Change in the Mode of Transmission, 13-067
Television Baird, 5-045
Television Broadcasting Practice in America: 1927–1944, 12-091
Television Broadcasting: Production, Economics, Technique, 12-095
Television by Electron Image Scanning, 12-088
Television Cyclopaedia, 1-036
Television Encyclopedia, 1-030
Television Engineering, 12-045
Television 50 Years Ago, 5-057
Television for All, 12-013
Television for the Amateur Constructor, 12-023
Television for the Home: The Wonders of "Seeing by Wireless," 12-004
Television for You, 12-030
Television Frequency Allocation Policy in the United States, 8-044
Television Handbook, 12-059
Television in America To-Day, 12-065
Television in Great Britain, 12-077
Television in Prospect, 1873–1927, 12-073
Television Man: The Story of John L. Baird, 5-056
Television Really Explained, 12-031
Television Reception, 1925–1975, 12-149
Television Reception Technique, 12-044
Television Reception: Construction and Operation of a Cathode-Ray Tube Receiver, 12-035
Television Standards and Practice: Selected Papers from the Proceedings of the NTSC and Its Panels. 12-090
Television Technical Terms and Definitions, 1-031
Television To-Day and To-Morrow, 12-012
Television Today and Tomorrow, 12-085
Television Today: Practice and Principles Clearly Explained, 12-029
Television Up-to-Date, 12-028
Television with Cathode-Rays, 12-038
Television, Theory and Practice, 12-024
Television: A Guide for the Amateur, 12-039
Television: A Practical Treatise on the Principles upon Which the Development of Television Is Based, 12-021
Television: A Series of Articles, 12-058
Television: A Struggle for Power, 12-051
Television: A World Survey and Supplement, 12-107
Television: An Account of the Development and General Principles of Television..., 12-043
Television: An International History, 12-114
Television: Its Methods and Uses, 12-017

Television: Present Methods of Picture Transmission, 12-008
Television: Seeing by Wire or Wireless, 12-002
Television: The Electronics of Image Transmission, 12-109
Television's Bright New Technology, 13-128
Telling the World, 3-054
Telstar Experiment, 14-054
Ten Founding Fathers of the Electrical Science, 5-020
Ten Years of Broadcasting, 9-118
Tent Life in Siberia, 6-136
Terrestrial Digital Audio Broadcasting in Europe, 13-100
Tesla 1984: Proceedings of the Tesla Centennial, 5-221
Tesla: Man out of Time, 5-214
Testing Television Sets, 12-050
Theater Television: A History, 12-092
Theodore N. Vail and the Role of Innovation in the Modern Bell System, 4-100
Theory and Practice of the Electric Telegraph, 6-052
Theory of the Submarine Telegraph and Telephone Cable, 7-149
Thermionic Tubes in Radio Telegraphy and Telephony, 11-021
Thermionic Vacuum Tube, 11-024
Thermionic Valve and Its Developments in Radiotelegraphy and Telephony, 11-019
Thermionic Valves, 1904–1954: The First Fifty Years, 11-008
Thirty Years in Cable TV: Reminiscences of a Pioneer, 13-063
This Fascinating Radio Business, 9-119
This Great Contrivance: The First Hundred Years of the Telephone in Rochester, 7-104
Thomas Alva Edison: An American Myth, 5-107
Three Degrees above Zero: Bell Labs in the Information Age, 4-116
Three Miles Deep: The Story of the Transatlantic Cables, 6-240
Through to 1970: Royal Signal Corps Golden Jubilee, 3-108
Time for Innovation, 7-074
Timetable of Technology, 1-053
Timetables of Technology, 1-043
Titanic: Signals of Disaster, 3-150
Tivador Puskas–A Forerunner, 7-124
Toll Telephone Practice, 7-067
Tomorrow's TVs: A Review of New TV Set Technology, Related Video Equipment and Potential Market Impacts, 1987–1995, 13-038
Tracing the Telephone in Western Massachusetts, 1877–1930, 7-114

Trail Blazers to Radionics and Reference Guide to Ultra High Frequencies, 9-134
Trailblazer to Television, 5-146
Transatlantic Communications..., 7-156
Trans-Atlantic Submarine Telegraph, 6-210
Transatlantic Telephone Cable, 7-148
Transcendent Cable: An Atlantic Telegraph, 6-248
Transfer of Telegraph Technologies in the Nineteenth Century, 6-113
Transistor, 11-084
Transistor 1948–1958: Ten Years of Progress at Bell Telephone Laboratories, 11-088
Transistor as an Industrial Research Episode, 11-070
Transistor Issue, 11-090
Transistor Radios, 1954–1968, 9-179
Transistor Radios: A Collector's Encyclopedia and Price Guide, 9-170
Transistor: Two Decades of Progress, 11-089
Transistors and Their Applications: A Bibliography, 1948–1953, 11-075
Transmission and Reception of Photodiagrams, 13-031
Transmission of Military Information, 3-145
Transmission of Photographs and Drawings by Wireless Telegraphy, 13-026
Transmission of Pictures by Radio, 13-009
Transmission of Pictures over Telephone Lines, 13-023
Transmission of Television Images, 5-113
Transmitters, Exciters, and Power Amplifiers, 1930–1980, 9-172
Transmitting World News: A Study of Telecommunications and the Press, 3-098
Transoceanic Communication by Means of Satellite, 14-046
Trans-Oceanic Radio Communication, 9-045
Traveling Wave Tube–A Record of Its Early History, 11-068
Traveling-Wave Tubes, 11-058
Treatise on Telephony, 7-032
Treatise on the Construction and Submersion of Deep-Sea Electric Telegraph Cables, 6-190
Trends in Telephone Services, 1-075
Trends in the International Communications Industry, 1975–1989, 1-074
Trip to Newfoundland...With an Account of the Laying of the Submarine Telegraph Cable, 6-169
Tropospheric Propagation: A Selected Guide to the Literature, 8-019
Tropospheric Scatter Communications: Past, Present, and Future, 8-018
True Alliance; or, The History of the Transatlantic Cable Uniting Britain with America, 6-180
Tube Lore: A Reference for Users and Collectors, 11-005a

Tube: The Invention of Televison, 12-116
Tubes and Transistors: A Comparative Study, 11-036
Turning Points in American Electrical History, 3-003
TV Is King: Exhibition at Sotheby's, 12-146
Twentieth Century Professional Institution: The Story of the Brit. I.R.E., 4-081
20 Years of Audio, 10-067
Twenty Years of Video Tape, 10-130
Twenty-Fifth Anniversary Historical Chronology, 4-026
25th Anniversary of Pulse Code Modulation, 8-016
Twenty-Five Years of BBC Television, 4-191
25 Years of Radio Progress with RCA, 4-161
Twenty-Five Years of World Television, 12-129
Two Hundred Meters and Down: The Story of Amateur Radio, 4-076
2MT Writtle: The Birth of British Broadcasting, 9-085
Two Paths to the Telephone, 5-130
Two-Way Television and a Pictorial Account of Its Background, 12-016

U.S. Industrial & Trade Outlook, 1-071
U.S. Information Technology Industry Trade Analysis, 1960–1991, 1-077
U.S. Microelectronics Industry, 11-073
U.S. Postal History Documents No. 2: Congress & the Telegraph, 6-111
U.S. Spectrum Management Policy: Agenda for the Future, 8-056
Ultrafax, 13-007, 13-035
Ultrasonics and Their Scientific and Technical Application, 10-002
Ultrasonics, 10-003, 10-015
Under the Glare of a Thousand Suns–The Pioneering Work of Sir J.C. Bose, 5-067
Understanding DAB, 13-101
Union List of Periodicals on Electronics and Related Subjects, 2-004
Union List of Periodicals: Science-Technology-Economics, 2-010
United States Corporate Histories: A Bibliography, 1965–1990, 4-007
United States Radio Development, 9-075
Untold Story of the Telephone, 7-066
Up in the Air–New Wireless Communications, 14-018
Use of Steel Tape Magnetic Recording Media in Broadcasting, 10-063

Vacuum Tubes in Wireless Communication, 11-014
Value of the Frequency Spectrum Allocated to Specific Uses, 8-062
Variations of Conductivity under Electrical Influence, 9-145

Title Index

VCR Age: Home Video and Mass Communication, 10-139
Victorian Internet: The Remarkable Story of the Telegraph and the Nineteenth Century's On-Line Pioneers, 6-121
Victory of Television, 12-055
Video Age: Television Technology and Applications in the 1980s, 13-055
Video Cassette: 1928–1971, 10-140
Video Discs: The Technology, the Applications and the Future, 10-145
Video Recording Technology: Its Impact on Media and Home Entertainment, 10-124
Video Recording: A History, 10-127
Video Telephone: Impact of a New Era in Telecommunications—A Preliminary Technology Assessment, 7-159
Video Worldwide: An International Study, 10-131
Videocassette Recorders in the Third World, 10-132
Videodisc, 10-146
Videodisc: The Next Step in the Communications Evolution, 10-147
Videotape Recorder: Its Evolution and the Present State of the Art of VTR Technology, 10-128
Videotex Systems and Services, 13-075
Viewdata and the Information Society, 13-078
Viewdata Revolution, 13-077
Vintage Crystal Sets, 1922–1927, 9-161
Vintage Telephones of the World, 7-176
Vision by Radio, Radio Photographs, 13-024
Visual Communication Systems: Trials and Experiences, 7-160
Vital Link: The Story of Royal Signals 1945–1985, 3-128
Voice across the Sea, 6-222
Voice by Wire and Post-Card, 7-009
Voice from Afar: The History of Telecommunications in Canada, 3-077
Voice of Generations: A History of Communications in Newfoundland, 3-079a
Voices in the Darkness: The European Radio War, 3-101
Voiceway to the Orient: First U.S.-Japan Telephone Cable, 7-154

War of the Black Heavens: The Battles of Western Broadcasting in the Cold War, 3-103
War on the Short Waves, 3-104
Watchers of the Waves, 3-154
Watching TV: Historic Televisions and Memorabilia from the MZTV Museum, 12-152
Wave Antenna, 8-027
Waveguides in Electrical Communication, 8-031

Waves of Sound and Speech As Revealed by the Phonograph, 10-036
We've Got Light! Building the Nation's First Coast-to-Coast Fiber-Optic Network, 14-082
Weaving the Web: The Original Design and Ultimate Destiny of the World Wide Web by Its Inventor, 14-085a
Web without a Weaver: How the Internet Is Shaping Our Future, 14-086
Werner von Siemens, 4-235
Western Electric and the Bell System: A Survey of Service, 4-169
Western Electric's First 75 Years: A Chronology, 4-170
Western Union, 4-179
Western Union Telegraph Company: 1851–1901, A Retrospect, 4-173
Western Union Telegraph Company: Its Past, Present and Future, 4-174
What Happened after Bell Spilled the Acid? Telecommunications History: A View Through the Literature, 1-019
What Hath God Wrought?, 6-114
What Hath God Wrought: The Story of the Queensland Telegraph Service from 1861, 6-156
What Telecommunications Did in the War, 3-115
Wheatstone and Cooke's Electric Telegraph, 5-090
When Old Technologies Were New: Thinking about Communications in the Late 19th Century, 3-044
When Telecom and ITT Were Young, 4-142
Where Do We Go from Here? The FCC Auctions and the Future of Radio Spectrum Management, 8-052
Where Wizards Stay Up Late: The Origins of the Internet, 14-087
Whisper in the Air: Marconi–The Canada Years, 1902–1946, 5-165
Who Invented the Telephone?, 7-045
Who Owns the Rainbow? Conserving the Radio Spectrum, 8-039
Who Really Invented the Telephone?, 7-085
William Henley, Pioneer Electrical Instrument Maker and Cable Manufacturer, 4-211
Wire and Wireless Telegraphy, 9-023
Wire and Wireless: A History of Telecommunications in New Zealand, 3-083
Wire Song: An Illustrated History of the Telephone in British Columbia, 1880–1930, 7-129
Wire Wars: The Canadian Fight for Competition in Telecommunications, 7-144
Wired Nation Continent: the Communication Revolution and Federating Australia, 3-079

Wired Nation: A Special Issue, 13-071
Wired Nation: Cable TV, the Electronic Communications Highway, 13-072
Wired Society: A Challenge for Tomorrow, 14-009
Wireless, 9-102
Wireless: The Revolution in Personal Telecommunications, 14-013
Wireless–The Crucial Decade: History of the British Wireless Industry, 1924–1934, 4-054
Wireless Access and the Local Telephone Network, 14-015
Wireless and Satellite Telecommunications, 14-010
Wireless and Shipping, 3-160
Wireless at Sea: The First Fifty Years, 3-156
Wireless Communication in the United States: The Early Development of American Radio Operating Companies, 4-063
Wireless Communication over Sea, 3-153
Wireless Constructor's Encyclopedia, 1-024
Wireless for the Warrior, 3-120
Wireless Man: His Work & Adventures on Land & Sea, 9-051
Wireless of To-Day, 9-063
Wireless over Thirty Years, 9-115
Wireless Personal Communications, 14-023
Wireless Pictures and Television, 13-004
Wireless Possibilities, 9-068
Wireless Radio: A Brief History, 9-096
Wireless Technologies and the National Information Infrastructure, 14-024
Wireless Telegraphic Communication, 9-071
Wireless Telegraphy, 9-005, 9-008, 9-031, 9-088, 9-101
Wireless Telegraphy and Hertzian Waves, 9-006
Wireless Telegraphy and Telephony Popularly Explained, 9-042
Wireless Telegraphy and Telephony without Wires, 9-063
Wireless Telegraphy and Telephony, 9-012, 9-034, 9-038
Wireless Telegraphy and Telephony: A Handbook of Formulae, Data and Information, 9-057
Wireless Telegraphy and Telephony: An Elementary Treatise, 9-033
Wireless Telegraphy and Wireless Telephony, 9-046
Wireless Telegraphy circa 1898–99: The Untold South African Story, 3-109
Wireless Telegraphy for Amateurs, 9-040
Wireless Telegraphy for Amateurs and Students, 9-036
Wireless Telegraphy for Army Purposes, 3-110
Wireless Telegraphy Popularly Explained, 9-002
Wireless Telegraphy: Its History, Theory and Practice, 9-027
Wireless Telegraphy: Its Origin, Development, Inventions, and Apparatus, 9-017
Wireless Telegraphy: Report of the Interdepartmental Board, 9-026
Wireless Telegraphy; Instruction Paper, 9-030
Wireless Telegraphy–Its Past and Present Status and Its Prospects, 9-016
Wireless Telephony, 9-060
Wireless Telephony and Broadcasting, 9-056
Wireless Telephony in Theory and Practice, 9-044
Wireless Transmission of Photographs, 13-029
Wireless World at War: 1939–1945, 3-125
Wires West: The Story of the Talking Wires, 6-102
Wiring a Continent: The History of the Telegraph Industry in the United States, 1832–1866, 6-101
Wiring the World: The Explosion in Communications, 13-049
Without the Valve: Some Alternatives to Thermionic Devices, 9-077
Wizard of the Wires: A Boy's Life of Samuel F.B. Morse, 5-193
Wizard: The Life and Times of Nikola Tesla, 5-222
Wonders of Wireless Telegraphy Explained in Simple Terms, 9-061
Wood's Plan of Telegraphic Instruction, 6-037
Words and Waves, 3-025
Work of Hertz and Some of His Successors, 8-010
Working of Long Submarine Cables, 6-211
Workshop of Engineers: The Story of the General Engineering Laboratory of the General Electric Company, 1895–1952, 4-134
World at Their Fingertips: The Story of Amateur Radio in the United Kingdom and a History of the Radio Society of Great Britain, 4-094
World Communication Report: The Media and the Challenge of the New Technologies, 3-073
World Communication: Threat or Promise? A Socio-Technical Approach, 3-085
World Communications: Press, Radio, Film, Television, 3-072
World Directory of Moving Image and Sound Archives, 1-039
World List of Scientific Periodicals, 2-011
World Telecommunications, 3-074
World's Greatest Industrial Laboratory, 4-115
World's Telephones, 1-076

Worldwide Satellite Communications, 14-066
WWW: Past, Present, and Future, 14-085

Year of the Transistor, 11-069
Yearbook of Common Carrier Telecommunication Statistics, 1-064
Year-Book of Wireless Telegraphy and Telephony, 9-070
You Have Been Listening...The Early History of Radio in South Africa, 9-126

Zenith Radio: The Early Years, 1919–1935, 4-182
Zenith Story: A History from 1919, 4-183
Zenith Trans-Oceanic: The Royalty of Radios, 9-158
Zworykin: Pioneer of Television, 5-225

About the Authors

Christopher H. Sterling is a professor of media and public affairs, and of telecommunication, at George Washington University, where he also serves as Associate Dean for Graduate Affairs in the arts and sciences. He has been on the GWU faculty since 1982, following two years at the FCC and a decade on the Temple University faculty. He holds undergraduate and graduate degrees from the University of Wisconsin. Sterling has authored or edited fifteen books, edits *Communication Booknotes Quarterly*, and served as editor of *Journal of Broadcasting* from 1972 through 1976. His own collection of historical works in telecommunication is reflected in the previous pages.

George Shiers (1908–1983) authored the Scarecrow book on which this one is based: *Bibliography of the History of Electronics* (1972). Born and raised in England, he worked in the electrical industry before World War I. After military service, he came to the United States in 1948, moving to California in 1954. He served as a freelance technical writer while teaching at a number of Santa Barbara area colleges. Over the years, he authored a number of articles on television and telecommunications history and the definitive *Early Television: A Bibliographic Guide to 1940* (Garland 1997), published after his death.